W0051311

NEUROMETHODS

Series Editor
Wolfgang Walz
University of Saskatchewan
Saskatoon, SK, Canada

For further volumes:
http://www.springer.com/series/7657

Neuromethods publishes cutting-edge methods and protocols in all areas of neuroscience as well as translational neurological and mental research. Each volume in the series offers tested laboratory protocols, step-by-step methods for reproducible lab experiments and addresses methodological controversies and pitfalls in order to aid neuroscientists in experimentation. *Neuromethods* focuses on traditional and emerging topics with wide-ranging implications to brain function, such as electrophysiology, neuroimaging, behavioral analysis, genomics, neurodegeneration, translational research and clinical trials. *Neuromethods* provides investigators and trainees with highly useful compendiums of key strategies and approaches for successful research in animal and human brain function including translational "bench to bedside" approaches to mental and neurological diseases.

Methods for Analyzing Large Neuroimaging Datasets

Edited by

Robert Whelan

School of Psychology, Trinity College Dublin, Dublin, Ireland

Hervé Lemaître

Institut des Maladies Neurodégénératives, UMR 5293, CNRS, University of Bordeaux, Bordeaux, France

Editors
Robert Whelan
School of Psychology
Trinity College Dublin
Dublin, Ireland

Hervé Lemaître
Institut des Maladies Neurodégénératives, UMR 5293
CNRS, University of Bordeaux
Bordeaux, France

Robert Whelan Institute of Neuroscience, Trinity College Dublin
This work was supported by Robert Whelan and Institute of Neuroscience, Trinity College Dublin.

ISSN 0893-2336 ISSN 1940-6045 (electronic)
Neuromethods
ISBN 978-1-0716-4259-7 ISBN 978-1-0716-4260-3 (eBook)
https://doi.org/10.1007/978-1-0716-4260-3

This Humana imprint is published by the registered company Springer Science+Business Media, LLC, part of Springer Nature.
The registered company address is: 1 New York Plaza, New York, NY 10004, U.S.A.

Foreword: New Methods for a New Approach to Neuroimaging

Neuroimaging finds itself at a thrilling point in its evolution. From its early days with small sample sizes, the neuroimaging field has evolved to embrace the scientific advantages of large datasets, an approach that is personally resonant. Collaborations are generating datasets incorporating thousands of images, and open science movements are succeeding in the synthesis and aggregation of data. However, this evolution means that researchers working with large neuroimaging datasets must, at a minimum, have a working knowledge of neuroscience, computer science, and best practices in conducting reproducible research. Therefore, there is a pressing need for a practical guide to navigating the terrain of population neuroimaging. This book provides such a guide: editors Whelan and Lemaître have curated a set of chapters that address the full spectrum of challenges and opportunities posed by large neuroimaging datasets. Chapters delve into the intricacies of cloud computing, EEG data processing at scale, and the application of deep learning to raw MRI images, to name just a few.

Designed to assist both novices and experienced researchers, each chapter offers a comprehensive guide to a particular topic, including practical examples. Dedicated chapters in the book focus on best practices for efficiency, reproducibility, preprocessing, and statistical analyses, ensuring researchers are well-equipped to seamlessly navigate the evolving landscape of neuroimaging. This book stands as a testament to the adaptability and forward-thinking nature of the neuroimaging community. Combining the insights of experts in the field with the latest technological advancements, this volume is positioned to be a guiding beacon for researchers tackling the complexities of large datasets.

Beyond data handling and analysis, the book champions the ethos of open science, collaboration, and equity, particularly concerning the use of open-source tools. The emphasis on transparent and reproducible research is a core value. It is hoped that an open-access book such as this will increase the diversity of researchers working in human neuroimaging around the globe, and ultimately increase the diversity of training samples for important applications (e.g., to Alzheimer's disease). Ultimately, it is through cooperation, across labs and countries, that will advance our science in the field of neuroimaging.

Paul Thompson

Imaging Genetics Center, Mark & Mary Stevens Institute for Neuroimaging & Informatics, Keck School of Medicine, University of Southern California, Los Angeles, CA, USA

Contents

Contributors

NICHOLAS ALLGAIER • *Department of Psychiatry, University of Vermont, Burlington, VT, USA*

GEORGIOS ANTONOPOULOS • *Institute of Neuroscience and Medicine, Brain and Behaviour (INM-7), Research Centre Jülich, Jülich, Germany; Institute of Systems Neuroscience, Medical Faculty, Heinrich Heine University Düsseldorf, Düsseldorf, Germany*

JÜRGEN BIRKLBAUER • *Centre for Cognitive Neuroscience and Department of Psychology, Paris-Lodron-University of Salzburg, Salzburg, Austria; Sport and Exercise Science, Paris-Lodron-University of Salzburg, Salzburg, Austria*

RORY BOYLE • *Department of Neurology, Massachusetts General Hospital, Harvard Medical School, Boston, MA, USA*

MICHAEL CONNAUGHTON • *Department of Psychiatry, School of Medicine, Trinity College Dublin, Dublin, Ireland; Trinity College Institute of Neuroscience, Trinity College Dublin, Dublin, Ireland*

ARNAUD DELORME • *Swartz Center for Computational Neuroscience, Institute for Neural Computation, University of California San Diego, La Jolla, CA, USA*

MONIQUE J. M. DENISSEN • *Centre for Cognitive Neuroscience and Department of Psychology, Paris-Lodron-University of Salzburg, Salzburg, Austria; SCCN, University of California San Diego, La Jolla, CA, USA*

OSCAR ESTEBAN • *Department of Radiology, Lausanne University Hospital and University of Lausanne, Lausanne, Switzerland*

HUGH GARAVAN • *Department of Psychiatry, University of Vermont, Burlington, VT, USA*

MÉLANIE GARCIA • *Trinity College Institute of Neuroscience, Trinity College, Dublin, Ireland; Department of Psychiatry at the School of Medicine, Trinity College Dublin, Dublin, Ireland*

CHRISTIAN GASER • *Department of Psychiatry and Psychotherapy, Jena University Hospital, Jena, Germany; Department of Neurology, Jena University Hospital, Jena, Germany; German Center for Mental Health (DZPG), Jena, Germany*

SAGE HAHN • *Department of Psychiatry, University of Vermont, Burlington, VT, USA*

NICOLE A. HIMMELSTOß • *Centre for Cognitive Neuroscience and Department of Psychology, Paris-Lodron-University of Salzburg, Salzburg, Austria*

FELIX HOFFSTAEDTER • *Institute of Neuroscience and Medicine, Brain and Behaviour (INM-7), Research Centre Jülich, Jülich, Germany; Institute of Systems Neuroscience, Medical Faculty, Heinrich Heine University Düsseldorf, Düsseldorf, Germany*

FLORIAN HUTZLER • *Centre for Cognitive Neuroscience and Department of Psychology, Paris-Lodron-University of Salzburg, Salzburg, Austria*

AGUSTIN IBAÑEZ • *Latin American Brain Health Institute (BrainLat), Universidad Adolfo Ibáñez, Santiago, Chile; Cognitive Neuroscience Center (CNC), Universidad de San Andrés, Buenos Aires, Argentina; National Scientific and Technical Research Council (CONICET), Buenos Aires, Argentina; Global Brain Health Institute, University of California, San Francisco, CA, USA; Trinity College Institute of Neuroscience (TCIN), Trinity College Dublin, Dublin, Ireland*

CLARE KELLY • *School of Psychology, Trinity College Dublin, Dublin, Ireland; Trinity College Institute of Neuroscience, Trinity College, Dublin, Ireland; Department of Psychiatry at the School of Medicine, Trinity College Dublin, Dublin, Ireland*

JOHANN D. KRUSCHWITZ • *Department of Psychology, MSB Medical School Berlin, Berlin, Germany*

ALEXANDER LEEMANS • *Image Sciences Institute, University Medical Center Utrecht, Utrecht, Netherlands*

HERVÉ LEMAÎTRE • *Institut des Maladies Neurodégénératives, CNRS UMR 5293, Université de bordeaux, Centre Broca Nouvelle-Aquitaine, Bordeaux, France*

CHRISTOPHER R. MADAN • *School of Psychology, University of Nottingham, Nottingham, UK*

TARA MADHYASTHA • *RONIN, Seattle, WA, USA; Department of Psychology, University of Washington, Oakland, CA, USA*

SCOTT MAKEIG • *Swartz Center for Computational Neuroscience, Institute for Neural Computation, University of California San Diego, La Jolla, CA, USA*

ANDRE F. MARQUAND • *Department of Cognitive Neuroscience, Radboud University Nijmegen Medical Centre, Nijmegen, Netherlands; Donders Institute, Radboud University Nijmegen, Nijmegen, Netherlands*

JANE MCGRATH • *Department of Psychiatry, School of Medicine, Trinity College Dublin, Dublin, Ireland; Trinity College Institute of Neuroscience, Trinity College Dublin, Dublin, Ireland*

SEBASTIAN MOGUILNER • *Latin American Brain Health Institute (BrainLat), Universidad Adolfo Ibáñez, Santiago, Chile; Cognitive Neuroscience Center, Universidad de San André s & CONICET, Buenos Aires, Argentina; Department of Neurology, Massachusetts General Hospital and Harvard Medical School, Boston, MA, USA*

ERIK O'HANLON • *Royal College of Surgeons in Ireland, Dublin, Ireland*

MATEUSZ PAWLIK • *Centre for Cognitive Neuroscience and Department of Psychology, Paris-Lodron-University of Salzburg, Salzburg, Austria*

JHONY ALEJANDRO MEJÍA PEREZ • *Latin American Brain Health Institute (BrainLat), Universidad Adolfo Ibáñez, Santiago, Chile; Memory and Aging Center, Department of Neurology, University of California San Francisco, San Francisco, CA, USA*

PAVEL PRADO • *Latin American Brain Health Institute (BrainLat), Universidad Adolfo Ibáñez, Santiago, Chile; Escuela de Fonoaudiología, Facultad de Odontología y Ciencias de la Rehabilitación, Universidad de San Sebastián, Santiago, Chile*

DECLAN QUINN • *School of Psychology, Trinity College Dublin, Dublin, Ireland*

JIVESH RAMDUNY • *School of Psychology, Trinity College Dublin, Dublin, Ireland; Trinity College Institute of Neuroscience, Trinity College, Dublin, Ireland*

ANNA N. RAVENSCHLAG • *Centre for Cognitive Neuroscience and Department of Psychology, Paris-Lodron-University of Salzburg, Salzburg, Austria*

FABIO RICHLAN • *Centre for Cognitive Neuroscience and Department of Psychology, Paris-Lodron-University of Salzburg, Salzburg, Austria*

KAY ROBBINS • *Department of Computer Science, University of Texas at San Antonio, San Antonio, TX, USA*

SAIGE RUTHERFORD • *Department of Cognitive Neuroscience, Radboud University Nijmegen Medical Centre, Nijmegen, Netherlands; Donders Institute, Radboud University Nijmegen, Nijmegen, Netherlands; Department of Psychiatry, University of Michigan-Ann Arbor, Ann Arbor, MI, USA*

AGUSTÍN SAINZ-BALLESTEROS • *Latin American Brain Health Institute (BrainLat), Universidad Adolfo Ibáñez, Santiago, Chile; Cognitive Neuroscience Center, Universidad de San Andrés & CONICET, Buenos Aires, Argentina*

EMIN SERIN • *Charité – Universitätsmedizin Berlin, Einstein Center for Neurosciences Berlin, Berlin, Germany; Bernstein Center for Computational Neuroscience, Berlin, Germany; Division of Mind and Brain Research, Department for Psychiatry, Charité-Universitätmedizin Berlin, Berlin, Germany*

DUNG TRUONG • *Swartz Center for Computational Neuroscience, Institute for Neural Computation, University of California San Diego, La Jolla, CA, USA*

NILAKSHI VAIDYA • *Centre for Population Neuroscience and Stratified Medicine (PONS), Department of Psychiatry and Psychotherapy, Charité-Universitätmedizin Berlin, Berlin, Germany*

HENRIK WALTER • *Division of Mind and Brain Research, Department for Psychiatry, Charité-Universitätmedizin Berlin, Berlin, Germany*

YIHE WENG • *School of Psychology, Trinity College Dublin, Dublin, Ireland*

ROBERT WHELAN • *School of Psychology, Trinity College Dublin, Dublin, Ireland; Global Brain Health Institute, Trinity College Dublin, Dublin, Ireland; Trinity College Institute of Neuroscience, Trinity College Dublin, Dublin, Ireland*

Chapter 1

Introduction to Methods for Analyzing Large Neuroimaging Datasets

Robert Whelan and Hervé Lemaître

Abstract

There is a recognition in the field of neuroimaging that sample size must drastically increase to achieve adequate statistical power and reproducibility. Several large neuroimaging studies and databases, such as OpenNeuro and the Adolescent Brain and Cognitive Development project, have emerged, offering open access to vast amounts of data. However, there is a dearth of practical guidance for working with large neuroimaging datasets, a deficit that this book seeks to address. With the emphasis on providing hands-on instruction, chapters contain worked examples using open-access data.

Key words Neuroimaging, Electroencephalography, Standardization, Toolboxes, Machine learning, Artificial intelligence

1 Structure of the Book

This book on methods for analyzing large neuroimaging datasets is organized as follows. In Subheading 4.1, the reader is shown how to access and download large datasets, and how to compute at scale. In Subheading 4.2, chapters cover best practices for working with large data, including how to build reproducible pipelines, the use of Git for collaboration, and how to make electroencephalographic and functional magnetic resonance imaging data sharable and standardized. In Subheading 4.3, chapters describe how to do structural and functional preprocessing data at scale, incorporating practical advice on potential trade-offs of standardization. In Subheading 4.4, chapters describe various toolboxes for interrogating large neuroimaging datasets, including those based on machine learning and deep learning approaches. These methods can be applied to connectomic and region-of-interest data. Finally, the book contains a glossary of useful terms.

Robert Whelan and Hervé Lemaître (eds.), *Methods for Analyzing Large Neuroimaging Datasets*, Neuromethods, vol. 218, https://doi.org/10.1007/978-1-0716-4260-3_1,

2 Why This Book Is Needed: Neuroimaging Datasets Are Getting (Much) Bigger

Human brain imaging is in a period of profound change. The sample sizes of the first neuroimaging studies were relatively low, perhaps with dozen subjects [1] or even single-subject analyses [2]. Such studies were incredibly valuable, facilitating non-invasive exploration of human brain structure and function. Those early studies paved the way for a new generation of neuroscientists. A few decades later, we have more complete and increasingly detailed maps of the human brain, but there is a growing recognition that sample size must drastically increase to achieve adequate statistical power and reproducibility. Indeed, it has been suggested that brain-wide association studies using neuroimaging may be unreliable without very large samples [3]. Fortunately—with added impetus from the fields of imaging genetics and neuroepidemiology—there has been a radical increase in the number of subjects within neuroimaging studies. For imaging genetics, the main reason was statistical power because the effects of common individual genetic variants are small and thus required large samples of subjects [4]. For neuroepidemiology, there was a paradigm shift from a goal of identifying robust diagnostic biomarkers to a prevention/prediction orientation, which necessarily requires large samples of healthy subjects, some of whom may subsequently develop a disease [5].

Given the scientific benefits of increased sample sizes, several large neuroimaging studies have been established. For example, the Adolescent Brain and Cognitive Development study (ABCD: https://abcdstudy.org/) is a 10-year neuroimaging project that will recruit over 10,000 people. Importantly, ABCD data are open access, and available with minimal restrictions. The UK Biobank (https://www.ukbiobank.ac.uk/) has neuroimaging data from 100,000 people, and these data are available to researchers for an access fee. Similar open-access magnetic resonance imaging (MRI) databases include the Alzheimer's Disease Neuroimaging Initiative (ADNI: http://adni.loni.usc.edu/) and Open Access Series of Imaging Studies (OASIS; https://www.oasis-brains.org/). Open access electroencephalographic (EEG) databases include Two Decades-Brainclinics Research Archive for Insights in Neurophysiology (TDBRAIN) [6] and the Child Mind Institute Multimodal Resource for Studying Information Processing in the Developing Brain (MIPDB). There are also datasets that belong to large consortia, such as IMAGEN (https://imagen-europe.com) [7] and EuroLADEEG [8].

As part of the effort to address poor reproducibility in neuroimaging research, the open science movement has fostered neuroimaging research to share data, codes, and publications [9]. In addition to the single studies with very large sample sizes described above, advances in open science have provided the conditions to allow researchers to access large neuroimaging datasets, or to combine several datasets to conduct meta- or mega-analyses [10, 11]. OpenNeuro (https://openneuro.org) [12] is a popular resource that hosts a variety of brain data, shared according to FAIR principles. The NeuroElectroMagnetic data Archive and tools Resource (NEMAR; https://nemar.org/) [13], contains EEG, magnetoencephalography, and intracranial EEG from OpenNeuro.

3 Why Focus Specifically on Analysis Methods for Large Neuroimaging Datasets?

We proposed to develop this book because several additional challenges arise when analyzing large—rather than small or medium—neuroimaging datasets. These challenges include the need for greater standardization, importance of good code management, use of scalable methods to process large volumes of data, and use of appropriate methods to uncover between-group or individual differences. The following chapters will bring the reader systematically through the essentials of working with large neuroimaging datasets, from downloading and storing data; to best practices for ensuring reproducibility; to preprocessing functional and structural data; to toolboxes for statistical analysis. Each chapter has comprehensive step-by-step instructions on a particular method (including examples of code where appropriate). We have not included any chapters in this book on neuroimaging acquisition. This is because, with respect to very large neuroimaging datasets, the data have either already been collected or will be collected according to a consensus protocol.

4 Overview of Chapters in This Book

4.1 Section 1: Accessing and Computing at Scale

Chapter 2 is titled "Getting Started, Getting Data" (Lemaître et al.) and illustrates different methods for downloading datasets using command lines (wget, curl), data management tools such as Datalad, Amazon Web Services, and graphical user interface options (e.g., Cyberduck). Chapter 2 demonstrates how to download data from OpenNeuro for a range of operating systems, which is important for using the worked examples later in the book. In general, after reading Chapter 2, researchers will be equipped with the knowledge and tools to download large neuroimaging datasets.

Analysis of large neuroimaging datasets requires scalable computing power and storage, plus methods for secure collaboration and for reproducibility. For example, data preprocessing is unlikely to be possible on a single computer. In Chapter 3 (Madhyastha)—Neuroimaging Workflows in the Cloud"—the theory and practice of using cloud computing to address many of these requirements is presented. Cloud computing offers a highly flexible model that is typically more cost-effective than a single laboratory investing in computer equipment to accommodate its peak demand. Chapter 3 describes the various considerations and options related to cloud-based neuroimaging analyses, including cost models and architectures. In this chapter, you will learn how to run a neuroimaging workflow in order to leverage cloud-computing capabilities. Using data from the AOMIC-PIOP2 project hosted on OpenNeuro, this chapter shows how to use Nextflow to create a very simple skull stripping and tissue segmentation workflow using FSL's *bet* and *fast* programs installed on a local computer. Nextflow allows scalability from a laptop to a cluster to cloud-native services with no code changes.

4.2 Section 2: Best Practices for Working with Large Data

Unlike neuroimaging datasets with ~30 participants, where it could be possible to manually or individually apply processing steps or statistical tests, a key part of working with large neuroimaging datasets involves controlling all steps with code. Therefore, we devote Chapter 4, "Establishing a Reproducible and Sustainable Analysis Workflow" (Ramduny et al.), to best practices for producing reproducible pipelines. In Chapter 4, you will also learn about FAIR principles (data should be Findable, Accessible, Interoperable, and Reusable). The BIDS (Brain Imaging Data Structure) format, which provides a common structure for data organization, is an extremely important tool for working with neuroimaging data (including M/EEG). Chapter 4 contains a worked example with Docker, using fMRIprep on an open-access dataset. There is also a section on working with Python notebooks and invaluable advice on writing sharable and reusable code, including commenting and debugging tips. Finally, perhaps the most impactful advice in Ramduny et al. is to write code and process data in the most efficient way possible to minimize the energy burden.

Continuing the emphasis on efficient code management, Chapter 5—"Optimising Your Reproducible Neuroimaging Workflow with Git" (Garcia and Kelly)—demonstrates the use of Git, which is a very important tool for collaboration and scalable neuroimaging. In Chapter 5, you will learn via a worked example of a cluster analysis on open-access data: version control, branching (especially useful when collaborating), and conflict resolution. Garcia and Kelly also describe how to use GitHub, and again the benefits of collaboration are outlined.

In addition to very large, centrally coordinated studies (e.g., ABCD), data aggregation is an efficient way to build large datasets. "Mega" analyses can provide insights that are otherwise not afforded by smaller studies. Making data sharable and standardized is therefore crucially important, and we include chapters here relevant to both EEG and functional MRI (fMRI). With respect to EEG, in Chapter 6—"End-to-End Processing of M/EEG Data with BIDS, HED, and EEGLAB"—Trong and colleagues introduce a combined BIDS and Hierarchical Event Descriptors (HED) approach that addresses a notable gap in the methods landscaper: namely, a standardized approach to characterize events during time series data. HED is a vocabulary designed to describe experiment events in a structured human-readable and machine-actionable way and HED metadata can enable intelligent combining of event-related data from different recordings and studies. HED can be accessed in several ways: as online tools, Python-based command line scripts and notebooks, and MATLAB scripts and plug-in tools for EEGLAB. Chapter 6 is focused on neuroelectric approaches and, as with all chapters, there is a worked example of end-to-end processing of EEG data using standardized BIDS and HED format to organize and describe information about the dataset. With respect to fMRI, in Chapter 7—"Actionable Event Annotation and Analysis in fMRI: A Practical Guide to Event Handling"—Denissen et al., again building on BIDS, describe tools for efficiently generating event files from experimental logs. These event restructuring tools (remodelers) allow users to modify a dataset's event files by specifying a series of operations in a JSON text file, improving reproducibility, and reducing the need for bespoke coding solutions. An example of using HED remodeling tools is given via a simple analysis of two datasets, working through the required event restructuring.

4.3 Section 3: Preprocessing

Preprocessing of large MRI and M/EEG datasets can be computationally expensive. Here, we describe methods for preprocessing data derived from functional MRI (Chapter 8), from structural MRI data—both gray (Chapter 9) and white matter (Chapter 10)—and from EEG (Chapter 11).

Chapter 8 (Esteban), using NiPreps as a foundation, focuses on the preprocessing stage of neuroimaging pipelines, exploring the rationale, benefits, and potential tradeoffs of standardization. It explores dimensions such as standardizing inputs and outputs, and modularization using tools such as NiPreps and NiReports. Emphasis is placed on version control, software engineering practices, and the use of TemplateFlow for standardizing spatial mappings. Challenges in implementation choice are discussed. Finally, Chapter 8 delves into the integration of artificial intelligence, including the importance of developing transparent, interpretable deep-learning models, trained on openly available data.

Voxel-based morphometry (VBM) is a widely used method for structural MRI analysis, quantifying local gray matter volume (GMV) by segmenting whole brain scans into tissue classes. Applying VBM to detect structural brain-behavior associations in moderate-to-large-sized samples faces challenges, with findings prone to overestimation and limited replicability. Chapter 9 (Hoffstaedter, Antonopoulos, and Gaser)—"Structural MRI and Computational

Neuroanatomy"—is a demonstration of the fully automatic processing of a public dataset with CAT12 in a fully reproducible workflow [14]. Chapter 9 emphasizes the importance of methodological transparency, public data sharing, and the availability of analysis code to enhance reproducibility and facilitate replications. A practical demonstration of fully automated processing using CAT12 on a public dataset is presented, showcasing a reproducible workflow.

Chapter 10—(Connaughton et al.) begins with an introduction to the concepts and techniques of diffusion MRI data processing used in the field and a step-by-step guide for processing diffusion imaging data and for generating tractography. Chapter 10 demonstrates the usage of the popular diffusion imaging toolbox, *ExploreDTI* [15]. Working with BIDS-formatted data, Chapter 10 contains advice specific to very large datasets, plus several helpful recommendations (especially for the novice), and identifies common pitfalls. Chapter 10 describes steps that can be taken to reduce processing time through resource optimization, and options are given to find the optimal balance between reconstruction accuracy and processing time.

In Chapter 11 (Sainz-Ballesteros et al.), the ConneEEGtome toolbox is introduced. Although this toolbox also contains methods for between-group comparisons, we included it in this section because users will encounter the preprocessing features first. Relative to MRI, multicentric high-density EEG studies are less common, even though EEG is much more scalable than MRI [16]. EEG presents extra challenges with data harmonization, not least because there are many different hardware configurations and montages. ConneEEGtome offers an elegant and open-access solution to these challenges, including the option of a graphical user interface. Using data in EEG-BIDS format, the authors bring the reader step-by-step through from preprocessing to classification. Notably, an automatic artifact rejection approach based on independent components is included, as is bad channel interpolation. Recommendations are provided for optimizing storage needs.

4.4 Section 4: Toolboxes for Statistical Analysis

In the next section, we introduce several toolboxes for statistical analysis of large datasets. The ConneEEGtome described in Chapter 11 is a toolbox that includes a classifier, with feature selection followed by Gradient Boosting Machines and a feature importance report. Chapter 12 (Hahn et al.) describes the *Brain Predictability toolbox* (BPt), which is a Python-based, cross-platform, toolbox. BPt can run analyses on single personal computers, with the option to scale up to be used in a cloud-computing environment, and has many user-friendly features, with inbuilt safeguards to prevent the many common errors that beginner users make. BPt provides support for several common data preparation steps: data organization, exploratory data visualization, transformations such as k-binning and binarization, automatic outlier detection, information on missing data, and other summary measures. A very useful aspect is the ability to correctly impute missing data (i.e., without data leakage). Recommendations are included for when to tune model hyperparameters. Understanding your model is made easier because feature importances from BPt can be easily visualized through the related Python package. As with the other chapters, a step-by-step example is included.

Chapter 13 (Serin et al.) describes the *NBS-Predict* toolbox, which builds on network-based statistics (NBS) to produce connectome-based predictions. In this way, connected graph components are used as features, thus incorporating the topological structure of features into account. NBS-Predict is a particularly user-friendly tool that allows the user to easily create models via a graphical user interface and enables automatic generation of training and test datasets for cross-validation purposes. There are two worked examples in Chapter 13: a linear regression and a classification.

The Predictive Clinical Neuroscience (PCN) Toolkit is described in Chapter 14 (Rutherford & Marquand), and offers an easy way to apply the powerful tool that is Normative Modeling. With Normative Modeling, the overarching aim is to define a reference range for a certain given structural or functional brain measurement in a certain sample and to create a reference standard. This reference standard can then be used as a comparator for individuals living with neurological or psychiatric conditions. The PCN toolbox is Python based. In the step-by-step tutorial, you will make predictions for a multi-site transfer dataset, derived from open-access data. The PCN toolkit contains a very useful facility to create an "adaptation dataset" to account for confounding variables such as site effects. The PCN toolkit outputs several evaluation metrics, which can be saved for later plotting/interrogation, and advice and recommendations for further post hoc analyses are included.

In Chapter 15 (Boyle and Weng) presents a flexible method, optimized for large datasets, for implementing *connectomic predictive modeling* (a popular approach for predicting phenotypes from fMRI connectivity data). Notably, this chapter includes an option to use leave-site-out validation. A schematic overview is provided plus all code necessary to conduct an analysis: only a beginner level of coding is needed. A helpful schematic of all decision points is included. Recommendations—such as doing global signal regression and for handling missing data—are provided, as are resources for plotting the output and for implementing computational lesions.

Turning to toolboxes that can be used to interrogate structural MRI, in Chapter 16, Moguilner and Ibañez describe in detail their application of *DenseNet* (a convolutional neural network) to MRI images. Notably, these images had not been preprocessed: the ability to utilize raw MR images expands the possibilities for the use of medical images, which are acquired with a variety of sequences from a range of manufacturers and include heterogeneous samples. All of the relevant code is open source, based on MONAI's PyTorch-based tools. A step-by-step example in Google Colab, using structural data from the AOMIC database, is provided including starting parameters for the DenseNet. Code is included for plotting model performance and metrics such as area under the curve of the receiver operating characteristic from the test set are included, as is code for an "occlusion sensitivity" map, which shows the brain regions that contribute to the prediction.

At the end of this book, we have added a list of resources. These include a list of tools described in the book. We also refer the reader to an excellent compilation of ~300 resources by Niso et al. [17]. There is also a list of open-access neuroimaging databases, and we refer the reader to Madan [18] Table 1 for a comprehensive list. The list of resources also includes information on resources for learning coding languages such as Python. There is also a glossary at the back with explanations of technical terms used in this book.

5 Concluding Comments

In summary, as neuroimaging datasets continue to grow in size, the need for standardized methodologies, efficient code management, and scalable data processing becomes increasingly crucial. This book seeks to address these challenges, and it is our goal to equip researchers with the practical knowledge necessary for conducting robust and reproducible analyses of large neuroimaging datasets.

References

1. Szucs D, Ioannidis JPA (2020) Sample size evolution in neuroimaging research: an evaluation of highly-cited studies (1990–2012) and of latest practices (2017–2018) in high-impact journals. NeuroImage 221:117164. https://doi.org/10.1016/j.neuroimage.2020.117164
2. Boddaert N, Barthélémy C, Poline J-B, Samson Y, Brunelle F, Zilbovicius M (2005) Autism: functional brain mapping of exceptional calendar capacity. Br J Psychiatry 187(1):83–86. https://doi.org/10.1192/bjp.187.1.83
3. Marek S, Tervo-Clemmens B, Calabro FJ, Montez DF, Kay BP, Hatoum AS, Donohue MR, Foran W, Miller RL, Hendrickson TJ, Malone SM, Kandala S, Feczko E, Miranda-Dominguez O, Graham AM, Earl EA, Perrone AJ, Cordova M, Doyle O, Moore LA, Conan GM, Uriarte J, Snider K, Lynch BJ, Wilgenbusch JC, Pengo T, Tam A, Chen J, Newbold DJ, Zheng A, Seider NA, Van AN, Metoki A, Chauvin RJ, Laumann TO, Greene DJ, Petersen SE, Garavan H, Thompson WK, Nichols TE, Yeo BTT, Barch DM, Luna B, Fair DA, Dosenbach NUF (2022) Reproducible brain-wide association studies require thousands of individuals. Nature 603(7902): 654–660. https://doi.org/10.1038/s41586-022-04492-9
4. The Alzheimer's Disease Neuroimaging Initiative, EPIGEN Consortium, IMAGEN Consortium, Saguenay Youth Study (SYS) Group, Thompson PM, Stein JL, Medland SE, Hibar DP, Vasquez AA, Renteria ME, Toro R, Jahanshad N, Schumann G, Franke B, Wright MJ, Martin NG, Agartz I, Alda M, Alhusaini S, Almasy L, Almeida J, Alpert K, Andreasen NC, Andreassen OA, Apostolova LG, Appel K, Armstrong NJ, Aribisala B, Bastin ME,

Bauer M, Bearden CE, Bergmann Ø, Binder EB, Blangero J, Bockholt HJ, Bøen E, Bois C, Boomsma DI, Booth T, Bowman IJ, Bralten J, Brouwer RM, Brunner HG, Brohawn DG, Buckner RL, Buitelaar J, Bulayeva K, Bustillo JR, Calhoun VD, Cannon DM, Cantor RM, Carless MA, Caseras X, Cavalleri GL, Chakravarty MM, Chang KD, CRK C, Christoforou A, Cichon S, Clark VP, Conrod P, Coppola G, Crespo-Facorro B, Curran JE, Czisch M, Deary IJ, EJC DG, Den Braber A, Delvecchio G, Depondt C, De Haan L, De Zubicaray GI, Dima D, Dimitrova R, Djurovic S, Dong H, Donohoe G, Duggirala R, Dyer TD, Ehrlich S, Ekman CJ, Elvsåshagen T, Emsell L, Erk S, Espeseth T, Fagerness J, Fears S, Fedko I, Fernández G, Fisher SE, Foroud T, Fox PT, Francks C, Frangou S, Frey EM, Frodl T, Frouin V, Garavan H, Giddaluru S, Glahn DC, Godlewska B, Goldstein RZ, Gollub RL, Grabe HJ, Grimm O, Gruber O, Guadalupe T, Gur RE, Gur RC, Göring HHH, Hagenaars S, Hajek T, Hall GB, Hall J, Hardy J, Hartman CA, Hass J, Hatton SN, Haukvik UK, Hegenscheid K, Heinz A, Hickie IB, Ho B-C, Hoehn D, Hoekstra PJ, Hollinshead M, Holmes AJ, Homuth G, Hoogman M, Hong LE, Hosten N, Hottenga J-J, Hulshoff Pol HE, Hwang KS, Jack CR, Jenkinson M, Johnston C, Jönsson EG, Kahn RS, Kasperaviciute D, Kelly S, Kim S, Kochunov P, Koenders L, Krämer B, JBJ K, Lagopoulos J, Laje G, Landen M, Landman BA, Lauriello J, Lawrie SM, Lee PH, Le Hellard S, Lemaître H, Leonardo CD, Li C, Liberg B, Liewald DC, Liu X, Lopez LM, Loth E, Lourdusamy A, Luciano M, Macciardi F, Machielsen MWJ, MacQueen GM, Malt UF, Mandl R, Manoach DS, Martinot J-L, Matarin M, Mather KA, Mattheisen M, Mattingsdal M, Meyer-Lindenberg A, McDonald C, McIntosh AM, McMahon FJ, McMahon KL, Meisenzahl E, Melle I, Milaneschi Y, Mohnke S, Montgomery GW, Morris DW, Moses EK, Mueller BA, Muñoz Maniega S, Mühleisen TW, Müller-Myhsok B, Mwangi B, Nauck M, Nho K, Nichols TE, Nilsson L-G, Nugent AC, Nyberg L, Olvera RL, Oosterlaan J, Ophoff RA, Pandolfo M, Papalampropoulou-Tsiridou-M, Papmeyer M, Paus T, Pausova Z, Pearlson GD, Penninx BW, Peterson CP, Pfennig A, Phillips M, Pike GB, Poline J-B, Potkin SG, Pütz B, Ramasamy A, Rasmussen J, Rietschel M, Rijpkema M, Risacher SL, Roffman JL, Roiz-Santiañez R, Romanczuk-Seiferth N, Rose EJ, Royle NA, Rujescu D, Ryten M, Sachdev PS, Salami A, Satterthwaite TD, Savitz J, Saykin AJ, Scanlon C, Schmaal L, Schnack HG, Schork AJ, Schulz SC, Schür R, Seidman L, Shen L, Shoemaker JM, Simmons A, Sisodiya SM, Smith C, Smoller JW, Soares JC, Sponheim SR, Sprooten E, Starr JM, Steen VM, Strakowski S, Strike L, Sussmann J, Sämann PG, Teumer A, Toga AW, Tordesillas-Gutierrez D, Trabzuni D, Trost S, Turner J, Van Den Heuvel M, Van Der Wee NJ, Van Eijk K, Van Erp TGM, Van Haren NEM, Van 'T Ent D, Van Tol M-J, Valdés Hernández MC, Veltman DJ, Versace A, Völzke H, Walker R, Walter H, Wang L, Wardlaw JM, Weale ME, Weiner MW, Wen W, Westlye LT, Whalley HC, Whelan CD, White T, Winkler AM, Wittfeld K, Woldehawariat G, Wolf C, Zilles D, Zwiers MP, Thalamuthu A, Schofield PR, Freimer NB, Lawrence NS, Drevets W (2014) The ENIGMA consortium: large-scale collaborative analyses of neuroimaging and genetic data. Brain Imaging Behav 8(2):153–182. https://doi.org/10.1007/s11682-013-9269-5

5. Vernooij MW, De Groot M, Bos D (2016) Population imaging in neuroepidemiology. In: Handbook of clinical neurology. Elsevier, pp 69–90

6. Van Dijk H, Van Wingen G, Denys D, Olbrich S, Van Ruth R, Arns M (2022) The two decades brainclinics research archive for insights in neurophysiology (TDBRAIN) database. Sci Data 9(1):333. https://doi.org/10.1038/s41597-022-01409-z

7. Mascarell Maričić L, Walter H, Rosenthal A, Ripke S, Quinlan EB, Banaschewski T, Barker GJ, Bokde ALW, Bromberg U, Büchel C, Desrivières S, Flor H, Frouin V, Garavan H, Itterman B, Martinot J-L, Martinot M-LP, Nees F, Orfanos DP, Paus T, Poustka L, Hohmann S, Smolka MN, Fröhner JH, Whelan R, Kaminski J, Schumann G, Heinz A, IMAGEN consortium, Albrecht L, Andrew C, Arroyo M, Artiges E, Aydin S, Bach C, Banaschewski T, Barbot A, Barker G, Boddaert N, Bokde A, Bricaud Z, Bromberg U, Bruehl R, Büchel C, Cachia A, Cattrell A, Conrod P, Constant P, Dalley J, Decideur B, Desrivieres S, Fadai T, Flor H, Frouin V, Gallinat J, Garavan H, Briand FG, Gowland P, Heinrichs B, Heinz A, Heym N, Hübner T, Ireland J, Ittermann B, Jia T, Lathrop M, Lanzerath D, Lawrence C, Lemaitre H, Lüdemann K, Macare C, Mallik C, Mangin J-F, Mann K, Martinot J-L, Mennigen E, De Carvahlo FM, Mignon X, Miranda R, Müller K, Nees F, Nymberg C, Paillere M-L, Paus T, Pausova Z, Poline J-B, Poustka L, Rapp M, Robert G, Reuter J, Rietschel M, Ripke S, Robbins T, Rodehacke S, Rogers J, Romanowski A,

Ruggeri B, Schmäl C, Schmidt D, Schneider S, Schumann M, Schubert F, Schwartz Y, Smolka M, Sommer W, Spanagel R, Speiser C, Spranger T, Stedman A, Steiner S, Stephens D, Strache N, Ströhle A, Struve M, Subramaniam N, Topper L, Walter H, Whelan R, Williams S, Yacubian J, Zilbovicius M, Wong CP, Lubbe S, Martinez-Medina L, Fernandes A, Tahmasebi A (2020) The IMAGEN study: a decade of imaging genetics in adolescents. Mol Psychiatry 25(11):2648–2671. https://doi.org/10.1038/s41380-020-0822-5

8. Parra-Rodriguez MA, Prado P, Moguilner S, Herzog RA, Birba A, Santamaría-García HA, Tagliazucchi E, Reyes PA, Cruzat JA, García AM, Whelan R, Lopera F, Ochoa JF, Anghinah R, Ibáñez A (2022) The EuroLaD-EEG consortium: towards a global EEG platform for dementia, for seeking to reduce the regional impact of dementia. Alzheimers Dement 18(S6):e059944. https://doi.org/10.1002/alz.059944
9. Gorgolewski KJ, Poldrack RA (2016) A practical guide for improving transparency and reproducibility in neuroimaging research. PLoS Biol 14(7):e1002506. https://doi.org/10.1371/journal.pbio.1002506
10. Boccia M, Piccardi L, Guariglia P (2015) The meditative mind: a comprehensive meta-analysis of MRI studies. Biomed Res Int 2015:1–11. https://doi.org/10.1155/2015/419808
11. Kochunov P, Jahanshad N, Sprooten E, Nichols TE, Mandl RC, Almasy L, Booth T, Brouwer RM, Curran JE, De Zubicaray GI, Dimitrova R, Duggirala R, Fox PT, Elliot Hong L, Landman BA, Lemaitre H, Lopez LM, Martin NG, McMahon KL, Mitchell BD, Olvera RL, Peterson CP, Starr JM, Sussmann JE, Toga AW, Wardlaw JM, Wright MJ, Wright SN, Bastin ME, McIntosh AM, Boomsma DI, Kahn RS, Den Braber A, De Geus EJC, Deary IJ, Hulshoff Pol HE, Williamson DE, Blangero J, Van'T Ent D, Thompson PM, Glahn DC (2014) Multi-site study of additive genetic effects on fractional anisotropy of cerebral white matter: comparing meta and mega-analytical approaches for data pooling. NeuroImage 95:136–150. https://doi.org/10.1016/j.neuroimage.2014.03.033
12. Markiewicz CJ, Gorgolewski KJ, Feingold F, Blair R, Halchenko YO, Miller E, Hardcastle N, Wexler J, Esteban O, Goncavles M, Jwa A, Poldrack R (2021) The OpenNeuro resource for sharing of neuroscience data. eLife 10:e71774. https://doi.org/10.7554/eLife.71774
13. Delorme A, Truong D, Youn C, Sivagnanam S, Stirm C, Yoshimoto K, Poldrack RA, Majumdar A, Makeig S (2022) NEMAR: an open access data, tools and compute resource operating on neuroelectromagnetic data. Database 2022:baac096. https://doi.org/10.1093/database/baac096
14. Wagner AS, Waite LK, Wierzba M, Hoffstaedter F, Waite AQ, Poldrack B, Eickhoff SB, Hanke M (2022) FAIRly big: a framework for computationally reproducible processing of large-scale data. Sci Data 9(1): 80. https://doi.org/10.1038/s41597-022-01163-2
15. Leemans A, Jeurissen B, Sijbers J, Jones DK (2009) ExploreDTI: a graphical toolbox for processing, analyzing, and visualizing diffusion MR data. p 3537
16. Whelan R, Barbey FM, Cominetti MR, Gillan CM, Rosická AM (2022) Developments in scalable strategies for detecting early markers of cognitive decline. Transl Psychiatry 12(1): 473
17. Niso G, Botvinik-Nezer R, Appelhoff S, De La Vega A, Esteban O, Etzel JA, Finc K, Ganz M, Gau R, Halchenko YO (2022a) Open and reproducible neuroimaging: from study inception to publication. NeuroImage 263:119623
18. Madan CR (2022) Scan once, analyse many: using large open-access neuroimaging datasets to understand the brain. Neuroinformatics 20(1):109–137

Chapter 2

Getting Started, Getting Data

Hervé Lemaître, Christopher R. Madan, Declan Quinn, and Robert Whelan

Abstract

This chapter explores the availability and accessibility of open-access neuroimaging datasets. It describes how to download datasets using command-line tools (e.g., wget, curl), data management tools such as Datalad, Amazon Web Services (i.e., AWS CLI), and graphical user interface options (e.g., CyberDuck). The chapter emphasizes the importance of accessibility and of documentation for improved research reproducibility. After reading this chapter, researchers will be equipped with the knowledge and tools to download large neuroimaging datasets, including those utilized in this book. We also demonstrate how to download data from OpenNeuro for a range of operating systems.

Key words Neuroimaging, Open access, Download, Data management

1 Introduction

There has been a remarkable increase in the availability of neuroimaging datasets through open access on the Internet (*see* Madan, 2022 for a comprehensive overview of these datasets, including their significance and diversity [1]). Open-access data allow researchers to conduct neuroimaging studies on over a thousand subjects without the need for scanning them anew. Moreover, this accessibility promotes research reproducibility by enabling the reanalysis of the same data. Notwithstanding the advantages of open-access data, it is important to consider financial or legal agreement issues before downloading these ostensibly "open-access" data. For instance, should the researchers who initially collected the data be included as authors? Should someone coordinate what projects are in progress, in the event that more than one group are working on the same data, and one group might "scoop" another?

One of the pioneering datasets accessible to researchers was the International Consortium for Brain Mapping (ICBM) dataset, which emerged in the late 1990s as a collaborative effort among multiple research institutions [2]. The field of neuroimaging has

Robert Whelan and Hervé Lemaître (eds.), *Methods for Analyzing Large Neuroimaging Datasets*, Neuromethods, vol. 218, https://doi.org/10.1007/978-1-0716-4260-3_2,

since witnessed substantial growth in initiatives offering open access to data. Some noteworthy examples include the Human Connectome Project (HCP), involving large-scale data collection of many imaging modalities from over 1000 young adults [3], UK Biobank study, aiming to include 100,000 scanned subjects [4], and the ABCD study, which is following more than 10,000 adolescents over 10 years [5].

The nature of the data that can be accessed varies depending on the neuroimaging dataset at hand. For example, data may be either "raw" or "derivative". Raw MRI data require subsequent preprocessing, which can be time intensive (cf. Chapter 16). Conversely, access may be limited to derivative images, eliminating the need for individual preprocessing but also preventing any modification, and limiting control over preprocessing steps. Another distinction concerns individual versus group-level data. Certain platforms provide access to individual subject data, allowing researchers to perform primary analysis at the group level according to their preferences. OpenNeuro (formerly openfMRI) is an example of such a platform [*see* Resources] [6, 7]. On the other hand, some platforms focus on granting access solely to group-level images, facilitating meta-analysis studies. NeuroVault [*see* Resources] is an online platform specifically designed as a repository for sharing, visualizing, and analyzing statistical maps derived from an extensive collection of neuroimaging studies [8].

In this chapter, we will explore various solutions for downloading such datasets, using the AOMIC dataset stored on OpenNeuro as an illustrative example [9]. The complete AOMIC dataset, including all derivatives, occupies approximately 408 GB of storage space. These data can be downloaded via a browser (*see* instructions https://openneuro.org/datasets/ds003097/versions/1.2.1/download); however, this is not recommended for larger datasets if the connection is not stable. Therefore, we demonstrate how to download using a robust method, for Windows, macOS, and Unix. By the end of this tutorial, the reader will be equipped to access other available datasets as well (e.g., HCP, https://www.humanconnectome.org; ADNI, https://adni.loni.usc.edu).

This chapter will focus on importing data to your local machine. It is worth noting that the reverse process also exists. For instance, Coinstac [*see* Resources] is a framework and platform that enables computation to be conducted locally on each participant's machine, while the data remains securely stored at its original source [10]. This approach can be viewed as exporting your analysis without the need to import the actual data, thereby addressing concerns related to data privacy, legal restrictions, and data-sharing agreements.

Throughout this tutorial, command lines will be predominantly employed for Unix-based systems (e.g., Linux, macOS), specifically Ubuntu, thus requiring basic familiarity with the Unix operating system. If you are using a different Unix operating system, please ensure the availability and installation of the required tools. For those unfamiliar with the Unix operating system, a good explanation of its structure and key components can be found at https://www.javatpoint.com/unix-operating-system. If you are a Windows or Mac user, you can download and use Ubuntu directly (https://ubuntu.com/desktop) for free, or you can try it without committing to major changes to your PC by using a virtual machine (https://ubuntu.com/tutorials/how-to-run-ubuntu-desktop-on-a-virtual-machine-using-virtualbox#1-overview). The following Unix/Ubuntu sections will use the "shell [*see* Glossary]", or command-line/terminal, to download open-source neuroimaging data files, and a good explainer/tutorial can be found at https://ubuntu.com/tutorials/command-line-for-beginners#1-overview. You will also find in the Annex section the specific DataLad instructions for the different chapters of the book.

2 Cyberduck (Windows, macOS)

If you prefer a graphical user interface (GUI) for your data transfer needs, there are several tools available (e.g., Filezilla, WinSCP). However, for the purpose of this tutorial, we will specifically use Cyberduck, which is compatible with both macOS and Windows operating systems. At the time of writing, there are no freely available file transfer clients with a GUI for Ubuntu that can establish a connection to the OpenNeuro repository.

Cyberduck [*see* Resources] is a popular file transfer client that supports various protocols, including FTP, SFTP, WebDAV, Amazon S3, and more. It provides a GUI that allows users to connect to different servers and transfer files between their local machine and remote servers. Once you have successfully downloaded and installed the suitable version of Cyberduck for your specific operating system, proceed to launch the Cyberduck application. Locate and click on the "Open Connection" option, as illustrated in Fig. 1, and configure the connection settings as follows:

- Select "Amazon S3"
- In the "Server" field, enter: **s3.amazonaws.com**
- In the "Port" field, enter: **443**
- In the "Access Key ID" field, enter: **anonymous**
- In the "More options" panel and in the "path" field, enter: **/openneuro.org/ds003097/**
- Click on "Connect"

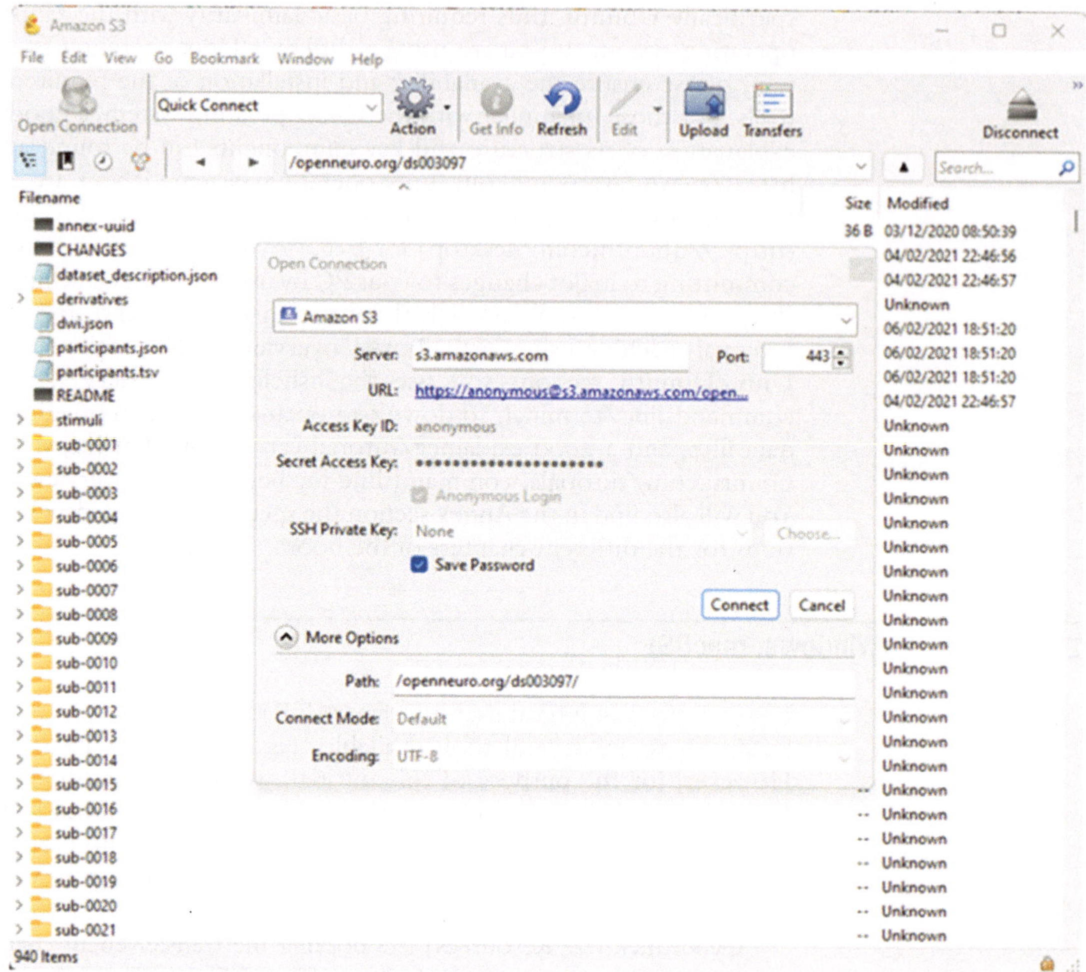

Fig. 1 Connect to the AOMIC dataset using Cyberduck

Following the aforementioned setup, you should now have the capability to navigate through the entirety of the AOMIC dataset and proceed with downloading ID 1000 data.

3 DataLad (Windows, macOS, Unix)

DataLad [*see* Resources] is an open-source data management tool designed to facilitate the management, sharing, and version control of large-scale datasets [11]. The name "DataLad" stands for "Data Lightweight Access and Distribution". It combines the features of data versioning systems, such as Git [*see* Chapter 5], with data distribution capabilities, making it easier to track changes, collaborate on datasets, and ensure reproducibility in scientific research and data analysis workflows.

First, you need to install DataLad on your system (https://www.datalad.org/#install).

In Ubuntu:

```
# Install Datalad
sudo apt-get install datalad
```

Notes

- *For Unix, DataLad can also be installed using pip or conda if you are more familiar with these tools.*
- *For Windows, ensure you have Git and Python with pip installed for successful download of DataLad.*

Then, you can use DataLad to download the AOMIC dataset:

```
# Install the AOMIC dataset
datalad install https://github.com/OpenNeuroDatasets/
ds003097.git
# Note that this command does not download data per se on your
local system but only the data structure
# Download the entire dataset. The get command will actually
download and store data on your local system
cd ds003097
datalad get.
# get: actually download and store data on your local system
# Download a subpart of the dataset (the raw data for the first
subjects)
cd ds003097
datalad get sub-000*
# You can use the data structure to download any kind of
subpart of the dataset
# example:
# sub-0001/anat
# derivatives/freesurfer/sub-0001
```

4 AWS (Windows, macOS, Unix)

AWS stands for Amazon Web Services [*see* Resources]. It is a comprehensive cloud computing platform provided by Amazon. AWS offers a wide range of cloud services, including computing power, storage, databases, networking, analytics, machine learning, artificial intelligence, security, and more. Some neuroimaging datasets are stored on the Amazon Simple Storage Service (S3) for object storage. The AWS CLI (Amazon Web Services Command Line Interface) is a unified command-line tool that can be used to download such neuroimaging datasets.

First, you need to install AWS CLI on your system (https://aws.amazon.com/cli/).

In Ubuntu:

```
# Install awscli
sudo apt-get install awscli
```

Then, you can use AWS CLI to download the AOMIC dataset:

```
# Download the entire AOMIC dataset
aws s3 sync --no-sign-request s3://openneuro.org/ds003097
ds003097

# Download one subfolder (one subject's raw data)
aws s3 cp --no-sign-request s3://openneuro.org/ds003097/sub-
0053 sub-0053 --recursive
# Select and download several subfolders (the raw data for the
first subjects)
aws s3 ls --no-sign-request s3://openneuro.org/ds003097/ --
recursive | \
awk '$NF ~ /^ds003097\/sub-000/ { print $NF }' | \
xargs -I {} aws s3 sync --no-sign-request s3://openneuro.org/
ds003097/{} {}

# the first aws command lists the files
# the awk command filters the lines that matches the pattern
# the xargs command passes the output to the second aws command
for download
```

Alternatively, S3 storage can also be downloaded from using a Python package, Boto[*see* Resources]. This can be useful when you want to selectively download parts of a large dataset such as from the HCP.

5 wget and curl (Unix, macOS)

wget is a command-line utility for downloading files from the web. It stands for "web get". *wget* allows you to retrieve files from remote servers using various protocols such as HTTP, HTTPS, and FTP. It is a versatile tool that supports recursive downloading, resuming interrupted downloads, following links on web pages, and downloading multiple files simultaneously.

curl is a command-line tool and a library for transferring data to or from a server using various protocols, including HTTP, HTTPS, FTP, SFTP, and more. The name “curl” stands for “client URL”. With curl, you can send requests to a server and retrieve responses, making it a versatile tool for interacting with web services, downloading files, and performing various network-related tasks.

Tools such as *wget* and *curl* offer the advantage of being readily available on Unix-like operating systems without the need for additional installation. If you wish to download files from the AOMIC dataset on your system using these command lines, you can follow the instructions below:

```
# with wget
wget https://s3.amazonaws.com/openneuro.org/ds003097/sub-0001/anat/sub-0001_run-1_T1w.nii.gz
# with curl
curl -O https://s3.amazonaws.com/openneuro.org/ds003097/sub-0001/anat/sub-0001_run-1_T1w.nii.gz
# -O: saves the downloaded file with the same name as the original file.
```

The method described above only allows for downloading one file at a time, which is not convenient when attempting to download an entire dataset. However, the OpenNeuro website offers a script specifically designed for downloading the complete AOMIC dataset (https://openneuro.org/datasets/ds003097/versions/1.2.1/download). This script navigates through all the files using the curl command.

If you have direct access to the remote directory [*see* Glossary], an alternative option is to employ the wget command for recursive downloads (i.e., to download everything in a folder, including files in subfolders), as curl does not support this functionality.

```
# with wget
wget -r -np http://WEBSITE/DIRECTORY
# -r: enabled recursive retrieval
# -np: avoids ascending to the parent directory when downloading recursively.
```

Note
OpenNeuro does not allow recursive access.

6 Conclusion

As demonstrated in this chapter, there are various options available for downloading your dataset to your local machine. It is recommended to choose the method that aligns with your operating system and personal experience with either graphical user interfaces (GUIs) or command-line interfaces. It is worth noting that while GUIs generally provide a more user-friendly experience, command-line interfaces offer greater automation potential through scripting. This aspect becomes particularly significant if you intend to re-download the same dataset and document the complete analysis process for your research.

Annexes

Using Datalad, the subsequent instructions facilitate the retrieval of data for the following chapters to which they are applicable:

Chapter 4: Establishing a Reproducible and Sustainable Analysis Workflow

```
# Install the AOMIC dataset
datalad install https://github.com/OpenNeuroDatasets/
ds002790.git
# Download the necessary files
cd ds002790
datalad get sub-0001 sub-0002
```

Chapter 5: Optimizing Your Reproducible Neuroimaging Workflow with Git

```
# Install the AOMIC dataset
datalad install https://github.com/OpenNeuroDatasets/
ds002790.git
# Download the necessary filescd ds002790datalad get deriva-
tives/fs_stats/data-cortical_type-aparc_measure-area_hemi-lh.
tsv
```

Chapter 6: End-to-End Processing of M/EEG Data with BIDS, HED, and EEGLAB

```
# Install the osf extension for datalad
pip install datalad-osf
# setting up OSF credential as a token (https://osf.io/
settings/tokens)
datalad osf-credentials
# Install the OSF repository
datalad install osf://p43rq/
# Download the necessary files
cd p43rq
datalad get *
```

Chapter 7: Actionable Event Annotation and Analysis in fMRI: A Practical Guide to Event Handling

```
# Install the osf extension for datalad
pip install datalad-osf
# setting up OSF credential as a token (https://osf.io/settings/tokens)
datalad osf-credentials
# Install the OSF repository
datalad install osf://u5w4j/
# Download the necessary files
cd u5w4j
datalad get *
```

Chapter 8: Standardized Preprocessing in Neuroimaging: Enhancing Reliability and Reproducibility

```
# Install the AOMIC dataset
datalad install https://github.com/OpenNeuroDatasets/ds002790.git
# Download the necessary files
cd ds002790
datalad get sub-0021
```

Chapter 9: Structural MRI and Computational Anatomy

```
# Clone the AOMIC dataset
datalad clone https://github.com/OpenNeuroDatasets/ds002790.git AOMIC-PIOP2
# Download the necessary files
datalad get -d AOMIC-PIOP2 AOMIC-PIOP2/sub-0111/anat/sub-0111_T1w.nii.gz
# Create an outputs directory and copy the T1w file there
mkdir -p CAT12_derivatives/TEST_sub-0111
cp AOMIC-PIOP2/sub-0111/anat/sub-0111_T1w.nii.gz CAT12_derivatives/TEST_sub-0111/
# delete/drop the local version of the file as we can get it back anytime
datalad drop --what filecontent --reckless kill -d AOMIC-PIOP2 AOMIC-PIOP2/sub-0111
```

Chapter 10: Diffusion MRI Data Processing and Analysis: A Practical Guide with ExploreDTI

```
# Install the AOMIC dataset
datalad install https://github.com/OpenNeuroDatasets/ds002790.git
# Download the necessary files
cd ds002790
datalad get sub-*/dwi/
# NICAP data are only accessible through their website
```

Chapter 13: NBS-Predict: An Easy-to-Use Toolbox for Connectome-Based Machine Learning

```
# Install the AOMIC dataset
datalad install https://github.com/eminSerin/NBSPredict_-SpringerNature.git
# Download the necessary files
cd NBSPredict_SpringerNature
datalad get *
```

Chapter 14: Normative Modeling with the Predictive Clinical Neuroscience Toolkit (PCNtoolkit)

```
# Install the braincharts data
datalad install https://github.com/predictive-clinical-neuroscience/braincharts.git
# Download the necessary files
cd braincharts
datalad get *
```

Chapter 15: Studying the Connectome at a Large Scale

```
# Install the AOMIC dataset
datalad install https://github.com/eminSerin/NBSPredict_-SpringerNature.git
# Download the necessary files
cd NBSPredict_SpringerNature
datalad get *
```

Chapter 16: Deep Learning Classification Based on Raw MRI Images

```
# Install the AOMIC dataset
datalad install https://github.com/OpenNeuroDatasets/ds003097.git
# Download the necessary files
cd ds003097
datalad get participants.tvs
datalad get sub-*/anat/
```

References

1. Madan CR (2022) Scan once, analyse many: using large open-access neuroimaging datasets to understand the brain. Neuroinformatics 20: 109–137. https://doi.org/10.1007/s12021-021-09519-6
2. Mazziotta JC, Woods R, Iacoboni M, Sicotte N, Yaden K, Tran M, Bean C, Kaplan J, Toga AW, Members of the International Consortium for Brain Mapping (ICBM) (2009) The myth of the normal, average human brain--the ICBM experience: (1) subject screening and eligibility. NeuroImage 44: 914–922. https://doi.org/10.1016/j.neuroimage.2008.07.062
3. Van Essen DC, Smith SM, Barch DM, TEJ B, Yacoub E, Ugurbil K, WU-Minn HCP Consortium (2013) The WU-Minn Human Connectome Project: an overview. NeuroImage 80: 62–79. https://doi.org/10.1016/j.neuroimage.2013.05.041
4. Miller KL, Alfaro-Almagro F, Bangerter NK, Thomas DL, Yacoub E, Xu J, Bartsch AJ, Jbabdi S, Sotiropoulos SN, Andersson JLR, Griffanti L, Douaud G, Okell TW, Weale P, Dragonu I, Garratt S, Hudson S, Collins R, Jenkinson M, Matthews PM, Smith SM (2016) Multimodal population brain imaging in the UK Biobank prospective epidemiological

study. Nat Neurosci 19:1523–1536. https://doi.org/10.1038/nn.4393

5. Casey BJ, Cannonier T, Conley MI, Cohen AO, Barch DM, Heitzeg MM, Soules ME, Teslovich T, Dellarco DV, Garavan H, Orr CA, Wager TD, Banich MT, Speer NK, Sutherland MT, Riedel MC, Dick AS, Bjork JM, Thomas KM, Chaarani B, Mejia MH, Hagler DJ, Daniela Cornejo M, Sicat CS, Harms MP, Dosenbach NUF, Rosenberg M, Earl E, Bartsch H, Watts R, Polimeni JR, Kuperman JM, Fair DA, Dale AM, Imaging Acquisition Workgroup ABCD (2018) The adolescent brain cognitive development (ABCD) study: imaging acquisition across 21 sites. Dev Cogn Neurosci 32:43–54. https://doi.org/10.1016/j.dcn.2018.03.001
6. Poldrack RA, Gorgolewski KJ (2017) OpenfMRI: open sharing of task fMRI data. NeuroImage 144:259–261. https://doi.org/10.1016/j.neuroimage.2015.05.073
7. Markiewicz CJ, Gorgolewski KJ, Feingold F, Blair R, Halchenko YO, Miller E, Hardcastle N, Wexler J, Esteban O, Goncavles M, Jwa A, Poldrack R (2021) The OpenNeuro resource for sharing of neuroscience data. eLife 10:e71774. https://doi.org/10.7554/eLife.71774
8. Gorgolewski KJ, Varoquaux G, Rivera G, Schwarz Y, Ghosh SS, Maumet C, Sochat VV, Nichols TE, Poldrack RA, Poline J-B, Yarkoni T, Margulies DS (2015) NeuroVault.org: a web-based repository for collecting and sharing unthresholded statistical maps of the human brain. Front Neuroinformatics 9(8). https://doi.org/10.3389/fninf.2015.00008
9. Snoek L, van der Miesen MM, Beemsterboer T, van der Leij A, Eigenhuis A, Steven Scholte H (2021) The Amsterdam Open MRI Collection, a set of multimodal MRI datasets for individual difference analyses. Sci Data 8:85. https://doi.org/10.1038/s41597-021-00870-6
10. Plis SM, Sarwate AD, Wood D, Dieringer C, Landis D, Reed C, Panta SR, Turner JA, Shoemaker JM, Carter KW, Thompson P, Hutchison K, Calhoun VD (2016) COINSTAC: a privacy enabled model and prototype for leveraging and processing decentralized brain imaging data. Front Neurosci 10:365. https://doi.org/10.3389/fnins.2016.00365
11. Halchenko YO, Meyer K, Poldrack B, Solanky DS, Wagner AS, Gors J, MacFarlane D, Pustina D, Sochat V, Ghosh SS, Mönch C, Markiewicz CJ, Waite L, Shlyakhter I, de la Vega A, Hayashi S, Häusler CO, Poline J-B, Kadelka T, Skytén K, Jarecka D, Kennedy D, Strauss T, Cieslak M, Vavra P, Ioanas H-I, Schneider R, Pflüger M, Haxby JV, Eickhoff SB, Hanke M (2021) DataLad: distributed system for joint management of code, data, and their relationship. J Open Source Softw 6:3262. https://doi.org/10.21105/joss.03262

Chapter 3

Neuroimaging Workflows in the Cloud

Tara Madhyastha

Abstract

Analysis of large neuroimaging datasets requires scalable computing power and storage, plus methods for secure collaboration and for reproducibility. The application of cloud computing can address many of these requirements, providing a very flexible model that is generally far less expensive than a lab trying to purchase the most computer equipment they would ever need. This chapter describes how researchers can change the way that they traditionally run neuroimaging workflows in order to leverage cloud-computing capabilities. It describes various considerations and options related to cloud-based neuroimaging analyses, including cost models and architectures. Next, using data from the AOMIC-PIOP2 project hosted on Open-NEURO, it shows how to use Nextflow to create a very simple skull stripping and tissue segmentation workflow using FSL's *bet* and *fast* programs installed on a local computer. Nextflow allows scalability from a laptop to a cluster to cloud-native services with no code changes.

Key words Neuroimaging, Cloud computing, Parallelization, Virtual machines, Containers

1 Introduction

Several drivers of modern neuroimaging research demand more scalable computing power, more storage, secure collaboration, and reproducibility. Recent papers have highlighted problems with small sample sizes and the need for greater numbers of subjects in studies [1]. At the same time, the Human Connectome Project has demonstrated the importance of higher resolution data both for structural and functional analysis, through gains in alignment and in more precise connectivity analysis [2]. Because it is often difficult to recruit sufficient subjects from specialized populations at a single site, and scanner throughput is limited, these forces necessitate multisite studies and sharing of data among researchers. Finally, reproducibility is of critical importance to the field, as workflows are incredibly complex, affected by subtle differences in operating systems and software package versions, and in our ongoing learnings about how different preprocessing steps may change or bias results [3].

Robert Whelan and Hervé Lemaître (eds.), *Methods for Analyzing Large Neuroimaging Datasets*, Neuromethods, vol. 218, https://doi.org/10.1007/978-1-0716-4260-3_3,

Cloud computing has several characteristics that directly address these drivers. Cloud computing offers virtually unlimited resources on-demand. Using cloud computing makes it possible to address statistical problems not only by scaling to analyze a larger number of subjects and higher resolution data, but by enabling more accurate statistical methods that are infeasible on a desktop or small cluster (such as permutation-based methods for correction of multiple comparisons). Further, cloud computing makes available a wide range of processor types and system architectures. This enables algorithms that can be accelerated through commodity Graphics Processing Units (GPU) [*see* Glossary] or perhaps through Field-Programmable Gate Arrays (FPGAs) [*see* Glossary] to take advantage of this hardware just for the duration of execution, without capital investment.

Storage is also scalable. Object-based storage [*see* Glossary] is an excellent fit for secure storage of raw and preprocessed images with fine-grained access control and auditing capabilities and a global footprint; these capabilities are a foundation for any large-scale data repository.

Finally, a fundamental characteristic of cloud computing is the ability to save computing infrastructure as code. This allows researchers to relaunch not just software pipelines on a new machine, but entire computational environments together with the operating system(s) and application libraries as code. This makes it easy to reproduce computations at scale with relatively low effort and technical knowledge.

These characteristics not only meet the demands of modern neuroimaging research, but they create new possibilities to develop massively parallel algorithms and analytical tools that leverage specialized hardware. However, cloud computing has very different cost models and architectures than the traditional on-premise computing resources that have been shaping the development of neuroimaging workflows. To leverage these cloud capabilities, researchers must change the way that they traditionally run neuroimaging workflows. This chapter describes these differences, and strategies for how researchers can leverage them.

2 Cloud Fundamentals

All cloud computing providers (e.g., Google Cloud Platform, Amazon Web Services, Microsoft Azure, Oracle Cloud Infrastructure, Alibaba Cloud) share some common characteristics and infrastructure services. All of the platforms provide "infrastructure as a service", which means that you can purchase virtual computers, disks, storage, and networking and assemble them to create computing architectures that are similar to computers you can build on-premises (e.g., workstations, high-performance clusters) but with the added benefit that because these are virtual components, you can create and manipulate them programmatically. This is called "infrastructure as code". We describe some features of infrastructure as a service that are generally common across providers, and their implications. However, when working with any specific provider, it is important to understand how the details of their implementation and the features of any cloud software that you use can impact how you work.

2.1 Pay-as-You-Go

A key characteristic of cloud providers is that you pay for the time that you use the infrastructure, rather than purchasing hardware upfront as you would in an on-premise lab (although there are often ways to pay in advance or reserve computing infrastructure to accommodate the bursty nature of grant funding). There are benefits and risks to this payment model. The strongest benefit is that it means that researchers have access to vast or specialized computing resources for short durations of time. This is a very flexible model and is generally far less expensive than a lab trying to purchase the most computer equipment they would ever need. Researchers can use the computers they need as their demands change and problems take them in new directions. The risk is that there is often no link between computing and research funding, so it is easy to run out of money if you cannot easily track and bound your spend. An e-commerce website that scales up to meet holiday demand spends more money on infrastructure in proportion to additional sales; however, a researcher who is able to obtain results faster on a large cluster does not necessarily obtain proportionately more funding. Thus, a critical implication of the pay-as-you-go cost model is that it rewards efficiency, a point we will discuss in terms of implications to neuroimaging workflows.

2.2 Computing (Virtual Machines)

Computing time is available in many forms from cloud providers. A basic characteristic is that it is possible to provision many different configurations of virtual computers, with different memory, CPU (Central Processing Unit) architectures and cores, and local storage footprints. The cost of virtual computers is closely related to the memory, storage, and compute resources that they have. Therefore, while there is no penalty to running a code on a dedicated on-premise workstation that only uses one core out of 32, to do so on a virtual cloud computer would be a waste of money. On the cloud, one selects a virtual computer with just enough resources to run an application efficiently. This can be a difficult optimization problem because it means selecting a virtual computer that can complete a job (perhaps with varying data) at the lowest cost in a reasonable time. A corollary of this is that when designing workflows, from the start, one should separate out components that have significantly different computational demands. Common workflows within popular neuroimaging packages such as Analysis of Functional Neuroimages (AFNI), FreeSurfer, FMRIB Software Library (FSL), and Statistical Parametric Mapping (SPM) are usually scripts that run different programs for specific processes. Some, such as probabilistic tractography, may be highly parallelizable and benefit from GPU acceleration. Others, such as an independent components analysis, may require a large amount of memory. Finally, simple image mathematical or transformational operations may be I/O bound (input/output bound; i.e., the time constraint in the workflow is the time taken to request the data, rather than the time needed to actually process the data). When these types of processes are run in a single script on a virtual computer, the computer will need to have sufficient resources to accommodate the highest demands from any process. From a cloud cost perspective, this is wasteful, because you are (for example) paying for a GPU when it is sitting idle, or paying for additional memory when it is not being used.

Another important concept across cloud providers is that of the "spot" or "preemptable" market. This is when extra cloud capacity is provided at a substantial discount, with some caveat that it can be reclaimed if needed, or after a specific amount of time (depending on provider). When a computer is reclaimed, it shuts down and any work that has not been saved to disk is lost. This is an excellent opportunity to obtain cloud computing resources at a fraction of the regular costs, but to take advantage of this, workflows must be written to save their state periodically (or when given a reclamation warning, if available) and restart from where they have left off.

2.3 Services, Serverless, and Containers

Cloud providers also offer services that are built on top of their infrastructure platform. For example, many business applications require a database, which is relatively complicated to manage. Offering a database as a managed service means that you can take care of such things as replicating and backing up the database and patching the underlying operating system on which it runs. These services incur an additional cost over the actual infrastructure costs but save on human time (from personnel that are often hard to hire) to manage the virtual servers. It is easier to create services that represent common IT or business functions than it is to create services for researchers; many services that seem appropriate for neuroimaging workflows such as Artificial Intelligence (AI), Machine-Learning (ML) services lock one into a particular cloud vendor, may leverage proprietary algorithms, and may not be reproducible. These concepts are not as important to a business analyst who is applying a machine-learning algorithm to their sales data to predict the impact of a specific marketing campaign. The analyst does not worry about sharing or reproducing the code, and if the results are easier to get from a service, that saves time and effort. However, taken to the limit, arbitrary services can run on computing infrastructure without the user having to actually start a virtual machine, secure the operating system, and perpetually make sure it is up to date. This abstraction is called "serverless computing", and it is another important cloud abstraction that also has implications for how one designs neuroimaging workflows.

Containers [*see* Glossary] are an important information technology development that have made strong inroads into neuroimaging, and simultaneously spurred the popularity of serverless computing. A container consists of an entire runtime environment: an application, plus all its dependencies, libraries and other binaries, and configuration files needed to run it, bundled into one package. A container is similar to a virtual machine but is lighter weight—multiple containers can run on a machine and share the underlying operating system. This allows multiple packages to be run efficiently within containers on the same underlying machine, without worrying about differences in operating system distributions or dependencies. Because the details of the underlying compute infrastructure are not critical, one does not need to manage a server simply to run a container. It is sufficient merely to specify what resources (GPU, memory, cores) a container needs to execute and what broad platform (Linux, Windows) and a cloud service can run it without requiring you to provision and maintain the underlying infrastructure. An additional advantage of containers is that—unlike virtual machines—files describing their contents can be written and kept under *version control* [*see* Glossary]. This makes it possible to recreate an environment from scratch to reproduce it, rather than merely interrogating it to find out what is in it.

Containers share a key characteristic with virtual machines: the ability to package up code and dependencies into a self-contained unit. For this reason, containers have been adopted by neuroimaging researchers [4] to encapsulate entire complex workflows bound together using Python code (e.g. Nipype [5]). Two popular containers are FMRIPrep [6] and MRIQC [7] (*see* Chapter 8). The idea of using Python both to connect workflow components that may be different applications and to program novel algorithms or in-line transformations within the same piece of code is appealing and simple. However, a major problem with this design is that different stages in such workflows require different computing resources, and so cannot take advantage of highly efficient cloud-based container services to execute them. In contrast, bioinformatics workflows often consist of smaller discrete containerized applications that are bound together with a dedicated workflow description language. Each step of the workflow can be accompanied by a specification of what resources are needed to run it. These workflow characteristics make it possible to cost-efficiently use many cloud services as well as on-premise resources, and this cross-discipline experience is starting to have an impact on neuroimaging workflows.

2.4 Security and Collaboration

An area where cloud architectures excel is in enabling global collaboration. Instead of copying data from one site to another, it is possible to architect secure research environments where researchers can come to access data and compute on it. Cloud providers can certify that their services adhere to compliance standards (such as Federal Risk and Authorization Management Program, General Data Protection Regulation) and architectures built on these services inherit these controls.

Depending on the sensitivity of the data, it can be controlled in many ways, ranging from restricting or enforcing access and maintaining an audit trail, to limiting exactly what flows in and out of a secure environment. The ability to protect data while maintaining environments where researchers anywhere can collaborate as though they were in the same lab is an enormous strength of cloud computing. Large datasets remain and are secured in one place while researchers come to the data to work on it. In particular, cloud computing enables large-scale collaborative studies and fine-grained control over dissemination of data with different levels of protected health information.

In most infrastructure-as-service platforms, security of data and the infrastructure is up to the user to configure. In a research context, this configuration would typically be designed and provided by research IT staff, leveraging third party products to avoid reinventing infrastructure where sensible.

2.5 Architectural Considerations

Cloud abstractions have been developed to serve business applications, which have different characteristics than research applications. For example, business applications are often critical; if an e-commerce website failed it could result in huge loss of revenue and damage to the company reputation. In contrast, if a research pipeline fails sporadically, it can generally be restarted without serious repercussions. Designing for high availability means creating architectures that build in failover (i.e., switching to a redundant system) to multiple independent data centers. They must also be elastic, so that if a single compute element becomes overwhelmed, it can be replicated to accommodate the bursty load. Often these characteristics are not important in a neuroimaging research context.

Cloud infrastructure components have specific service level agreements: *availability* (can you get to them), *reliability* (do they work correctly), and *durability* (is your data intact). These are normally much higher than what can be provided by on-premise components. For example, the default S3 object storage on AWS will maintain several copies of your data in distinct data centers that are unlikely to fail for the same cause unless there is a large geographic disaster. This will probably be safer than network attached storage in a lab machine room, so your backup plan may look different. It is important to reassess strategies for maintaining availability and durability of your workflows and data in the context of the cloud.

3 Where Traditional Neuroimaging Workflows Fall Short

There are a variety of packages commonly used in neuroimaging, such as AFNI, FSL, FreeSurfer, SPM, Advanced Normalization Tools (ANTs) [8], HCP Informatics Infrastructure [9], that contain programs and graphical user interfaces (GUI) [*see* Glossary] to make it possible to process neuroimaging data. There are many steps involved in taking structural or functional imaging data from raw DICOM (digital imaging and communications in medicine) format to processed images and group results; the details of these steps are not the topic of this paper. However, our main focus in the context of cloud computing is how they are connected together.

There are different schools of thought on the flexibility one should have to "mix and match" different algorithms (embodied in programs) from different packages. Some neuroimaging researchers prefer to stick to workflows created within a package, to make sure that subtle differences in how individual programs work do not cause errors. In this case, the different programs are typically connected using scripts provided by the packages that govern their execution. Another school of thought suggests that some packages are better at some capabilities than others, so they mix and match programs from different packages in a single workflow.

Regardless of which approach is used, one characteristic of neuroimaging workflows is that they can take advantage of parallelism, often at multiple levels. Different brains can be preprocessed at the same time (for example, on different cores within a cluster). This is called *coarse-grained parallelism* [*see* Glossary]. Often analysis of a single image can be parallelized (for example, probabilistic tractography lends itself well to fine-grain parallelism). At the level of course-grained parallelism, several approaches have been used to take advantage of multiple cores or a cluster where available. Several workflows (e.g., many scripts involving multiple tools in FSL, FreeSurfer) can automatically submit independent jobs to a cluster where available. Some tools can readily take advantage of multiple cores where available. Although parallelism may exist under the hood, the general approach is to abstract that from users within a single package so that they do not need to worry about how their code is being parallelized. Figure 1 shows an illustration of how this may be accomplished. To the user, processing appears to be conducted by a single tool that executes on a traditional computer architecture (a single node, or a cluster).

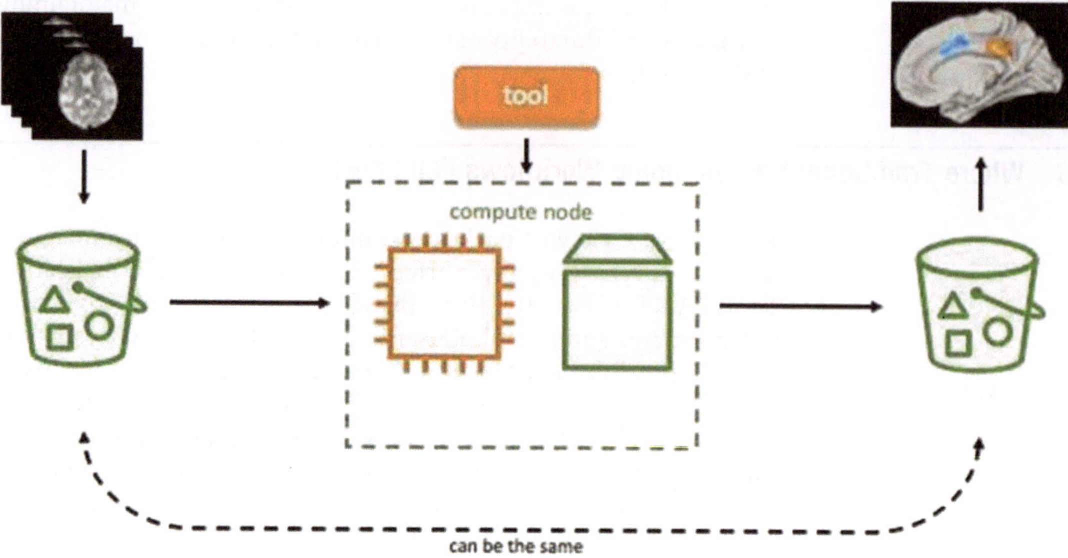

Fig. 1 A single tool approach to cloud-based workflow

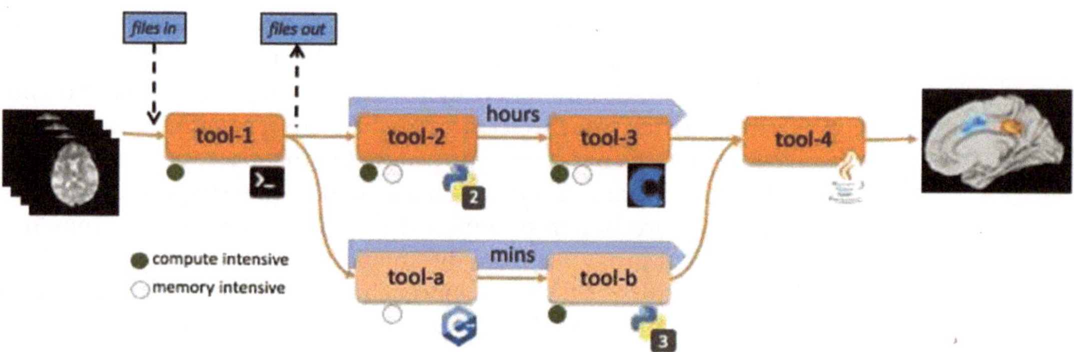

Fig. 2 An example of "mix and match" workflow

Another approach to take advantage of parallelism is used more in the "mix and match" scenario. It is common to express the course-grained parallelism of a workflow as a directed acyclic graph (DAG) [*see* Glossary], with dependencies clearly spelled out. This tells an execution engine which programs need to be run before others. Where dependencies do not exist between programs, they can all be run simultaneously. A dependency graph can also provide information about conditions for successful step completion, so if the workflow is interrupted, it can be restarted without replication of work. A typical sequential programming script does not express this dependency information, and so is an inefficient way to structure a workflow from a performance perspective. Figure 2 shows an example of a "mix and match" workflow. Each tool has different software and hardware requirements, and some steps take minutes and some take hours. However, the dependencies between tools and their requirements (even if they are all from the same neuroimaging package and designed to work together) is articulated so that they can automatically be parallelized.

Two programs that have been used to describe parallel workflows are GNU Make [10] and Nipype [5] [*see* Resources]. GNU Make has the advantage that it is a robust and relatively simple tool and can be used to connect scripts written in multiple different languages. Its main disadvantage is that it is well grounded in traditional UNIX conventions (such as files, standard input and output), and has its own unique syntax. For this reason, it is difficult to describe dependencies on very complex output directory structures, and researchers need to learn Make syntax. Nipype has gained a lot of traction in the neuroimaging community as a Python-based library that can be used to express dependencies among neuroimaging workflows, and is used by several other packages. It has the strong advantage that researchers can work exclusively in Python, and the Nipype wrappers handle the complexity of neuroimaging outputs. The disadvantage is that components need to be wrapped and as an open-source project, changes can cause problems with backward compatibility.

Both GNU Make and Nipype approaches fail to leverage modern aspects of cloud computing. To take full advantage of serverless and elastic computing, we would want to be able to separate the compute needs of each component that makes up a workflow, and express dependencies among them. We would want to be able to execute containers as well as code. To take advantage of cloud storage we would want to be able to natively indicate that our input and output data live on object storage in the cloud. And to avoid lock-in to a specific cloud platform or architecture, we would want to be able to run the same workflow on on-premise architectures and multiple cloud platforms.

4 Futureproofing Neuroimaging Workflows

To leverage the capabilities of cloud computing in neuroimaging workflows, enabling scaling to larger, more compute-intensive datasets, we need to rethink how to write workflows. Below are some basic principles and how to implement them.

4.1 Portability and Reproducibility

Researchers will have access to different computing resources at different points in their careers, and so must not be locked into a specific cloud platform or architecture. Moreover, their colleagues, who may need to be able to reproduce their work, cannot be expected to have access to the same platforms or architectures.

Note

To improve portability and reliability of neuroimaging workflows. First, avoid using proprietary algorithms embedded in cloud services (e.g., AI/ML services) that impact research results. At best, these can lock researchers into a specific vendor that they or their colleagues lose access to in the future. At worst, services can change or be retired, making it impossible to reproduce results. Second, stick to the "lowest common denominator" of services when using cloud computing for running workflows. For example, virtual machines, object storage, and container execution platforms are available on all platforms and have on-premise analogs. Failure to do so makes it difficult or impossible to change platforms.

4.2 Workflow and Serverless

To take advantage of serverless cloud computing and spot market offerings, it is critical to structure your applications from the start to be parallelizable, architecture-aware, and fault tolerant.

1. Separate time-consuming and resource-intensive components from other parts of the workflow. Distribute these components as containers so that they can be run unmodified on cloud services.
2. Create workflows from components within the same package or across packages by using a cloud-native workflow description language that can be easily ported to on-premises clusters and cloud-native architectures across multiple platforms. As cloud computing grows more prevalent, more software for workflows (e.g., Nextflow, snakemake [*see* Resources]) is written to make code portable across platforms.
3. Avoid loops to process multiple subjects, and instead use a workflow description language, parallel job submission, or multiple cores (e.g., via GNU parallel [11]) to run them.
4. Write long-running applications so that they checkpoint their work and can resume upon interruption. Choose a workflow description language that permits resuming a workflow after it has been interrupted or when a step needs to be modified.

4.3 Data Management

Storage options in the cloud are incredibly powerful but form a model that is more complex than most on-premises storage systems and cost models. Cloud object storage is scalable and highly reliable. However, pricing for object storage is typically based on the size of the data stored, the storage tier (how readily accessible and/or available is the data), and access charges. There are also potentially charges for data egress from the cloud or between regions. Entire machines can be saved along with everything necessary to create an analysis. Finally, object storage is not suitable as a file system; to use data on object storage one needs to stage it to a file system. To use these features cost-effectively requires thinking about data management at the start of an analysis.

1. Save virtual machines, containers, code and data together at the end of a completed analysis so that you can reproduce the analysis.
2. Use directory and file naming conventions consistently so that you can create automatic rules for creating versions of objects for backup, deleting versions of objects, moving objects to less expensive tiers of storage, or archiving objects.
3. Automatically migrate important objects to lower-cost archival storage when appropriate.
4. Save data products for completed analyses when the cost to store them for an appropriate timeframe is less than the cost to reproduce them, and delete them otherwise.
5. Write workflows to copy data from object storage to file system storage and write back results to minimize the size of a working disk.

5 Step-by-Step Example: Nextflow and AWS

Nextflow is a workflow description language that has these characteristics and is the basis of Tractoflow [12] and several other workflows from the Sherbrooke Connectivity Imaging lab (https://scil.usherbrooke.ca). Nextflow is well-established in bioinformatics workflows, which share a lot in common with neuroimaging workflows, but because of larger data sizes and sharing requirements have migrated to the cloud earlier. Key characteristics of Nextflow are the ability to take advantage of cloud native storage and batch computing services to execute containers, an extremely flexible language to describe expected inputs and outputs, and the ability to configure multiple engines. A language such as Nextflow allows scalability from a laptop to a cluster to cloud native services with no code changes. Other bioinformatics workflow systems such as snakemake [13] and WDL/CWL [14] share similar characteristics with Nextflow, and there is some effort to introduce these into the neuroimaging community. AWS is a major cloud provider, and may be familiar to neuroimagers because data from the Human Connectome Project and OpenNeuro are stored on AWS S3 object storage.

In this example, we use data from the AOMIC-PIOP2 project [15] hosted on OpenNEURO (https://openneuro.org/datasets/ds002790) and Nextflow to create a very simple skull stripping and tissue segmentation workflow using FSL's bet and fast programs installed on a local computer. We will stage data from where it is stored on AWS S3 object storage.

To follow along with this example, you will need a Linux or MacOS terminal environment (commands assume bash), and you will need to have FSL installed (if you do not use FSL, feel free to substitute any other simple neuroimaging commands that you prefer).

5.1 Installing Nextflow

Installation of Nextflow is very simple (see directions for most recent information). You must have a recent version of Java installed, which you can check by typing:

```
java -version
```

If you do not have Java installed then click here (https://www.java.com/en/download/help/download_options.html) for more installation instructions.

Then, install the Nextflow software:

```
curl -s https://get.nextflow.io | bash
```

This will create the executable program called nextflow in your current working directory. You can test that this program works by running a canned workflow.

```
./nextflow run hello
```

If everything works, move the nextflow program to your ~/bin directory and add this directory to your path in your .bashrc so it will be there every time you log in:

```
mkdir -p ~/bin
mv nextflow ~/bin
cat << EOF >> ~/.bashrc
export PATH=$PATH:~/bin/nextflow
EOF
source ~/.bashrc
```

Note

Create a configuration file. The files in the AOMIC-PIOP2 repository are on S3, and require no specific permissions, but if you have not configured your environment with valid AWS credentials, you will obtain an error when Nextflow attempts

(continued)

to stage the S3 files locally. To get around this, you can create a file called nextflow.config with the following contents.

```
aws {
 client {
 anonymous='true'
 }
 }
```

This tells Nextflow that the request will not be authenticated, so you do not need any credentials.

5.2 Create a Small Workflow

Add the following code in a text file: *script.nf*.

```
#!/usr/bin/env nextflow

nextflow.enable.dsl=1

t1 = Channel.of(["0001_bet.nii.gz", "s3://openneuro.org/ds002790/sub-0001/anat/sub-0001_T1w.nii.gz"], ["0002_bet.nii.gz", "s3://openneuro.org/ds002790/sub-0002/anat/sub-0002_T1w.nii.gz"])

/* perform skull stripping */
process skull_strip {
 input:
 tuple val(bet), path(t1) from t1

 output:
 path(bet) into betout
"""
/home/ubuntu/fsl/bin/bet $t1 $bet
"""
}

process fast {
 input:
 path bet from betout

 output:
 path '*_bet_*' into fastout

 """
```

```
  /home/ubuntu/fsl/bin/fast $bet
  """
}

fastout
.flatMap()
 .subscribe{ println "File: ${it.name}" }
```

This workflow has two processes. The first, skull_strip executes the bet skull stripping command, and the second executes the fast tissue segmentation command. The execution order is determined by the inputs and outputs. The skull_strip process takes inputs from a set of tuples (anatomical T1-weighted files on S3, and the friendly name we would like to give to the skull stripped images). The second, fast, executes the fast tissue parcellation command using the output from skull_strip.

We have specified multiple files as input. Unlike a script, each process runs independently and in a separate working directory. This enables us to run all the processes as quickly as possible—as soon as prerequisites have completed—without worrying about files with the same name being overwritten. Nevertheless, here we show how we can pass in friendly file names to help keep our outputs straight.

5.3 Run the Workflow

To run the workflow, type:

```
nextflow run script.nf
```

This command will run the workflow locally, which is great for testing, and publish the output in the directory work. Note that each process is stored separately, so file names do not conflict with each other. There are many settings you can use to move output to other directories.

5.4 To Infinity: Going Cloud Native

A workflow system such as Nextflow covers the basics of portable neuroimaging workflow that can scale. By defining processes separately from each other and clearly specifying the parallel structure of the steps and the input data, you now have the capability to run at scale and use scalable object storage effectively. This is the most important precursor to neuroimaging in the cloud. When resources are limited, there is little to be gained by structuring your workflow in this way; a script that cannot exploit parallel computing will be slow if there is no extra capacity to be had. This will allow you to effectively use even an autoscaling cluster in the cloud, or a multicore server. To leverage serverless computing and optimize your use of resources further, you will need to replace each process with a container and describe the compute resources necessary to run each step. With this work done, you can move to a serverless container-based platform.

References

1. Marek S, Tervo-Clemmens B, Calabro FJ, Montez DF, Kay BP, Hatoum AS, Donohue MR, Foran W, Miller RL, Hendrickson TJ, Malone SM, Kandala S, Feczko E, Miranda-Dominguez O, Graham AM, Earl EA, Perrone AJ, Cordova M, Doyle O, Moore LA, Conan GM, Uriarte J, Snider K, Lynch BJ, Wilgenbusch JC, Pengo T, Tam A, Chen J, Newbold DJ, Zheng A, Seider NA, Van AN, Metoki A, Chauvin RJ, Laumann TO, Greene DJ, Petersen SE, Garavan H, Thompson WK, Nichols TE, Yeo BTT, Barch DM, Luna B, Fair DA, Dosenbach NUF (2022) Reproducible brain-wide association studies require thousands of individuals. Nature 603:654–660. https://doi.org/10.1038/s41586-022-04492-9
2. Elam JS, Glasser MF, Harms MP, Sotiropoulos SN, Andersson JLR, Burgess GC, Curtiss SW, Oostenveld R, Larson-Prior LJ, Schoffelen J-M, Hodge MR, Cler EA, Marcus DM, Barch DM, Yacoub E, Smith SM, Ugurbil K, Van Essen DC (2021) The human connectome project: a retrospective. NeuroImage 244:118543. https://doi.org/10.1016/j.neuroimage.2021.118543
3. Poldrack RA, Baker CI, Durnez J, Gorgolewski KJ, Matthews PM, Munafò MR, Nichols TE, Poline J-B, Vul E, Yarkoni T (2017) Scanning the horizon: towards transparent and reproducible neuroimaging research. Nat Rev Neurosci 18:115–126. https://doi.org/10.1038/nrn.2016.167
4. Gorgolewski KJ, Auer T, Calhoun VD, Craddock RC, Das S, Duff EP, Flandin G, Ghosh SS, Glatard T, Halchenko YO, Handwerker DA, Hanke M, Keator D, Li X, Michael Z, Maumet C, Nichols BN, Nichols TE, Pellman J, Poline J-B, Rokem A, Schaefer G, Sochat V, Triplett W, Turner JA, Varoquaux G, Poldrack RA (2016) The brain imaging data structure, a format for organizing and describing outputs of neuroimaging experiments. Sci Data 3:160044. https://doi.org/10.1038/sdata.2016.44
5. Gorgolewski K, Burns CD, Madison C, Clark D, Halchenko YO, Waskom ML, Ghosh SS (2011) Nipype: a flexible, lightweight and extensible neuroimaging data processing framework in python. Front Neuroinformatics 5. https://doi.org/10.3389/fninf.2011.00013
6. Esteban O, Markiewicz CJ, Blair RW, Moodie CA, Isik AI, Erramuzpe A, Kent JD, Goncalves M, DuPre E, Snyder M, Oya H, Ghosh SS, Wright J, Durnez J, Poldrack RA, Gorgolewski KJ (2019) fMRIPrep: a robust preprocessing pipeline for functional MRI. Nat Methods 16:111–116. https://doi.org/10.1038/s41592-018-0235-4
7. Esteban O, Birman D, Schaer M, Koyejo OO, Poldrack RA, Gorgolewski KJ (2017) MRIQC: advancing the automatic prediction of image quality in MRI from unseen sites. PLOS ONE 12:e0184661. https://doi.org/10.1371/journal.pone.0184661
8. Avants B, Tustison NJ, Song G (2009) Advanced normalization tools: V1.0. Insight J. https://doi.org/10.54294/uvnhin
9. Marcus D, Harwell J, Olsen T, Hodge M, Glasser M, Prior F, Jenkinson M, Laumann T, Curtiss S, Van Essen D (2011) Informatics and data mining tools and strategies for the human connectome project. Front Neuroinformatics 5. https://doi.org/10.3389/fninf.2011.00004
10. Askren MK, McAllister-Day TK, Koh N, Mestre Z, Dines JN, Korman BA, Melhorn SJ, Peterson DJ, Peverill M, Qin X, Rane SD, Reilly MA, Reiter MA, Sambrook KA, Woelfer KA, Grabowski TJ, Madhyastha TM (2016) Using make for reproducible and parallel neuroimaging workflow and quality-assurance. Front Neuroinform 10:2. https://doi.org/10.3389/fninf.2016.00002
11. Tange O (2018) GNU parallel 2018. Ole Tange
12. Theaud G, Houde J-C, Boré A, Rheault F, Morency F, Descoteaux M (2020) TractoFlow: a robust, efficient and reproducible diffusion MRI pipeline leveraging Nextflow & Singularity. NeuroImage 218:116889. https://doi.org/10.1016/j.neuroimage.2020.116889
13. Mölder F, Jablonski KP, Letcher B, Hall MB, Tomkins-Tinch CH, Sochat V, Forster J, Lee S, Twardziok SO, Kanitz A, Wilm A, Holtgrewe M, Rahmann S, Nahnsen S, Köster J (2021) Sustainable data analysis with Snakemake. F1000Research 10:33. https://doi.org/10.12688/f1000research.29032.2
14. Amstutz P, Crusoe MR, Tijanić N, Chapman B, Chilton J, Heuer M, Kartashov A, Leehr D, Ménager H, Nedeljkovich M, Scales M, Soiland-Reyes S, Stojanovic L (2016) Common workflow language, v1.0. 5921760 Bytes
15. Snoek L, van der Miesen MM, Beemsterboer T, van der Leij A, Eigenhuis A, Steven Scholte H (2021) The Amsterdam Open MRI Collection, a set of multimodal MRI datasets for individual difference analyses. Sci Data 8:85. https://doi.org/10.1038/s41597-021-00870-6

Chapter 4

Establishing a Reproducible and Sustainable Analysis Workflow

Jivesh Ramduny, Mélanie Garcia, and Clare Kelly

Abstract

Getting started on any project is often the hardest thing—and when it comes to starting your career in research, just figuring out *where* and *how* to start can seem like an insurmountable challenge. This is particularly true at this moment—when there are so many programming languages, programs, and systems that are freely available to neuroimaging researchers, and even more guides, tutorials, and courses on how to use them. This chapter is intended to set you off on the right foot as you get stuck into the task of learning to work with large neuroimaging data. We will cover a number of processes, systems, and practices that you should adopt to help ensure that your work is efficient, your processing steps traceable and repeatable, your analyses and findings reproducible, and your data and processing scripts amenable to sharing and open science. While this chapter is aimed at those getting started, it will also be of use to established researchers who want to streamline their processes and maximize robustness and reproducibility of their neuroimaging analyses. Finally, this chapter is also intended to help make neuroimaging work practices and processes more environmentally sustainable by reducing demands on computational resources through better planning, efficiency, and awareness of resource use.

Key words Reproducibility, BIDS, Docker, Python, Sustainability

1 Why Establish a Reproducible Workflow?

In the wake of the "replication crisis" in science [1, 2] reproducibility has become a cornerstone of neuroimaging research. The term *reproducibility* [*see* Glossary] refers to the ability to obtain the same results as a prior study, using procedures that are closely matched with those used in the original research [3]. As a result of this increased emphasis on reproducibility, a whole set of disciplinary norms have been instituted. It is now routinely expected that researchers will share not only their data and derivatives (to the extent that data protection regulations permit) but also their analysis code, so that others may reproduce their findings with the same

Authors Jivesh Ramduny and Mélanie Garcia have equally contributed to this chapter.

Robert Whelan and Hervé Lemaître (eds.), *Methods for Analyzing Large Neuroimaging Datasets*, Neuromethods, vol. 218, https://doi.org/10.1007/978-1-0716-4260-3_4, © The Author(s) 2025

data, or attempt to replicate findings using different data. Disorganized or idiosyncratically named data and esoterically written and uncommented code are of little use to anyone, including yourself, when you inevitably return to a project after a break (e.g., to address peer reviewers' comments). Putting a reproducible workflow in place from the outset will help ensure that your data and code are both reproducible and useful to yourself and to other researchers.

The practices we outline in this chapter may seem like a considerable investment at first—particularly when students are also just beginning to learn about their research topic itself. There can be an understandable urge to "just get stuck in", but, we can confirm—from personal experience—that investments made now will reap many benefits in the future. In contrast, cutting corners now may lead to heartache (and extra work) down the line. Not only will other researchers thank you for adopting these practices—but future "you" will also appreciate it!

Note
Time invested now in learning reproducible research practices will reap benefits throughout your career. Believe it or not, you'll never have more time to learn than as a PhD student!

1.1 FAIR Principles

Based on the recognition that data reuse is central not only to reproducibility but also to maximizing the value of research, a set of best-practice guidelines have been developed to maximize the usability of such data. These are known as the FAIR principles [*see* Resources]. The FAIR principles [*see* Glossary] prescribe characteristics of data and digital objects to maximize their reuse by the scientific community—that is, to maximize data sharing, exploration, reuse and deposition by parties other than the original researcher [4]. The application of the FAIR principles for neuroimaging data has been extensively and accessibly documented by ReproNIM [*see* Resources]—we recommend that you take time to explore their excellent module on Data and FAIR Principles. In brief, the FAIR principles require that data are:

- **F**indable: Data and supplementary materials have sufficiently rich metadata [*see* Glossary] and a unique and persistent identifier (PID).
- **A**ccessible: Data are deposited in a trusted and accessible repository [*see* Glossary]. Both the data and metadata are accessible and downloadable via platforms such as OpenNeuro, Open Science Framework, github, amongst others [*see* Resources].

- **I**nteroperable: Data and metadata use a formal, agreed-upon and shared language or format such as the BIDS standard (explained in more detail below).
- **R**eusable: Data are described with clear and understandable attributes, and there should be a clear and acceptable license for reuse (e.g., CC0 public domain).

In the rest of this chapter, we outline how you can build reproducibility into your workflow. Our guide is intended to get you started, rather than to be exhaustive, so we also recommend that you also build on these basics by exploring other guides [5, 6] and resources (e.g., ReproNIM). Much of what we outline is simply good practice for keeping your data organized and maximally reusable for yourself and your collaborators, but it will become very important when you reach the point of publishing your study. If you wish to share your data using a public platform such as OpenNeuro (increasingly the norm for the field), then you must curate your dataset to comply with the FAIR principles and the BIDS standard. Next, we'll take a look at what this means.

2 Working with the BIDS Ecosystem

BIDS stands *for Brain Imaging Data Structure* [*see* Glossary]. It is a standard for the organization of neuroimaging datasets that follows FAIR principles and facilitates both data reuse and automated processing by open science data analysis pipelines [5]. The BIDS standard specifies machine-readable directory structure, filenames, file formats, and metadata for various neuroimaging modalities. The need for these standards arose as a result of increasing demand for data sharing in the field, and the difficulties created when such data are idiosyncratically named, organized, and formatted. The BIDS specifications, initially developed by those working with the OpenfMRI (now OpenNeuro) data repository [*see* Resources], are consensus-based and community-driven, and leverage existing conventions in the neuroimaging community (Fig. 1). They emphasize **simplicity**, **readability**, and **accessibility**. BIDS is now widely accepted as the standard for the field and includes specifications for modalities beyond MRI, including EEG and MEG [*see* Chapter 6]. There is an active BIDS community that welcomes involvement [*see* Resources]. Databases such as OpenNeuro.org, LORIS, COINS, XNAT, SciTran, and others will accept and export datasets organized according to BIDS, and some open-source software such as fMRIPrep, C-PAC, and MRIQC, works only or optimally with BIDS data [*see* Resources]. Making the BIDS standard a key feature of your reproducible workflow is therefore a very important first step!

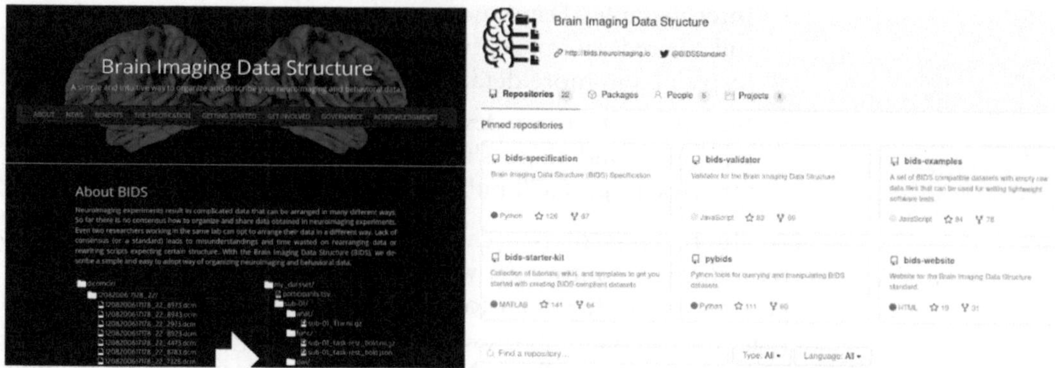

Fig. 1 BIDS

```
project/
└── subject
    └── session
        └── datatype
```

Fig. 2 BIDS dataset hierarchical organization

Let's take a more detailed look at how to build a BIDS dataset, and how to create and run compatible applications called BIDS apps [*see* Glossary]. From the outset, we should say that BIDS is extremely well documented (e.g., https://bids.neuroimaging.io/; https://github.com/bids-standard). Here, we highlight some of the most important features and resources for further information.

2.1 BIDS-Structured Data

BIDS datasets are hierarchically organized to contain both modality-agnostic and modality-specific files. The directory organization follows four main levels of hierarchy, shown in Fig. 2.

The main directory "project" contains all of the project files while sub-directories contain subject-level data, which is further organized according to session (if data were collected over multiple sessions) and modality (e.g., anatomical, diffusion-weighted, functional, etc.). The session folder is not necessary when the study contains only one session of data per participant.

Let's illustrate BIDS in more detail using an example: the AOMIC-PIOP2 dataset [7], which can be downloaded from the OpenNeuro database [see Chapter 2].

The AOMIC-PIOP2 dataset is already BIDS-structured. Once you have downloaded the data, if you look inside the main folder (*see* Fig. 3), you will find:

- **README**: The first file to open in order to quickly understand what the dataset contains; for instance, we can learn about the data types/modalities, and whether the dataset contains raw, preprocessed or processed data.

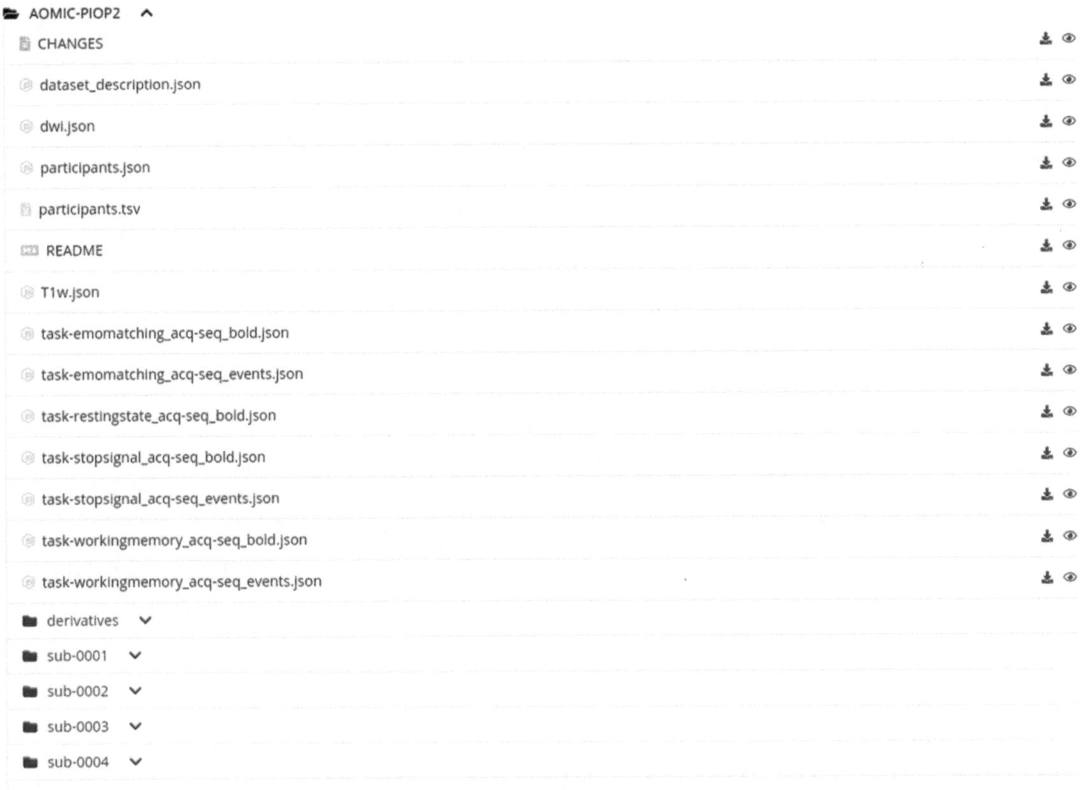

AOMIC-PIOP2
CHANGES
dataset_description.json
dwi.json
participants.json
participants.tsv
README
T1w.json
task-emomatching_acq-seq_bold.json
task-emomatching_acq-seq_events.json
task-restingstate_acq-seq_bold.json
task-stopsignal_acq-seq_bold.json
task-stopsignal_acq-seq_events.json
task-workingmemory_acq-seq_bold.json
task-workingmemory_acq-seq_events.json
derivatives
sub-0001
sub-0002
sub-0003
sub-0004

Fig. 3 Files and folders in the main folder of the AOMIC-PIOP2 dataset

- **dataset_description.json**: A json format text file [*see* Glossary] containing metadata on the dataset, including details such as the name, the authors, the version, a reference to a paper describing the data, etc.
- **participants.tsv**: A tab-delimited file (.tsv format—column separators are tabulations) that contains at least one column labelled "participant_id", which provides a unique identifier for each subject (i.e., person or animal) in the dataset. Further columns may contain other relevant information on each participant (e.g., s, gender, age, clinical status etc.). The first row of the file provides the column names.
- **participants.json**: A json text file that provides a legend for each column of participants.tsv. In the "Tree" section, the "Root Property" list can be expanded. Clicking on an item in the list (e.g., "Age" will expand the label to provide a Description and Units etc.). *See* Fig. 4.
- **CHANGES**: A timeline of all the updates/changes to the dataset.

PARTICIPANTS.JSON

Tree

Root Property-

Age-

Description: Age of participant in years

Units: Years, with one quantile precision

Sex-

Description: (Self-reported) biological sex of participant

Levels-

M: male

F: female

Fig. 4 Screenshot of an expanded Participants.json list

- There are also several modality-specific files:
  - dwi.json: Metadata on the diffusion-weighted MRI data,
  - T1w.json: Metadata on the T1-weighted MRI data.
  - There are several metadata files providing information on the fMRI, scans labelled according to experimental condition (emomatching, restingstate, stopsignal, etc.). Some of these (..._bold.json) provide information on the MRI scanning parameters themselves, such as slice timing information. For task-based fMRI there json files (and the experimental task events, such as response accuracy type ("correct", "incorrect", "miss") (..._events.json):
    - task-emomatching_acq-seq_bold.json,
    - task-emomatching_acq-seq_events.json,
    - task-restingstate_acq-seq_bold.json,
    - task-stopsignal_acq-seq_bold.json,
    - task-stopsignal_acq-seq_events.json,
    - task-workingmemory_acq-seq_bold.json,
    - task-workingmemory_acq-seq_events.json,
- **sub-X folders**: There is one folder per participant, containing raw data (and potentially participant metadata in json files) (*see* Fig. 5).
- a **derivatives** directory [*see* Glossary]: This contains processed data. For each type of processing, the organization of the corresponding subdirectory mirrors that of the main folder: there is a group.tsv file containing information on the participants whose files were processed, and participants' folders similarly named sub-X (*see* Fig. 6). The organization of each subject's folder and of the folder **derivatives** is specific to each study.

sub-0001
anat
dwi
fmap
func
sub-0002
sub-0003
sub-0001
anat
sub-0001_T1w.json
sub-0001_T1w.nii.gz

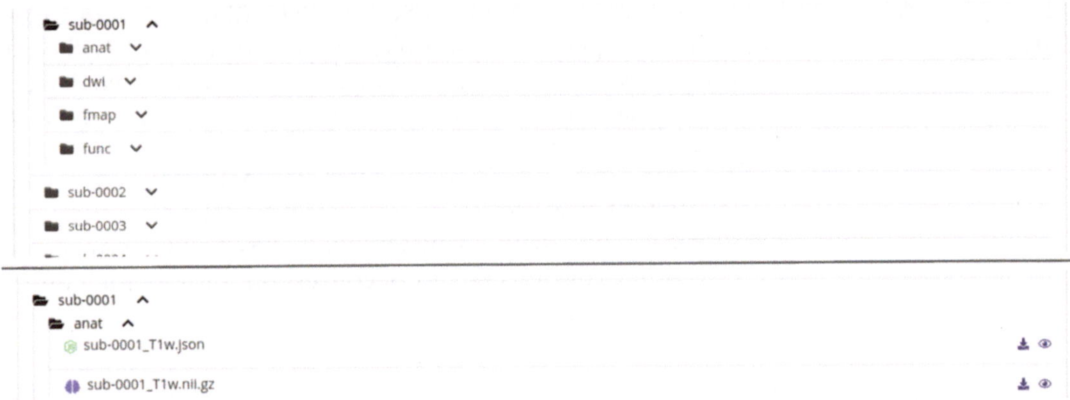

Fig. 5 The contents of each participant's folder "sub-X" of the AOMIC-PIOP2 dataset: (**a**) on the top panel, there are different subdirectories for every data modality (e.g., anat—anatomical; dwi—diffusion weighted imaging; func—functional); (**b**) the bottom panel shows the contents of the anat subdirectory

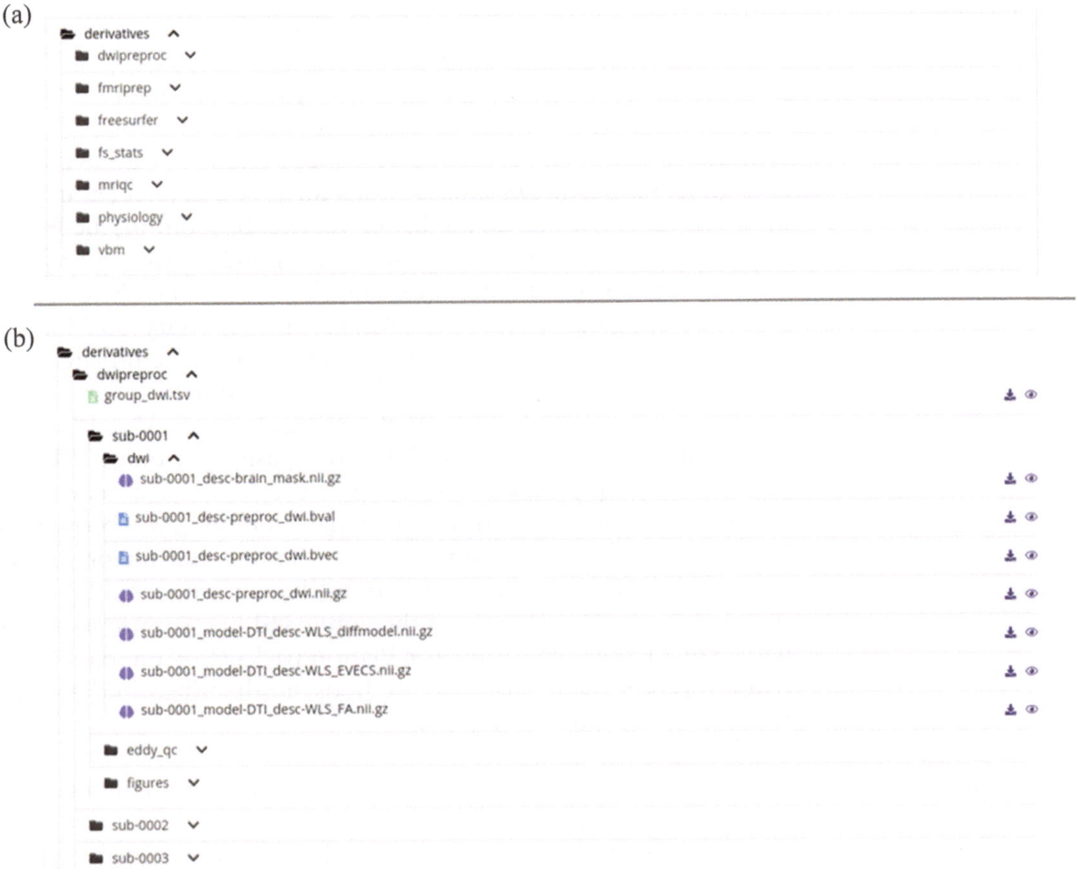

(a)
derivatives
dwipreproc
fmriprep
freesurfer
fs_stats
mriqc
physiology
vbm
(b)
derivatives
dwipreproc
group_dwi.tsv
sub-0001
dwi
sub-0001_desc-brain_mask.nii.gz
sub-0001_desc-preproc_dwi.bval
sub-0001_desc-preproc_dwi.bvec
sub-0001_desc-preproc_dwi.nii.gz
sub-0001_model-DTI_desc-WLS_diffmodel.nii.gz
sub-0001_model-DTI_desc-WLS_EVECS.nii.gz
sub-0001_model-DTI_desc-WLS_FA.nii.gz
eddy_qc
figures
sub-0002
sub-0003

Fig. 6 Contents of the "derivatives" folder of the AOMIC-PIOP2 dataset: (**a**) as the top-panel shows, preprocessed data derivatives are organized according to the process/program applied (e.g., dwipreproc); (**b**) the bottom panel shows the contents of one subdirectory of data processed using dwipreproc

2.2 Implementing BIDS

If you are working with Open Science data, obtained from a platform such as OpenNeuro, it is likely to already have a BIDS organization. However, if you are working with newly acquired data, or preexisting data in your lab, you may have to implement BIDS organization yourself.

This is easiest to do with new data. For MRI data, converting from DICOM (typical scanner output) to BIDS is relatively straightforward as there are BIDS converters [*see* Glossary]. The BIDS website provides a full list of conversion software for a variety of data modalities.

> **Note**
> Implement BIDS organization for your data from the beginning (ideally, when extracted/imported from the scanner). This is best practice and will save the harder work of having to convert your data later, when you want to share it.

If you have preexisting data in, for example, NIFTI format, complying with BIDS specifications is likely to be a matter of data organization/structure, relabelling, and the creation of the required .json files, tasks that can all be scripted (e.g., with python). A complete list of BIDS specifications for how a dataset should be organized are provided on the BIDS website. In addition, the BIDS Starter Kit [*see* Resources] provides a very good explanation of the folder and file names and formats, as well as pieces of code to create your own .tsv [*see* Glossary] or .json files [*see* Glossary] in Matlab, Python and R. The starter kit also includes a general template for a BIDS-dataset, and examples of BIDS-organized data.

Finally, once you have created your dataset, you should verify that it fully complies with BIDS structure, using the online **BIDS-validator** tool [*see* Resources]. To use the tool (which currently only supports Google Chrome and Mozilla Firefox), you browse to your data folders—the tool will check whether your file organization meets BIDS criteria. If you do not have a supported browser on your machine, you can use the command line version working with Node.js, the bids_validator Python package or the bids/validator Docker image. All these methods are described in detail on the **BIDS-validator** github pages.

2.3 Using BIDS Apps to Work with Data

Standardizing the organization of neuroimaging datasets has greatly accelerated the development of open-source data analysis packages and tools. A variety of tools have been developed to facilitate procedures from data handling and manipulation (e.g., PyBIDS and BIDS-Matlab) to fMRI data analysis (e.g., fMRIprep).

A particularly useful template of reproducible apps—called BIDS apps [*see* Resources]—has been developed by the BIDS community [8]. A **BIDS App** "*is an analysis pipeline that takes* ***a BIDS formatted dataset as input*** *and has* ***all of its dependencies packaged within a container*** *[see Glossary] (Docker or Singularity)*". This feature—the capture of all the dependencies of an app within a container—means that the app is not dependent on any software outside the container [*see* Chapter 3]. This removes what was traditionally a huge headache for users: the requirement to ensure that all dependencies were installed, that the version (i.e., release/revision number) installed was the correct one, etc.). Containers dramatically increase the reproducibility of studies by enabling users—and therefore reusers—to record and fix the version of an app (and all its dependencies) used in an analysis [*see* Chapter 3].

Before we dig in, let's quickly review some Docker usage. If you are not familiar with Docker, the BIDS community recommends that you view this tutorial (https://neurohackweek.github.io/docker-for-scientists/) and this video (https://www.slideshare.net/chrisfilo1/docker-for-scientists) on Docker and Singularity. The official "get started" Docker tutorial may help you too [*see* Resources]. **You will need to install Docker or Singularity before getting started with BIDS Apps**.

When you download an app using Docker (known as a *Docker image*), the entrypoint for interaction with that BIDS app is set to a preset script called /run.py. This script defines the required parameters for the BIDS app, which will vary by app but will include:

- **bids_dir**: The directory (full path) on your computer containing the input dataset organized according to BIDS standard.
- **output_dir**: The directory (full path) on your computer where the output files should be stored. If you are running a group level analysis, this folder should be prepopulated with the results of the participant-level analysis.
- **analysis_level**: Level of the analysis that will be performed (e.g., participant, group). Multiple participant level analyses can be run independently (in parallel).
- --**participant_label**: The label(s) of the participant(s) that should be analyzed. The label corresponds to sub-<participant_label> from the BIDS spec (so you should not include the "sub-" prefix). If this parameter is not provided, all the participants in the dataset will be analyzed. Multiple participants can be specified with a space-separated list.

In addition to these required parameters, you may add as many optional parameters as needed and appropriate to the app you are running. These parameters are visible in the script run.py of the BIDS app.

2.3.1 Worked Example

Let's illustrate how to run such a BIDS app with an example. In this example, we want to run **fMRIPrep**, which is a BIDS app that runs a customizable preprocessing pipeline for structural and functional MRI data analysis (*see*: https://fmriprep.org/en/stable/workflows.html). Here, we will run it on participants 0001 and 0002 of the AOMIC-PIOP2 dataset (which we downloaded in the previous example, Subheading 2.1).

Step 1 Get the fMRIPrep BIDS app.

In a shell on your laptop, computer, or server, run:

```
docker pull nipreps/fmriprep:latest
```

Here we call Docker's action `pull`, which finds the `latest` version of the Docker image `nipreps/fmriprep` on the DockerHub platform [*see* Resources] and copies it to your machine (this may take several minutes). The Docker image `nipreps/fmriprep` contains the scripts and installations for all the dependencies (specified in the Dockerfile) to be able to run the fMRIPrep pipeline under various conditions. For this reason, Docker images can be quite large (several GB).

If you want to use a specific version of a BIDS app (e.g., to reproduce analyses reported in a paper and performed with an older version of the BIDS app), you will find it on the DockerHub platform. For instance, if we want to use the version 1.5.10 of fMRIPrep, we can check if it is still openly shared in https://hub.docker.com/r/nipreps/fmriprep/tags. We can then adapt our docker command to pull this version instead of the latest: `docker pull nipreps/fmriprep:1.5.10`.

Step 2 Run fMRIPrep on specific participants.

The generic command line to run fMRIprep is as follows:

```
docker run -it --rm -v <freesurfer_licenses>:/opt/freesurfer/license.txt -v <bids_dir>:/data:ro -v <output_dir>:/out -v <temporary_dir>:/tmpnipreps/fmriprep:latest /data /out <analysis_level> -w /tmp --participant_label <label1 label2 label3>
```

Here, we are calling Docker's `run` action, which launches a container as a new instance of the *nipreps/fmriprep:latest* image. All fMRIPrep's dependencies are installed in the container.

Following `docker run`, we see several options are passed to the command:

`-it` is the combination of two options, -i and -t, that are the single character versions of --interactive (which keeps STDIN open even if not attached—here STDIN is the terminal window you are using to input the commands) and --tty (which allocates a pseudo-TTY connected to the container's STDIN). These two options are often passed when running `docker run` because most programs need to be run in -it mode.

`--rm` is another optional parameter that specifies that after the container exits (for instance, after the fMRIPrep pipeline is completed), the container will be automatically removed from your machine.

`-v` makes it possible for the container to interact with your machine filesystem. This is required because while a container uses computational and memory resources on your machine, it runs ***independently*** of the machine filesystem (it has its own filesystem which is defined in the Dockerfile). If we want the BIDS app to run on a BIDS dataset located on our machine, then we need to create "volumes" or "bind mounts" between the filesystems of our machine and of the container. In the command above, we use `-v` to specify three volumes:

- We link `<bids_dir>` on our machine with the directory `/data` in the container, specifying the mode `:ro` for read-only, to avoid saving files inside the BIDS directory. So everything in `<bids_dir>` will be accessible for reading from the container, in its directory `/data`.
- We link `<output_dir>` on our machine (a directory where the outputs will be saved) with the directory `/out` of the container. No mode is specified, so the default read-and-write mode is applied.
- We link `<temporary_dir>` on our machine (a directory where we want intermediate outputs of the pipeline to be stored) with the directory `/tmp` of the container. No mode is specified, so the default read-and-write mode is applied.

Note

To know what parameters you should specify when calling "docker run", check the documentation of the BIDS app you are using, or ask the creator!

For example, fmriprep uses freeSurfer tools, which require a license to run. To obtain a freeSurfer license, simply register for free at https://surfer.nmr.mgh.harvard.edu/registration.html.

Like all the BIDS apps, `nipreps/fmriprep` also takes parameters specific to the fMRIPrep pipeline like `-w`, a single character version of the parameter —work-dir, which specifies the directory where intermediate outputs should be stored. Here, we are passing */tmp* as a value for *-w*, which will save the intermediate results locally on our machine in *<temporary_dir>* thanks to the volume specified with the *-v* option.

A full list of the parameters you can pass to the BIDS app *nipreps/fmriprep* is given here: https://fmriprep.org/en/stable/usage.html#execution-and-the-bids-format.

Now we understand the basic ingredients of a call to a Docker container, let's look at a fully specified example, working with the AOMIC-PIOP2 dataset. In our case, the AOMIC-PIOP2 data is locally stored in /home/melanie/, so we replace the directory placeholders such as *<bids_dir>* with our specific directories, we specify "participant" for the analysis_level and provide our participant list. This gives us the command:

```
docker run -it --rm -v /home/melanie/AOMIC-PIOP2:/data:ro -v
-v $HOME/.licenses/freesurfer/license.txt:/opt/freesurfer/li-
cense.txt
/home/melanie/AOMIC-PIOP2/derivatives/fmriprep_test:/out -v /
home/melanie/AOMIC-PIOP2/derivatives/fmriprep_test/tmp:/tmp
nipreps/fmriprep:latest /data /out participant -w /tmp --
participant_label 0001 0002
```

Running this command will launch fMRIPrep on participants 0001 and 0002, store the intermediate results locally in the folder ***/home/melanie/AOMIC-PIOP2/derivatives/fmriprep_test/tmp*** and the final results in ***/home/melanie/AOMIC-PIOP2/derivatives/fmriprep_test***.

If you do not specify participants using the option—participant_label, the program will process all the subjects in the BIDS dataset.

Step 3 Run a group level analysis.

Once the participant-level analysis is completed for all participants (or the ones selected) in your dataset, you may need to run the analysis at the group-level:

```
docker run -it --rm -v /home/melanie/AOMIC-PIOP2:/data:ro -v /
home/melanie/AOMIC-PIOP2/derivatives/fmriprep_test:/out  -v /
home/melanie/AOMIC-PIOP2/derivatives/fmriprep_test/tmp:/tmp
nipreps/fmriprep:latest /data /out group -w /tmp
```

Notice that the command is identical, apart from the specification of analysis_level. Here, we omit a participant list, because we want all participants to be included.

This example shows how easy it is to work with your BIDS data using BIDS apps. There are many different BIDS apps [*see* Resources]. Once you have established your own analysis pipeline, you may even consider creating your own BIDS app. A template is provided here: https://github.com/BIDS-Apps/example.

More information on how to run BIDS apps, especially on HPC clusters, is provided (http://bids-apps.neuroimaging.io/tutorial/) and there is a very helpful FAQ (https://bids-standard.github.io/bids-starter-kit/apps.html). Don't forget to join the BIDS community (https://bids.neuroimaging.io/get_involved.html)!

3 Getting Started with Python Notebooks

If you have followed the steps above, you will now have some familiarity with the BIDS ecosystem, you'll have implemented BIDS structure for your data, and you'll have run your first analysis using a BIDS app. Now it's time to work with your data!

When performing reproducible research, one key recommendation is to use *open-source* tools, software, and programming languages, where possible. By using open-source tools, we remove the barrier of access to expensive, proprietary tools, and maximize the opportunity for any researcher to repeat our analyses, regardless of their resources. Python is an open-source programming language that has recently gained popularity across a broad spectrum of disciplines including neuroscience, psychology, biomedical science, and beyond, due to its intuitive platform and substantial computational capability for basic and complex analyses. While python can be used in traditional scripts, like other programming languages, **Python notebooks** offer a particularly useful way to write reproducible and portable code. Python notebooks make use of Markdown—a markup language that allows easy plain-text formatting (e.g., for adding comments, headings, bold, italics, hyperlinks, etc.). Notebooks are also commonly used with another open-source program for statistical analysis, R.

Python can be installed either alone or using a platform like Anaconda. Installing Jupyter Notebook will enable notebook use on your own local computer, but, alternatively, Google Colaboratory offers an easy web-based platform for creating and running Python notebooks, for which no local installation of Python is required. Google Colab also offers free access to (carbon neutral) GPUs and facilitates the storage and sharing of code using Google Drive. For large projects or analyses of large neuroimaging volumes, however, it is preferable to have your own local installation of Python, since Google Colab has resource limitations (12GB RAM) and may not be scalable for very large datasets. For analysis of smaller datasets, or less resource-demanding analyses of time series or behavioral data, this resource limitation should not be a problem. The example notebook included with this book chapter has been written using Jupyter Notebook (installed locally), but it can also easily be used in Google Colab.

Before we take a look at our example, there are several Python packages that you are likely to use frequently for reproducible neuroimaging and behavioral analyses. These are very well documented and supported by tutorials that are easily found online. These include [*see* Resources]:

- numpy: Numerical package for scientific and arithmetic computing,
- pandas: Supports data manipulation and analysis, particularly useful for reading data in csv/xls format,
- nilearn: Package supporting neuroimaging analyses of structural and functional volumetric data. Includes tools for voxelwise statistical analyses, multi-voxel pattern analysis (MVPA), GLMs, clustering and parcellation, etc.
- scikit-learn: Tools for machine learning, including classification, model selection, and dimensionality reduction,
- https://nltools.org: Neuroimaging package for fMRI data analyses (e.g., resting-state, task-based, movie-watching) that incorporates code from nilearn and scikit-learn,
- statsmodels: Package to build statistical models and perform statistical tests,
- pingouin: Simple statistical functions and graphics,
- matplotlib and seaborn: Tools for data visualization.

3.1 Writing and Sharing Code

As you begin your analysis, it is important to think about future **code reuse and sharing**—so that other researchers (including "future you") can reproduce your analyses and findings. Because neuroimaging analyses are often complex in nature, with many different parameters, each with a number of possible settings, it can be very difficult to reproduce a workflow or set of results if code/scripts are not provided with the published paper. For this reason, when writing code, one of the most important things to think about is whether someone else will be able to look at your code, understand what is being done, and reproduce or adapt the analyses themselves.

The idea of code sharing can be anxiety-inducing—researchers often fear that their code will contain errors that may be identified by more experienced researchers and programmers. While this is an understandable and common fear, practice helps. Learning to write reproducible code is just like learning a second language or a musical instrument—if you don't practice regularly, you will not master the ability to speak fluently or play coherently. If you have never written Python code before, it can be daunting to start. We are not going to cover the basics and mechanics here, since there are many excellent resources available online. If you are completely new to Python, you might want to consider some of the beginner's tutorials and guides (https://wiki.python.org/moin/BeginnersGuide/Programmers) linked to on the Python wiki (https://wiki.python.org/moin/), or one of the MOOCS offered by Coursera, edX, Udemy, and others. In what follows, we're going to highlight some of the most important things you should know. We have also created an **example notebook** (https://osf.io/fye48) to accompany this chapter, which illustrates some of the concepts and practices we outline below.

First, there are some basic good practices that all programmers can follow to ensure their code is well-structured and clean (avoid "spaghetti code"), clear (without ambiguous or confusing variable names), and concise (avoiding unnecessary loops). You should check out this excellent guide from the Alan Turing Institute (https://the-turing-way.netlify.app/reproducible-research/reproducible-research.html), python best practices (https://towardsdatascience.com/5-python-best-practices-every-python-programmer-should-follow-3c92971ed370), and, although very detailed, you should also familiarize yourself with the official Python Style Guide (https://peps.python.org/pep-0008/), including best practice naming conventions (https://peps.python.org/pep-0008/#naming-conventions).

Beyond following best practices for programming style, the best way to ensure usability of your code is to **#comment** it. Commenting means adding short (one-line) explanatory notes to your code, preceded by a designated comment symbol—the # symbol is used for this purpose in many coding languages, including Python. The comment notes explain the intent of the code, or summarize what the code does or refers to. Any time you write code, you should try to comment it as much as possible—not just for others but for "future you". We are all susceptible to the myth that we will successfully recall the "what" and "why" of our code at a later date, but the unfortunate truth is that even highly experienced programmers return to uncommented code after a period of time and can't decipher it. To save heartache down the line, you should carefully comment your code as you write it. One useful recommendation from Kirstie Whitaker of The Alan Turing Institute is that comments should represent about 40% of your code. In our example notebook (https://osf.io/fye48), we provide lots of examples of how to comment code. Because notebooks offer an easy means for separation of code and mark-up (plain-text formatting), they allow for detailed annotation of your analyses, which will maximize reproducibility and reuse.

Note Always comment your code, to make it accessible to your (future) self and others! By allowing for markdown plain-text sections as well as code, notebooks have made this easier and tidier than ever.

Another important coding skill is **debugging**—figuring out why your code is not working or is producing the wrong output. As with code-writing in general, debugging is a skill that develops over time and involves a lot of trial-and-error (trying something, seeing if it works, trying something else), combined with some intrepid Googling skills. Believe it or not, the default way for even quite experienced programmers to understand errors in their code is to copy/paste the error message into Google! This works because it is very likely that your error has been encountered and posted about before—a quick Google search often reveals the solution, on platforms such as Stack Overflow [*see* Resources].

Note
Even the best programmers use Google to help debug and find solutions to errors!

3.2 Planning and Implementing Your Analysis

When it comes to implementing a reproducible and sustainable analysis pipeline, it is important to carefully consider and select the preprocessing steps for your raw imaging data, so that this step can be run **only once**. This is because it is data preprocessing—steps such as motion correction and normalization to standard space—that consume the majority of computational resources and therefore energy consumption. Thankfully, the development of standardized preprocessing pipelines such as fMRIPrep [9], C-PAC [10]), and NiPreps, amongst others [*see* Resources], has the promise of not only harmonizing analyses and increasing their reproducibility, but also increasing the attractiveness of sharing and using preprocessed rather than raw data—a crucial next step in reducing the carbon footprint of neuroimaging.

Our example notebook (https://osf.io/fye48) works with data preprocessed using fMRIPrep, described above. The example works with time series extracted using a standard functional parcellation of the brain into Regions of Interest (ROIs). Working with ROI time series, rather than voxelwise data, saves both time and computational resources, because the dimensionality of the data is considerably reduced—from ~200,000 voxels × *n_timepoints* to 268 (ROIs) × *n_timepoints*. One drawback of this approach is the fine detail offered by high resolution data may be lost—the specific approach adopted and level of dimensionality required should be selected based on the goals of your analysis.

Here, we highlight some of the features of the notebook.

1. Notebook basics: For those who are completely new to Python notebooks, we introduce and demonstrate Markdown and code cells (Fig. 7).
2. Examples of good coding practice—commenting, variable names, and style (Fig. 8).
3. Using neuroimaging packages to work with voxelwise and time series data (Fig. 9).
4. An example functional connectivity analysis with data from the AOMIC-PIOP2 dataset (Fig. 10).

1. Getting Started with notebooks

Notebooks are composed of cells, each of which can contain either Python code or text.

Jupyter notebooks use a special kind of text known as Markdown, which allows formatting (such as headings or text styles like bold and italics). You can edit any of the text by double-clicking on the cell. Once you are finished editing, press Shift+Enter, and the editor view will close.

This is a markdown cell, for formatted text, and the following cell is a code cell, which we use for Python code. Code cells are denoted in Jupyter by shading in gray and square brackets to the left. In a code cell, we type Python commands. When we run the code (by pressing Shift+Enter) our commands will run and a result will print out in an output cell beneath.

```
In [2]: #Here is a little bit of simple code
        x = 3
        y = 2
        z = x + y
        print("The value of z is", z)

        The value of z is 5
```

Fig. 7 Code and markdown cells in a Python notebook

Naming Conventions, You will need to create variables to store the intermediate and/or final results. Variables should follow consistent naming conventions to maximise interpretability and readability. Use descriptive rather than temporary names. Here are some of the most common descriptive naming conventions:

- lowercase - `behav`
- uppercase - `BEHAV`
- underscore - `BEHAV_AOMIC_DATA`, `behav_df`, `BEHAV_DF`
- mixed - `behav_AOMIC_df`

Below are some examples of good and bad naming conventions:

```
In [4]: #good naming practice
        behav = pd.read_csv('ds002790/participants.tsv', index_col = 'participant_id', delimiter = '\t')
        behav_df = pd.read_csv('ds002790/participants.tsv', index_col = 'participant_id', delimiter = '\t')
        BEHAV_HBN_DATA = pd.read_csv('ds002790/participants.tsv', index_col = 'participant_id', delimiter = '\t')

        #exampls of bad naming practices
        #use of temp
        temp = pd.read_csv('ds002790/participants.tsv', index_col = 'participant_id', delimiter = '\t')
        #including a space
        final var = pd.read_csv('ds002790/participants.tsv', index_col = 'participant_id', delimiter = '\t')
        #a long yet undescriptive name
        veryimportantvariablethatiwontforget = pd.read_csv('ds002790/participants.tsv', index_col = 'participant_id', delimite
        #undescriptive name including a number
        final_temp1 = pd.read_csv('ds002790/participants.tsv', index_col = 'participant_id', delimiter = '\t')
```

It is not advisable to use single letter names for your variables as this will create a lot of confusion. In some Python platforms, single characters such as 'l,' 'o,' or 'I' (lowercase L and O and uppercase I) may produce an error as the compiler cannot distinguish these characters from numerals 1 and 0. Also, you cannot use numbers as variable names as Python will interpret these as numbers. Here are two examples which showcase these problems:

```
In [4]: l = pd.read_csv('ds002790/participants.tsv', index_col = 'participant_id', delimiter = '\t')
        1temp = pd.read_csv('ds002790/participants.tsv', index_col = 'participant_id', delimiter = '\t')

          File "<ipython-input-4-84f481fbf59c>", line 2
            1temp = pd.read_csv('ds002790/participants.tsv', index_col = 'participant_id', delimiter = '\t')
             ^
        SyntaxError: invalid syntax
```

Fig. 8 Some examples of good and bad coding practices

Once we have the resulting 3D nifti image, we can plot it with different functions such as `view_img` and `plot_epi`. While both functions plot segments of an EPI image, the former provides an interactive HML viewer of the image and the latter provides a static cut of the image as shown below.

```
In [13]: from nilearn.plotting import plot_epi

         #plot the 3D nifti image in a static format
         plot_epi(func_3D)
```

```
Out[13]: <nilearn.plotting.displays.OrthoSlicer at 0x7fd384fc97c0>
```

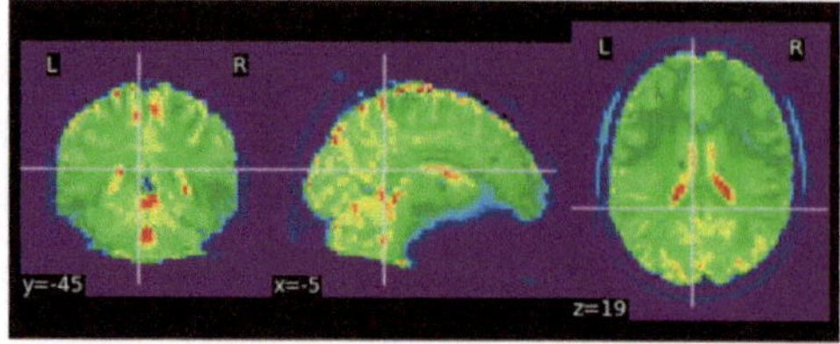

We can also create a brain mask from the 3D nifti image of a participant using the `compute_epi_mask` function as part of the `nilearn` package. Once you generate a brain mask, you can easily plot it as an ROI using `plot_roi`.

```
In [14]: from nilearn.masking import compute_epi_mask
         from nilearn.plotting import plot_roi

         #create a brain mask using the 3D nifti image of the first participant and plot it as an ROI
         mask_img = compute_epi_mask(func_3D)
         plot_roi(mask_img, func_3D)
```

Fig. 9 Working with neuroimaging packages and data

8. Computing Whole-Brain Functional Connectivity (Approach 1)

```
Now that we have the preprocessed fMRI timeseries using the Shen 268 parcellation, we can derive the FC matrix of
each participant by computing the Pearson's correlation coefficient (PCC) between all possible ROI pairs to construct
a 268 x 268 FC matrix. The resultant correlation values represent the connectivity strengths (i.e., edges) between
two ROIs (i.e., nodes).
```

```
In [34]: all_subs_RSFC = []

         for sub in range(len(clean_timeseries)):

             #compute FC matrix of each participant using np.corrcoef
             timeseries_2d = clean_timeseries[sub]
             RSFC = np.corrcoef(timeseries_2d.T)
             #append the FC matrix of all participants to a list
             all_subs_RSFC.append(RSFC)

         all_subs_RSFC_3D = np.array(all_subs_RSFC)
         #compute the mean correlation across the participant to produce a symmetrical 268 x 268 FC matrix
         mean_correlations = np.mean(all_subs_RSFC_3D, axis = 0).reshape(all_subs_RSFC_3D.shape[1], all_subs_RSFC_3D.shape[1])
         #plot the symmetrical 268 x 268 FC matrix with the colorbar showing the correlation ranging from -1 to 1
         plotting.plot_matrix(mean_correlations, vmax = 1, vmin = -1, colorbar = True, title = 'Whole-brain FC')
```

```
Out[34]: <matplotlib.image.AxesImage at 0x7f939881df10>
```

Whole-brain FC

1.00
0.75
0.50
0.25
0.00
−0.25
−0.50
−0.75
−1.00

Fig. 10 Performing an analysis with the AOMIC-PIOP2 dataset

4 Thinking About Sustainability for Your Workflow

The twin climate and biodiversity crises constitute the greatest challenge that we and our planet have ever faced. It is a challenge of our own making—the unprecedented changes in our climate and devastating destruction of ecosystems are a direct result of human behavior and the social, political, and economic systems and structures we have created. Addressing the planetary crisis—by mitigating our changing climate, stalling biodiversity loss, restoring nature, and learning to thrive within planetary boundaries—requires urgent and collective action at all levels of society.

The trouble is that most of us don't know where or how to start. As neuroimagers, neuroscientists, psychologists, researchers, academics, we doubt what contribution we could possibly make in our professional lives to addressing our planetary crisis. We worry that we don't have the right expertize, that we need to change too much about what we do, or that any small changes we make won't have an impact on such a huge problem. But in reality, there's lots we can do—as recent papers from Aron et al. [11], and Rae et al. [12] have compellingly outlined. The actions range from reducing air travel (e.g., for conferences), teaching or doing research on the crisis, being aware of the environmental impact and sustainability of helium extraction, engaging in advocacy and activism, including nonviolent direct action, to simply talking about the planetary crisis with your colleagues, friends, and family. Relevant to the goals of this chapter, one of the simplest and most immediate things we can all do is to reduce the energy usage and associated environmental impact of our data analyses. Another important action neuroimagers at all career stages can take is to better inform themselves about the origins and consequences of the planetary crisis, as well as potential solutions. In addition to the papers by Aron et al. [11], and Rae et al. [12], which review environmental issues in neuroimaging and neuroscience research, scientific computing, and our field's conferences, and propose practical steps towards sustainability, we recommend recent articles by Keifer & Summers [13] and Zak et al. [14]. Looking beyond our field to the broader picture, we strongly recommend Naomi Klein's *This Changes Everything* [15], Jason Hickel's *Less is More—How Degrowth Will Save the World* [16], and the podcasts Drilled (https://www.drilledpodcast.com/drilled-podcast/), Scene on Radio (fifth season—The Repair: http://www.sceneonradio.org/the-repair/), and Upstream (https://www.upstreampodcast.org/).

The advice contained in this chapter is intended to set you off on the right foot in terms of implementing an efficient and effective workflow that minimizes unnecessary resource use. In the next chapter, we cover Git, which can also help with this. You might also want to actively monitor your and your lab's energy use. Just a few years ago, the task of measuring the resource demands and carbon footprint of neuroimaging processing pipelines would have been exceedingly difficult. Luckily, in 2020, the Organization for Human Brain Mapping Sustainability and Environmental Action Special Interest Group (SEA-SIG, https://ohbm-environment.org/) was launched, in recognition of the need to reduce the environmental impact of the organization, its conference, and its members. As one of its first actions, the SEA-SIG established a working group focused on assessing the environmental impact of neuroimaging processing pipelines. The group has developed several carbon tracker toolboxes, based on existing utilities (Code Carbon, EIT) that monitor CPU and GPU resource use during data processing [*see* Resources]. For example, you can monitor the carbon footprint of your BIDS app by following this excellent tutorial (https://github.com/sebastientourbier/tutorial_carbonfootprint_neuropipelines). There are also guidelines (https://github.com/nikhil153/fmriprep/blob/carbon-trackers/singularity/carbon_trackers_readme.md) for measuring the impact of the way you use fMIPrep, and if you use Deep Learning models, this carbon tracker (https://github.com/lfwa/carbontracker/) can help you perform energy-optimized model training and selection procedures. Facing up to the planetary crisis is a challenging task that is made easier when you are part of a community. Finding like-minded folks to meet, talk, and take action with is one of the best things you can do to support yourself. Whether it's in your university, local community, you'll find that climate action is easier when you can draw inspiration, motivation, solidarity, support, information and resources from others, and give back in return. For neuroimagers, a global community like the SEA-SIG is particularly relevant. The SEA-SIG Neuroimaging Research Pipeline WG have shared their work on quantifying the carbon footprint of neuroimaging tools and their vision for making our processing pipelines greener by improving their reproducibility and sustainability (https://neuropipelines.github.io/)—a vision we support, share, and express throughout this chapter.

Note

We can all make concern for the climate and biodiversity crisis part of who we are and how we work. All actions matter, and those that scale to collective action and social norms are the most powerful.

References

1. Munafò MR, Nosek BA, Bishop DVM, Button KS, Chambers CD, du Sert NP, Simonsohn U, Wagenmakers E-J, Ware JJ, Ioannidis JPA (2017) A manifesto for reproducible science. Nat Hum Behav 1:0021. https://doi.org/10.1038/s41562-016-0021
2. Poldrack RA, Baker CI, Durnez J, Gorgolewski KJ, Matthews PM, Munafò MR, Nichols TE, Poline J-B, Vul E, Yarkoni T (2017) Scanning the horizon: towards transparent and reproducible neuroimaging research. Nat Rev Neurosci 18:115–126. https://doi.org/10.1038/nrn.2016.167
3. Goodman SN, Fanelli D, Ioannidis JPA (2016) What does research reproducibility mean? Sci Transl Med 8:341ps12. https://doi.org/10.1126/scitranslmed.aaf5027
4. Wilkinson MD, Dumontier M, Aalbersberg IJJ, Appleton G, Axton M, Baak A, Blomberg N, Boiten J-W, da Silva Santos LB, Bourne PE, Bouwman J, Brookes AJ, Clark T, Crosas M, Dillo I, Dumon O, Edmunds S, Evelo CT, Finkers R, Gonzalez-Beltran A, Gray AJG, Groth P, Goble C, Grethe JS, Heringa J, 't Hoen PAC, Hooft R, Kuhn T, Kok R, Kok J, Lusher SJ, Martone ME, Mons A, Packer AL, Persson B, Rocca-Serra P, Roos M, van Schaik R, Sansone S-A, Schultes E, Sengstag T, Slater T, Strawn G, Swertz MA, Thompson M, van der Lei J, van Mulligen E, Velterop J, Waagmeester A, Wittenburg P, Wolstencroft K, Zhao J, Mons B (2016) The FAIR guiding principles for scientific data management and stewardship. Sci Data 3:160018. https://doi.org/10.1038/sdata.2016.18
5. Gorgolewski KJ, Poldrack RA (2016) A practical guide for improving transparency and reproducibility in neuroimaging research. PLoS Biol 14:e1002506. https://doi.org/10.1371/journal.pbio.1002506
6. Sandve GK, Nekrutenko A, Taylor J, Hovig E (2013) Ten simple rules for reproducible computational research. PLoS Comput Biol 9: e1003285. https://doi.org/10.1371/journal.pcbi.1003285
7. Snoek L, van der Miesen MM, Beemsterboer T, van der Leij A, Eigenhuis A, Steven Scholte H (2021) The Amsterdam open MRI collection, a set of multimodal MRI datasets for individual difference analyses. Sci Data 8:85. https://doi.org/10.1038/s41597-021-00870-6
8. Gorgolewski KJ, Alfaro-Almagro F, Auer T, Bellec P, Capotă M, Chakravarty MM, Churchill NW, Cohen AL, Craddock RC, Devenyi GA, Eklund A, Esteban O, Flandin G, Ghosh SS, Guntupalli JS, Jenkinson M, Keshavan A, Kiar G, Liem F, Raamana PR, Raffelt D, Steele CJ, Quirion P-O, Smith RE, Strother SC, Varoquaux G, Wang Y, Yarkoni T, Poldrack RA (2017) BIDS apps: improving ease of use, accessibility, and reproducibility of neuroimaging data analysis methods. PLoS Comput Biol 13: e1005209. https://doi.org/10.1371/journal.pcbi.1005209
9. Esteban O, Markiewicz CJ, Blair RW, Moodie CA, Isik AI, Erramuzpe A, Kent JD, Goncalves M, DuPre E, Snyder M, Oya H, Ghosh SS, Wright J, Durnez J, Poldrack RA, Gorgolewski KJ (2019) fMRIPrep: a robust preprocessing pipeline for functional MRI. Nat Methods 16:111–116. https://doi.org/10.1038/s41592-018-0235-4
10. Craddock C, Sharad S, Brian C, Ranjeet K, Satrajit G, Chaogan Y, Li Q, Daniel L, Vogelstein J, Burns R, Stanley C, Mennes M, Clare K, Adriana D, Castellanos F, Michael M (2013) Towards automated analysis of connectomes: the configurable pipeline for the analysis of connectomes (C-PAC). Front Neuroinform 7. https://doi.org/10.3389/conf.fninf.2013.09.00042
11. Aron AR, Ivry RB, Jeffery KJ, Poldrack RA, Schmidt R, Summerfield C, Urai AE (2020) How can neuroscientists respond to the climate emergency? Neuron 106:17–20. https://doi.org/10.1016/j.neuron.2020.02.019
12. Rae CL, Farley M, Jeffery KJ, Urai AE (2022) Climate crisis and ecological emergency: why they concern (neuro)scientists, and what we can do. Brain Neurosci Adv 6: 23982128221075430. https://doi.org/10.1177/23982128221075430
13. Keifer J, Summers CH (2021) The neuroscience community has a role in environmental conservation. eNeuro 8.: ENEURO.0454-20.2021. https://doi.org/10.1523/ENEURO.0454-20.2021
14. Zak JD, Wallace J, Murthy VN (2020) How neuroscience labs can limit their environmental impact. Nat Rev Neurosci 21:347–348. https://doi.org/10.1038/s41583-020-0311-5
15. Klein N (2015) This changes everything: capitalism vs. the climate, First Simon&Schuster trade paperback edition. Simon & Schuster Paperbacks, New York
16. Hickel J (2020) Less is more: how degrowth will save the world. Penguin Random House

Chapter 5

Optimizing Your Reproducible Neuroimaging Workflow with Git

Mélanie Garcia and Clare Kelly

Abstract

As a neuroimager working with open-source software and tools, you will quickly become familiar with the website GitHub, which is a (for profit) platform for storing, managing, and sharing code, software, and projects. Many of the open-source tools discussed in this book are hosted on GitHub. Although people are generally very familiar with GitHub (or GitLab), they are often less familiar with its foundation—Git. Better understanding Git will help you better manage your own projects *and* will also help you to better understand GitHub and how to use it optimally. In this chapter, we will first explain Git and how to use it, then we will turn to GitHub. A worked example—a clustering analysis—with open access neuroimaging data is provided, demonstrating utilities such as version control, branching, and conflict resolution.

Key words Neuroimaging, Git, GitHub, Version control, Reproducibility

1 What Is Git?

> "Git is a free and open source distributed version control system designed to handle everything from small to very large projects with speed and efficiency." (https://git-scm.com/)

To understand what Git [*see* Resources] is, we first need to understand the concept of *version control* [*see* Glossary]. At its most basic, *version control* means tracking and managing changes to your code (scripts) and projects. Git will help you to do this in a structured and systematic way, which helps you to manage and share your code with others. The Git version control system means that parts of your project can be developed or edited, while your work is also protected from unwise or undesirable changes through the possibility of reverting to an earlier version. These features of Git facilitate the collaborative development of tools. Git is widely used in academia but also in most business environ-

Robert Whelan and Hervé Lemaître (eds.), *Methods for Analyzing Large Neuroimaging Datasets*, Neuromethods, vol. 218, https://doi.org/10.1007/978-1-0716-4260-3_5,

ments that involve IT programming. While using Git may seem cumbersome or complicated at first, the habits that it establishes, the version control system it implements, and the ease of collaboration it enables will, in the long run, make the investment worth it.

2 Version Control with Git

Git and version control are best explained through an example. As a prerequisite, you need to install Git (https://git-scm.com/book/en/v2/Getting-Started-Installing-Git) on your machine. Below, we work through one example, providing command lines you can type into a terminal on your machine. In this example, we will perform a clustering analysis on a file from the derivatives folder in the AOMIC-PIOP2 dataset (https://openneuro.org/datasets/ds002790/versions/2.0.0), using Python. To follow the tutorial, you will need to download the file *derivatives/fs_stats/data-cortical_type-aparc_measure-area_hemi-lh.tsv* from the Open Neuro repository [*see* Chapter 2].

2.1 Create a Git Repository for Your Project

If you want to version control a project, you will need to create a repository [*see* Glossary] (which is simply a directory or folder; often called a "repo") that will contain all the code related to the project.

Let's call this folder "my-analysis". Note that when using the command line, it is best to avoid spaces or special characters when naming folders.

```
mkdir my-analysis
```

Next, change directory into this folder.

```
cd my-analysis
```

Now, initialize a Git repository associated with this folder, by running the command:

```
git init
```

This creates a .git directory inside the folder "my-analysis". The .git directory is the only difference between a Git repository and an ordinary folder.

Note
Be very careful! Deleting the .git folder will mean that all the files in the folder "my-analysis" become *unversioned*—you will lose the history of all changes to those files.

2.2 Basic Configuration of the Git Repository

First, configure Git by telling it who you are:

```
git config --global user.name "Your Name"
git config --global user.email your.email@example.com
```

Note
In the first command, the inverted commas are required. For the second command, you should use the same email address used to set up your github account (this will be explained later). These steps have a very practical benefit in that they allow you to track *who* coded each part of a collaborative project.

2.3 Check the Status of the Git Repository

You should check the status of the Git repository frequently, using the command:

```
git status
```

In this example, this command should return:

```
On branch master

No commits yet

nothing to commit (create/copy files and use "git add" to track)
```

The first line indicates that you are on the branch "master". We will explain the notion of a branch later in this tutorial. The second line tells us that we have not yet begun to track ("commit") any files. The final line tells us that we do not yet have any files in the repository that we can track—there is "nothing to commit".

Users often prefer to rename the "master" branch as "main", in recognition of the fact that technology terminology such as "master" has its origins in slavery and colonialism. GitHub recently made the same change across its platform (https://github.com/github/renaming). This change cannot be done until you have made your first commit, however, so we will return to this below.

2.4 Create a File

Let's create a notebook to visualize the data, called visualization.ipynb. The notebook is available here for download (https://github.com/garciaml/my-analysis). Download the whole folder as a zipped folder (*see* Fig. 1)—then unzip it and move the file **visualization.ipynb** into the "my-analysis" folder. To run this notebook, you will need to have installed python and the libraries pandas, matplotlib and seaborn [*see* Resources].

You also need to move the data downloaded from the AOMIC-PIOP2 ("derivatives fs_stats data-cortical_type-aparc_measure-area_hemi-lh.tsv") into your "my-analysis" folder.

Next, verify the status of the Git repository again:

```
git status
```

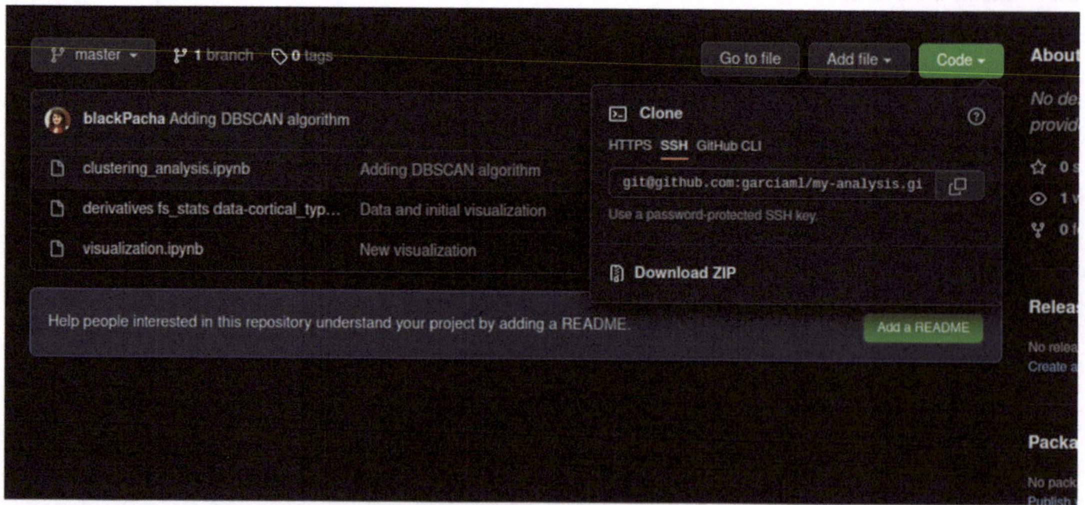

Fig. 1 Repository containing the files needed for this tutorial. You can download these files as a zipped folder

This command should return:

```
On branch master

No commits yet

Untracked files:
  (use "git add <file>..." to include in what will be committed)
    .ipynb_checkpoints/
    derivatives fs_stats data-cortical_type-aparc_measure-area_hemi-lh.tsv
    visualization.ipynb

nothing added to commit but untracked files present (use "git add" to track)
```

Now that you have some files inside your my-analysis folder, the *git status* command returns a lot more information. It now has a section, "Untracked files", which lists two files:

derivatives fs_stats data-cortical_type-aparc_measure-area_hemi-lh.tsv

and visualization.ipynb. These files are untracked, which means that no versioning has yet been applied to these files. The directory ".ipynb_checkpoints/" is a hidden folder containing cache memory related to the notebook. It should not be tracked.

2.5 Staging Files

The first step in versioning files is called *staging*—which involves letting Git know which files we intend to version. Stage the files "derivatives fs_stats data-cortical_type-aparc_measure-area_hemi-lh.tsv" and "visualization.ipynb" using the following command:

```
git add derivatives fs_stats data-cortical_type-aparc_measure-area_hemi-lh.tsv visualization.ipynb
```

This operation means that you intend to track these files, but have not yet started to save versions of them. If needed, you can go back and remove these files from being staged/tracked with no impact on the Git repository, using the command *git restore --staged <file>*.

You can verify the status with:

```
git status
```

The outcome should have changed now, and you should see the two stage files now included in the list of files to be committed.

2.6 Committing Files

In Git, a "commit" is a snapshot of the staged changes to the project. Each time you perform a "commit", you create a *version* of your project. Git keeps these versions safe, so we can always return to an earlier version of our project if needed.

Let's create your first commit. First, check the history of commits of the project.

```
git log
```

Because this is a new project and no commits have yet been made, it should return:

```
fatal: your current branch 'main' does not have any commits yet
```

One way to commit is to run:

```
git commit
```

This command will lead to a window where you can type a brief message related to the commit. The message should capture what the commit does—what changes the commit makes to your repository (here's a guide to writing Git commit messages: https://cbea.ms/git-commit/). Next, save the changes and exit. This will have created a new commit in the history of our Git repository.

Another way to commit is to use the command-line:

```
git commit -m "Data and initial visualization"
```

The part inside the inverted commas (quotes) is the commit message.

We can visualize this commit and obtain its identifier by running:

```
git log
```

Now that you have made your first commit, you can rename the "master" branch as "main" using the command:

```
git branch -m master main
```

2.7 Create and Track a New File in Your Project

Let's create a notebook that will contain our clustering analysis, called "clustering_analysis.ipynb". You'll find this notebook in the folder you previously downloaded from the repository. Move it into the "my-analysis" folder.

Stage and commit it:

```
git add clustering_analysis.ipynb

git commit clustering_analysis.ipynb -m "First clustering: PCA + k-Means"
```

Note

Your commit command is different this time, in that you specified the name of the file you wanted to commit. This may be necessary when you have several files that are staged but you only want to commit one of these.

Now you can make changes and edits to your project files, staging and committing those changes as often as needed (*see* Fig. 2).

2.8 Remove Modifications in a File Before Committing

Sometimes you will want to undo changes you have made to a file. For example, let's modify the file "visualization.ipynb" by adding a cell and writing some code inside. Save the changes. Next, run:

```
git status
```

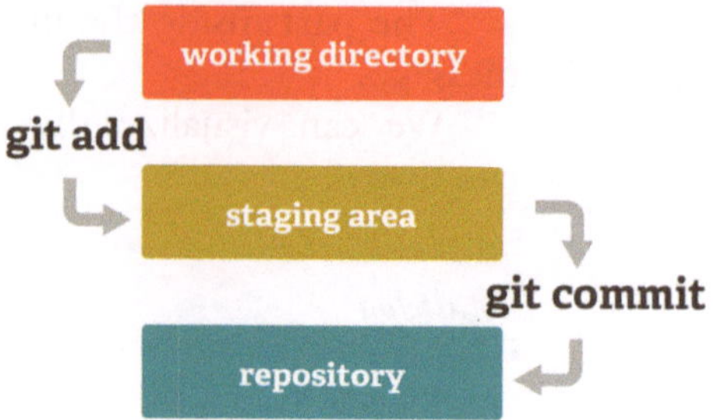

Fig. 2 Visualizing the staging and committing process

The name of the file should be in the section "Changes not staged for commit":

Now, let's say you regret the change you made—they didn't produce the output expected and you want to revert to an earlier version of your notebook.

To do this, run:

```
git restore visualization.ipynb
```

Open the notebook in Jupyter or another interface. You should find that you have restored the last version of the file before you made those modifications—the changes you made to the file have disappeared.

2.9 Check a Specific Old Version of a File in the Git Repository

Let's say you have developed a project quite a bit, but you'd like to look at a specific version of one of your files—for example, the first version of the file "visualization.ipynb". You can do this by:

- Finding the ID of the commit corresponding to the version of the project you want to check, by running: git log --oneline.
- The need for informative commit messages should now be clear! You need to be able to identify the commit containing the version of the project that you want. Here, you want the first commit, "Data and initial visualization", which the log indicates is commit 62901c0.
- To access this version, use the command `git checkout 62901c0`.
- You can now open the notebook "visualization.ipynb" in Jupyter or another editor. This is the old version of your file. Now, things can get complicated if you want to modify *this* version of the file. In that case, it is recommended that you start a *new branch* of your project, and commit any new changes within this

new branch. Otherwise you will run into trouble when you return to the most recent version of the project in the branch main. We will look at how to build a new branch in the next section.

- Finally, to return to the most recent version of your project in the branch main, run the command line: git checkout main.
- You can check that all the recent changes in "visualization. ipynb" are still there by opening that notebook.

2.10 Developing in Branches

Let's say you want to add a new type of algorithm to our analysis but are not sure it will work. One way to do this safely is to create a new "branch" of your project where you can develop this new analysis feature, and make commits, but without modifying the stable version of your project (which is by default on a branch named "main").

First, let's see what branches there currently are in your project:

```
git branch
```

Since we have not yet created any branches, we should see just one listed:

```
* main
```

You will now create a new branch named "dbscan" because this is the new feature to add to the notebook "clustering_analysis. ipynb".

```
git branch dbscan
```

Now, you need to go to this branch, so the work you do is staged and committed to the dbscan branch and not the main:

```
git checkout dbscan
```

You can verify that you are in the dbscan branch by running *git branch* (you will see an asterisk in front of the current branch).

```
* dbscan    main
```

Modify the file clustering_analysis.ipynb. You can add a cell with these lines of code:

```
# Perform a k-Median

clustering = cluster.DBSCAN(eps=10000, min_samples=3).fit(X)

print(clustering.labels_)

# Let's visualize the clusters

plt.figure(figsize=(16,9))

plt.scatter(X[:, 0], X[:, 1], c=clustering.labels_)

plt.figure()
```

Save your changes in the Git repository within the current branch:

```
git add clustering_analysis.ipynb

git commit -m "Adding DBSCAN algorithm"
```

By working within a branch, you can make sure that the changes you make are what you want. Once you have a version of a new feature that works and that you are fully satisfied with, you can *merge* the branch with the branch main to integrate our feature in the stable version of our project.

First, go back to the branch master:

```
git checkout main
```

Next, merge the branch dbscan:

```
git merge dbscan
```

Finally, in order to keep our Git repository clean and refined, you should delete the branch dbscan since we will not use it later in the project:

```
git branch -d dbscan
```

Fig. 3 Branches in Git projects

Sometimes you will need to manage conflicts while merging branches. Conflicts arise due to the line-by-line differences between two versions of the same file. To resolve conflicts, you can edit files in their respective branches and make the different versions compatible. Git will help in this task by showing us the conflicting parts of the files. For example:

```
<<<<<<< HEAD    What is written in the version in the branch you want to merge files into.
    =======
    What is written in the version in the branch where you want to get the version from.
    >>>>>>> dbscan
```

Branches (*see* Fig. 3) are particularly useful when you want to work in a collaborative way. Every contributor can create new branches where they can develop new features, and, when agreed, can later merge their branches with the stable version of the project. When you code alone, it is also a safe way to add new features to your projects, without breaking the stable version.

2.11 Link Your Project with a GitHub Repository

GitHub is a platform for storing projects in public or private repositories. It is widely used within the scientific community and uses Git as backbone. It makes it possible to access your projects from anywhere, and to share your projects in an open-source way, thanks to the "public repository" mode.

To work remotely on your laptop on a project that is stored in GitHub, and to keep the state of the project synchronized between your laptop and GitHub, you will need to perform several steps.

First, you'll need a GitHub account. Next, create an ssh connection that will serve as your secure authentication when synchronizing your project with your GitHub repository. You can follow this GitHub tutorial: https://docs.github.com/en/authentication/connecting-to-github-with-ssh/about-ssh

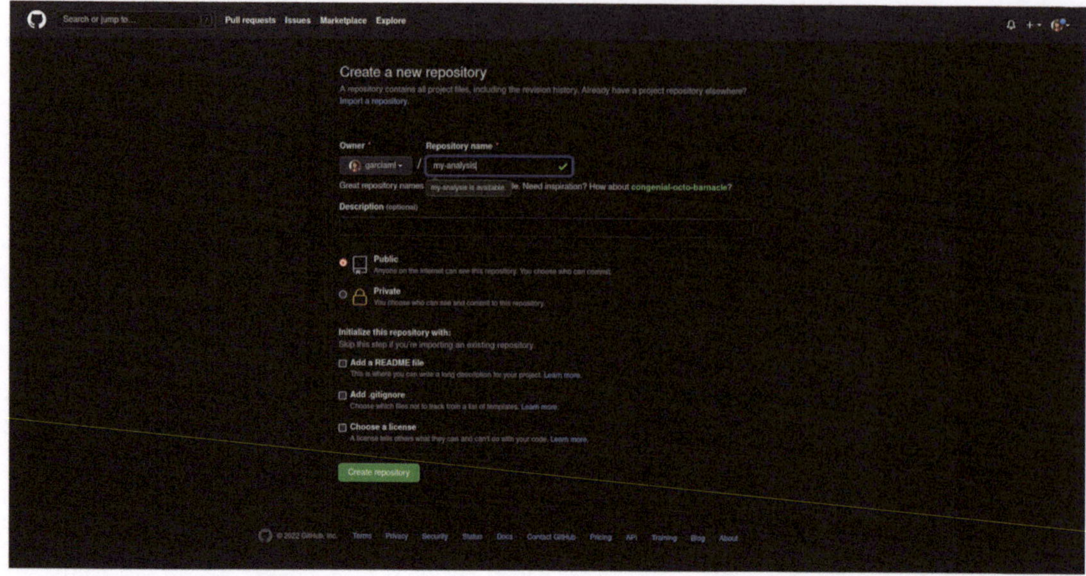

Fig. 4 Creating a GitHub repository

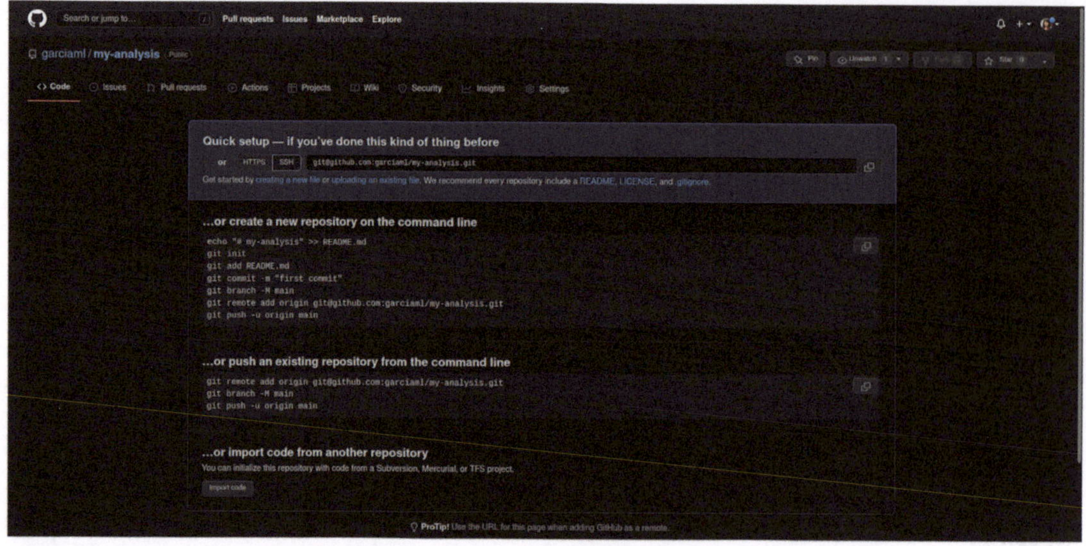

Fig. 5 Quick setup after creating your new repository on GitHub

Next, you can create a new GitHub repository, for instance with the same name: my-analysis (*see* Fig. 4).

To be able to connect your GitHub repository via ssh, select the ssh button. This will display the commands you'll need to launch (*see* Fig. 5).

Once you have run those setup commands, from the my-analysis folder, run:

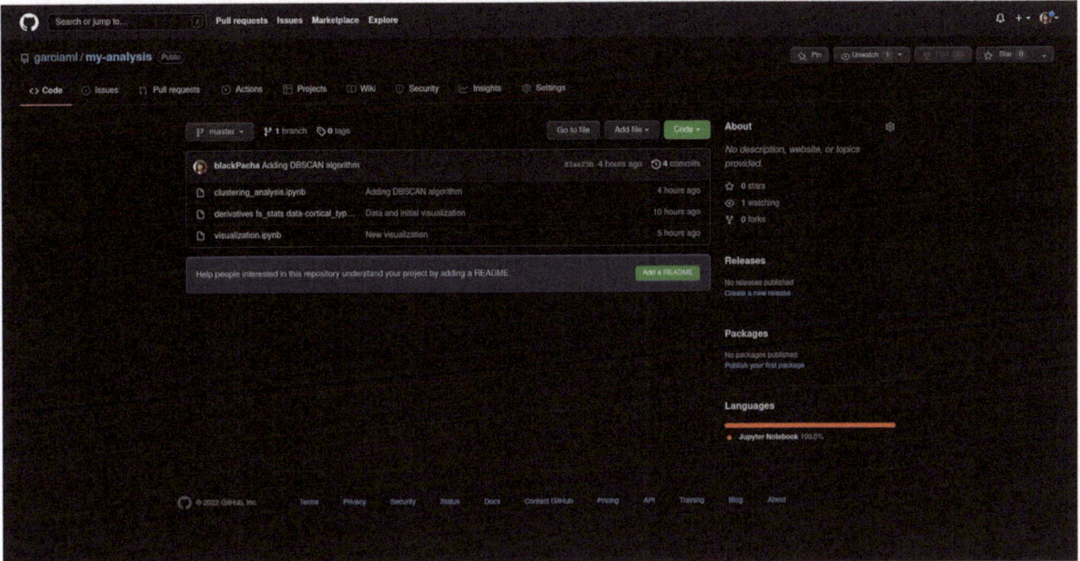

Fig. 6 GitHub repository after *git push*

```
git remote add origin git@github.com:garciaml/my-analysis.
git      git branch -M main
```

> **Note**
> Be sure to change git@github.com:garciaml/my-analysis.git to the name of your own GitHub repository!

Next, you can push everything you have coded and tracked in your Git repository to the GitHub repository by running:

```
git push -u origin main
```

If you update your GitHub webpage, you will see your files uploaded on the platform (*see* Fig. 6).

It is very important to regularly push your commits so that you keep an updated version of your project in your GitHub repository.

When you develop a project with other people or when you develop alone but using different machines, you will need to synchronize the version of the project on your machine with the one in the GitHub repository each time you connect to your machine to keep on working on the project. Do this using the command:

```
git pull
```

2.12 Collaborating with Git and GitHub

Part of what makes Git and GitHub so popular is that they are important tools for collaborative projects. In GitHub, you can add collaborators to your project. These collaborators will then be allowed to directly push their changes to the GitHub repository and to pull the project regularly to stay synchronized with changes committed by others.

Generally, collaborators develop on their own branches before merging with the main branch which is usually the branch of the "deployed" project (i.e., the code on the main branch should run well and correspond to a stable version of the project).

Note

It is useful to have a "develop" branch (*see* Fig. 2) that will allow the collaborators to develop new features of a project until the next stable version that will be merged with the main branch. It also prevents anyone from breaking a stable version of a project with a bad merge (because the stable version remains on the main branch!).

Each collaborator can contribute to developing new features in a project or solving issues, using separate branches. This can make the division of tasks amongst collaborators structured and clear.

It is important to commit often (code that runs well) and push these changes to the GitHub repository to make the evolution of the project visible for all and to avoid overlap or duplication of effort between contributors. There are many ways to establish a productive and efficient work environment. For instance, GitHub Flow [*see* Resources] can help with setting up "*a lightweight, branch-based workflow*" for collaborative projects in neuroscience. Trunk-based-development [*see* Resources] is another type of workflow that can help teams of developers to build a project or a software in an efficient way.

What we have just described is known as the "shared repository model", when team members directly share and collaborate on the source repository with others. Another way to contribute to projects is through the "Fork and pull model". Here, you can create a "fork" from any repository you have access to (e.g., public repositories, the authors of which you might not know or formally collaborate with). The "fork" is a copy of that repository, which exists on your own GitHub. Now you can develop and push your changes on the fork, *without the permission of the owner of the source repository.* If you want to suggest your changes to the owner of the source repository, you can create a "pull request" and the owner will be able to accept or decline your suggestions for changes to the source repository. The "fork and pull model" is often used for big collaborative projects, when the code is made public and open to be developed by the wider community. You can learn more about collaborative development models from the GitHub documentation (https://docs.github.com/en/pull-requests/collaborating-with-pull-requests/getting-started/about-collaborative-development-models).

2.13 Some Other Practical Tips for Git and GitHub Users

- You can tag specific versions of your project (i.e., commits) with tag names using the command *git tag -a <tag-name> -m "<description>"*.
- You can create a .gitignore file that will contain the names of the untracked files or folders that Git should ignore.
- You can easily clone a GitHub project on your machine by using git clone.
- You can undo committed changes with the command git revert.
- Using the GitHub CLI might be useful (https://cli.github.com/).
- Our example used command-lines directly typed into a terminal. Another option is to use a Graphical User Interface like Sourcetree or GitHub Desktop [*see* Resources]. Here's a list of good Git GUIs (https://dev.to/theme_selection/best-git-gui-clients-for-developers-5023).

2.14 Tutorials to Further Develop Your Git Skills

Kirstie Whitaker's intro: https://kirstiejane.github.io/friendly-github-intro/

The GitHub quickstart: https://docs.github.com/en/get-started/quickstart/hello-world

The official Git tutorial: https://git-scm.com/docs/gittutorial

Videos: https://git-scm.com/videos; https://johnmathews.is/rys-git-tutorial.html

Git Cheat Sheets: https://training.github.com/

2.15 Ten Simple Rules for Taking Advantage of Git and GitHub

Reproduced verbatim here from [1]:

Rule 1: Use GitHub to Track Your Projects

Rule 2: GitHub for Single Users, Teams, and Organizations

Rule 3: Developing and Collaborating on New Features: Branching and Forking

Rule 4: Naming Branches and Commits: Tags and Semantic Versions

Rule 5: Let GitHub Do Some Tasks for You: Integrate

Rule 6: Let GitHub Do More Tasks for You: Automate

Rule 7: Use GitHub to Openly and Collaboratively Discuss, Address, and Close Issues

Rule 8: Make Your Code Easily Citable, and Cite Source Code!

Rule 9: Promote and Discuss Your Projects: Web Page and More

Rule 10: Use GitHub to Be Social: Follow and Watch

Reference

1. Perez-Riverol Y, Gatto L, Wang R, Sachsenberg T, Uszkoreit J, da Leprevost FV, Fufezan C, Ternent T, Eglen SJ, Katz DS, Pollard TJ, Konovalov A, Flight RM, Blin K, Vizcaíno JA (2016) Ten simple rules for taking advantage of git and GitHub. PLoS Comput Biol 12:e1004947. https://doi.org/10.1371/journal.pcbi.1004947

Chapter 6

End-to-End Processing of M/EEG Data with BIDS, HED, and EEGLAB

Dung Truong, Kay Robbins, Arnaud Delorme, and Scott Makeig

Abstract

Reliable and reproducible machine-learning enabled neuroscience research requires large-scale data sharing and analysis. Essential for the effective and efficient analysis of shared datasets are standardized data and metadata organization and formatting, a well-documented, automated analysis pipeline, a comprehensive software framework, and a compute environment that can adequately support the analysis process. In this chapter, we introduce the combined Brain Imaging Data Structure (BIDS) and Hierarchical Event Descriptors (HED) frameworks and illustrate their example use through the organization and time course annotation of a publicly shared EEG (electroencephalography) dataset. We show how the open-source software EEGLAB can operate on data formatted using these standards to perform EEG analysis using a variety of techniques including group-based statistical analysis. Finally, we present a way to exploit freely available high-performance computing resources that allows the application of computationally intensive learning methods to ever larger and more diverse data collections.

Key words EEG, Neuroinformatics, BIDS, HED, EEGLAB

1 Introduction

As demonstrated in a recent study [1], the complex and varied landscape of individual and phenotypic brain differences may require studies of thousands of individuals to establish reliable and reproducible associations between brain dynamics and function, and personal experience and behavior. Machine-learning approaches can now facilitate these discoveries but typically require large collections of *diverse* input training data, including well-labelled data, to create generalizable models. Public data archives that use common data formatting standards are thus critical infrastructure for enabling large-scale neuroimaging data analysis (and "meta=" or "mega-analysis" [*see* Glossary]) applied within and across individual studies and data recordings. Standardized data formats enable tool interoperability [*see* Glossary]. Machine-actionable [*see* Glossary] metadata [*see* Glossary] supports the

Robert Whelan and Hervé Lemaître (eds.), *Methods for Analyzing Large Neuroimaging Datasets*, Neuromethods, vol. 218, https://doi.org/10.1007/978-1-0716-4260-3_6,

interpretation of the results [*see* Chapter 4]. Finally, complete specification and accurate documentation of the applied computational pipelines promote reproducibility and testing.

This chapter presents an end-to-end overview of an electroencephalography (EEG) data analysis process based on open community standards, beginning with the identification of suitable raw data and ending with the presentation of results suitable for discussion and publication. The process begins with the conversion of raw data (here, collected EEG plus sensory and behavioral event descriptions and timings) to standard BIDS (Brain Imaging Data Structure [*see* Glossary]) archival format [2]. The BIDS project, which initially focused on specification of file organization for functional magnetic resonance imaging (fMRI) datasets, is evolving into a widely adopted, community-driven set of data format specifications for a variety of imaging modalities including fMRI, positron emission tomography (PET), diffusion-weighted imaging (DWI), EEG, magnetoencephalography (MEG), intracranial EEG (iEEG), and microscopy data [3–5] [*see* Chapter 4]. Further BIDS specifications are under development for body motion capture, eye-tracking data, and multiple modality data. The in-common BIDS formatting and metadata annotation standards allow tools to be built using standard APIs (application program interfaces) enabling automated ingestion and processing of datasets representing either one or more than one study and experimental paradigm.

While initial BIDS efforts focused on raw data formats and layouts, recent efforts are also underway to standardize derivative datasets (i.e., containing results of computations performed on the raw data) for different imaging modalities including structural and functional fMRI, electrophysiology, MEG, and PET. Also underway is an initiative to develop a framework for statistical models (BIDS Stats Models) allowing computations to be specified using a JSON file [*see* Glossary] so that they can be more easily documented and reproduced. Finally, a BIDS-mega specification is being proposed to standardize the way that multiple BIDS datasets can be integrated for large-scale computations. (A full listing of BIDS extensions proposals under development can be found at https://bids.neuroimaging.io/get_involved.html.)

A second requirement for interpretable large-scale, cross-study analyses is a standardized metadata specification. While the BIDS conventions include specifications for basic metadata (who, what, when, where, etc.), one important annotation category has remained nearly unstandardized and very often underspecified—the nature of the *events* recorded during or later discovered to have occurred during time series recordings. That is, the answer to the question, "What exactly did the participant(s) experience and do during this recording?

The Hierarchical Event Descriptors (HED or 'H-E-D' system is a standardized method of capturing information about events in dataset in a common metadata format to produce event annotation ready for machine analysis [*see* Resources]. HED, first proposed by Nima Bigdely-Shamlo over a decade ago [6] and now in its third major version, has many tools and features that facilitate production and validation of standardized annotations that can be searched and used in analyses [7, 8]. HED was formally accepted as part of all BIDS modality standards in 2019 (BIDS v1.2.1-). Current and in-progress HED tools support data annotation, validation, search, summary, and analysis [https://www.hed-resources.org]. The HED framework includes a base standardized vocabulary and a tool suite supporting annotation, validation, and analysis in a combination of online tools, Python-based command line scripts and notebooks, and MATLAB scripts and plug-in tools for EEGLAB [*see* Resources]. See https://osf.io/8brgv/ for links to HED and other resources discussed in this chapter.

HED metadata can enable intelligent combining of event-related data from different recordings and studies in sophisticated analyses including those using machine learning. Because EEG, MEG, and iEEG data, in particular, have very fine time resolution (much quicker than our thoughts and actions), event-related analysis is crucial in processing of neuroelectromagnetic data. The dominant EEG analysis approaches including event-related potential (ERP), time/frequency, and dynamic connectivity averaging rely on event descriptions and time markers to isolate sets of similar data excerpts (or epochs) for meaningful comparison. HED annotation is equally applicable to any other time series data, including fMRI [*see* Chapter 7].

HED also now supports a 'library' mechanism using which specialized research communities can develop additional HED "library schemas" to specify HED term vocabularies for event description of neuroimaging research subfields including language, body movement control, clinical neurophysiology, and others. HED's first library schema, a library of terms used in interpretation of clinical EEG recordings (now available at https://www.hedtags.org/display_hed.html), incorporates the standardized SCORE (Standardized Computer-based Organized Reporting of EEG) for annotation of clinical EEG [9].

The past (and still most current) conception of experiment events conflates *event processes* (unfolding over time) with *event-phase markers* that point (typically) to the *onsets* or to other time points of interest in event processes on the experiment timeline. HED distinguishes in principle between *event processes* (having duration) and *event-phase markers* (pointing to a single moment on the experiment timeline). For many experiment events, the second most important *event phase* to mark is its *offset*. BIDS allows (but does not demand) that event onset markers include a measure (in seconds) of the *duration* of the event, from which the event *offset* moment can be calculated. HED allows other (inset) event phases to be annotated as well, facilitating detailed analysis of complex events.

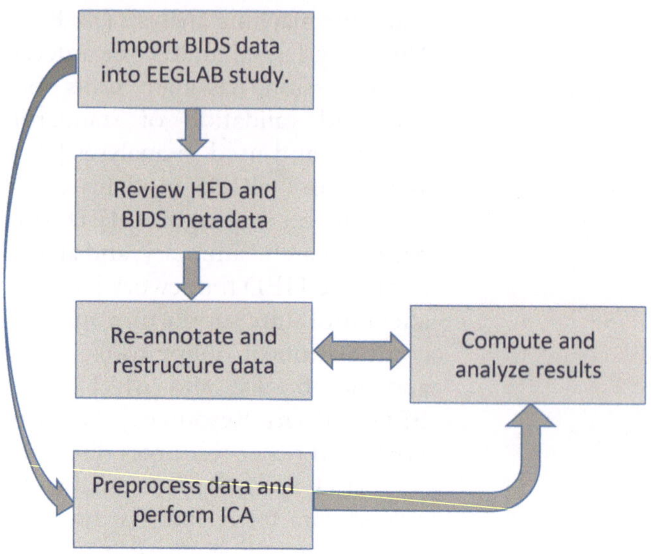

Fig. 1 Chapter overview of the end-to-end process of analyzing EEG data incorporating HED annotation

A further pillar supporting reproducible data analysis is the use of well-documented and automated analysis pipelines. This process is highly dependent on the tools and tool platform used in the analysis. In this chapter, we demonstrate an approach to such an analysis using tools from the EEGLAB environment [10], a widely used processing platform for M/EEG data that fully supports both BIDS and HED. Here, we provide practical guides on the input/output (I/O) workflow between BIDS, HED, and the EEGLAB environment, and show how researchers using EEGLAB can create and/or import BIDS-formatted, HED-annotated datasets. We also provide practical comments on event-related processing pipelines that can be applied *within* or *across* datasets, including those using different task paradigms.

Figure 1 gives a compact overview of the topics covered in this chapter. After reviewing some background material, we demonstrate how to import a BIDS dataset as an EEGLAB Study. We introduce tools available for reviewing and correcting HED annotation and other BIDS metadata, and perform re-annotation as necessary. Independent of metadata review, we demonstrate preprocessing the data including ICA decomposition [11], and as an example show how to test a simple hypothesis concerning source-resolved event-related potentials (ERPs) to sensory presentations of different classes of letters in a demonstration experiment.

2 Methods

2.1 Starting Point: Obtaining the Data

This chapter discusses end-to-end processing of shared or newly collected M/EEG data, from raw data to end results. We assume that the raw data have been made available in BIDS format from an open repository using BIDS formats such as OpenNeuro or its neuroelectromagnetic data portal NEMAR (https://nemar.org/). We also demonstrate tools for importing new or unannotated raw data into BIDS using EEGLAB tools.

2.2 Data Storage and Computing

We use data from one participant in the 24-participant Sternberg Modified Memory Task [12], available in BIDS format with HED annotation in OpenNeuro and NEMAR under accession number *ds004117*. Its paradigm has recently been chosen for replication as part of the EEGManyLabs reproducibility study [13]. The single-participant (~380 MB) demo dataset used here is available at https://osf.io/8brgv/ [*see* Chapter 2].

We assume for demonstration purposes that users will download the demo data on their local computer to experiment with using the tools we describe. Users should have at least 16 GB of memory on their local machine to comfortably run even the single-participant analysis. The Neuroscience Gateway (NSG, https://www.nsgportal.org) at the San Diego Supercomputer Center is a world neuroscience community resource that supports free use of high-performance computing resources to run user-defined analyses built on any of multiple analysis environments (e.g., MATLAB, python, R) and toolsets (EEGLAB, Freesurfer, Open Brain, TensorFlow, PyTorch, NEURON, etc.). NSG enables users to submit analysis scripts of their own design to process either their own or publicly shared (on NEMAR.org) data using either the nsgportal EEGLAB plug-in or the NSG web-browser interface. Data shared via the NEMAR resource (www.nemar.org) are immediately available for analysis via NSG without data download and re-upload. EEGLAB users can also use EEGLAB *nsgportal* plug-in tools to submit jobs to NSG directly from an EEGLAB MATLAB session, as discussed later in this chapter.

2.3 Software and Coding

Although some basic understanding of analysis scripting may be needed, the tools discussed here focus on those providing support for the use of GUIs (graphical user interface) based analysis pipelines. The demonstration analyses use EEGLAB running on MATLAB (The Mathworks, Inc.). To run the demos, users must have MATLAB installed and must download and install EEGLAB as described in https://eeglab.org/download/. An extensive EEGLAB tutorial is available at https://eeglab.org/tutorials/. Preliminary assessment of event structure and annotation using

Hierarchical Event Descriptors (HED) can be done entirely using online HED tools at https://hedtools.org/hed, with no coding or software installation required. We discuss two paths for running an EEGLAB analysis pipeline, one using EEGLAB GUI windows and/or command line scripts that run locally, and another using online high-performance computational resources freely available to neuroscience researchers via the Neuroscience Gateway (NSG).

2.4 Computational Requirements

Computational requirements for importing and annotating the data, as shown here, are minimal. In general, analysis time per data recording depends on the complexity of the analysis, while processing time on the entire dataset grows linearly with the number of processed recordings. When processing time is a limiting factor (making some analyses extend beyond the practical compute horizon), processing large bodies of data via NSG can take advantage of parallel processing across 128 or more cores, with GPU resources also available.

2.5 Background

In this section we review some useful background material about the demo data, BIDS data organization, EEG data formats, and some HED basics. If you are already familiar with these topics, you may skip this section or skip to Subheading 3.

2.5.1 Sternberg Working Memory Dataset

The example data recordings used here are from the Modified Sternberg Working Memory dataset [12], which can be examined on NEMAR (https://nemar.org) and downloaded from NEMAR (or from OpenNeuro.org) as dataset *ds004117*. The single-participant demonstration dataset used here is available separately at https://osf.io/8brgv/. Figure 2 shows a schematic timeline of each experiment task trial. Under BIDS, onset markers for each event are stored in an *events.tsv* file for each recording, as well as, here, in the EEG data itself in EEGLAB format. In both cases the event record has a tabular structure with each row representing an event onset marker and columns recording event aspects. One column always records the time of the event marker relative to the experiment timeline.

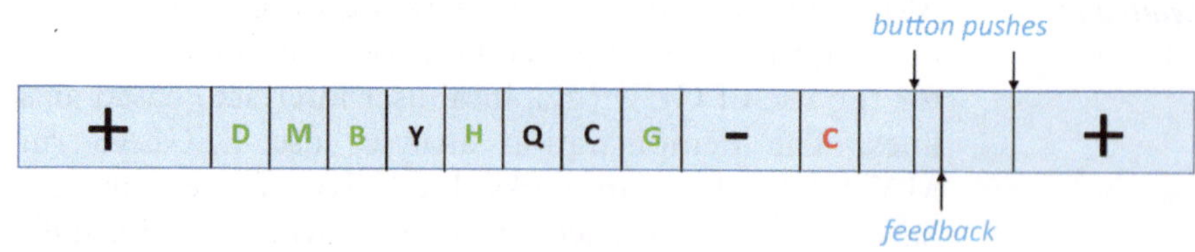

Fig. 2 Schematic timeline of the sequence of sensory presentation and participant action events in each trial of the Modified Sternberg Working Memory experiment. See the text for details including the meanings of the letter colors

In this experiment, each task trial begins with the display of a central fixation cross for 5 s followed by a sequence of 8 centered single letter presentations, each displayed for 306 ms, followed by a 1.4-s empty-screen delay. Each 8-letter sequence includes between 3 and 7 black letters "to be remembered" as well as (5 to 1) green letters "to be ignored". A central dash is then displayed for between 2 s and 4 s (the "memory maintenance" period). A red probe letter then appears, prompting the participant to click either a right-hand controller button (using their dominant hand index finger) if the probe letter was presented in the preceding sequence as a black (to be remembered) letter, or otherwise a left button (with their thumb). All participants were right-handed. The participant response was followed by a feedback sound—a "beep" for "correct" or a "buzz" for "incorrect". Thereafter the participant pressed either controller button to indicate their readiness to proceed to the next trial.

Note

As in nearly all actual experiment datasets, participants were not always able to perform each pair of required trial button press actions appropriately. Across the dataset, there were a few trials in which the participant pressed a button multiple times prior to receiving the trial feedback sound, a few trials in which they did not press a response button, and one session in which for some reason the nature and timing of the auditory feedback was not recorded. The raw dataset also included event marker sequences indicating incomplete trial presentations (e.g., trials in which no letters were presented).

Trial irregularities make it more difficult to perform automated analysis without building specialized programs to handle such irregularities. To help downstream users minimize required special handling, we recommend that, where appropriate, a trial-number column be included in the *events.tsv* file to allow analysis scripts to identify and then efficiently process valid trial events in the data.

If you are annotating a dataset new to you, it is useful to study the experimental event sequence and check, for each participant, whether all its recording segments (runs) and run elements (trials, if any) conform to the expected syntax. The Modified Sternberg Working Memory dataset paradigm is well-structured, producing a well-defined sequence of event onsets that should be present and accounted for in each data trial, making it easy to check for and exclude "bad" (non-standard) trials. If an experiment paradigm is not trial-structured or does not have an easily defined trial

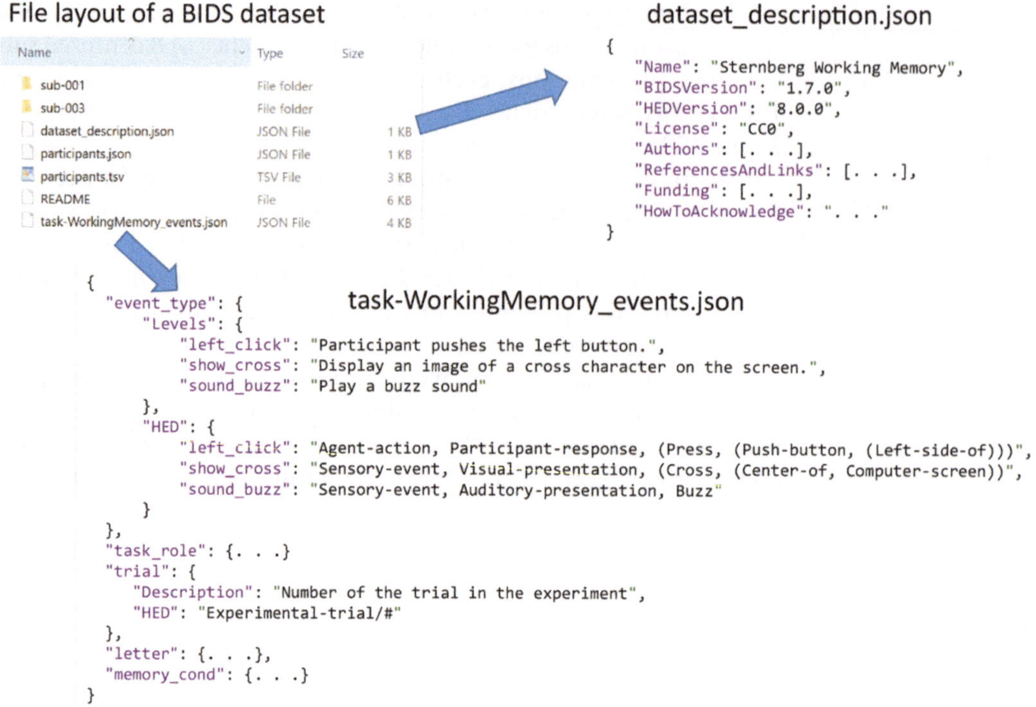

Fig. 3 The Sternberg Working Memory Demo dataset in BIDS format. Top left: the top-level file organization. Top right: The *dataset_description.json* file providing basic dataset identification. Lower: The *task-WorkingMemory_events.json* file giving event-related metadata for the dataset. In particular, it gives text descriptions and HED strings for events whose event types (here using any convenient titles) are indicated in a column of the event list file

sequence, this may require custom coding. The chapter by Denissen et al. introduces some tools that support event file restructuring without programming.

2.5.2 BIDS Dataset Format

BIDS specifications specify data and metadata formats and dataset disk file organization. Dataset metadata are stored in a variety of *JSON* (*.json*) and *tab-separated value* (*.tsv*) ASCII files. The top-level file organization for the Modified Sternberg Working Memory demo dataset is shown in the upper left of Fig. 3. Contents of two important top-level metadata files, *dataset_description.json* and *task-WorkingMemory_events.json*, are shown in the upper right and bottom center respectively.

BIDS datasets are organized by participant (subject), with each participant's data organized in a unique folder (named '*sub-xxx*'). The Dataset EEG data are stored in a participant *eeg/* subdirectory. Optionally, intervening sub-folders may provide organization of the participant's data into session subfolders, as in our example. Several excellent tutorials in the *BIDS Starter Kit* explain this file organization and its implications in more detail (see https://bids-standard.github.io/bids-starter-kit/).

The required top-level files (by name *dataset_description.json*, *participants.tsv*, and *README*) contain some overall information about the data. Of immediate interest in the *dataset_description.json* are the versions of BIDS and HED formatting used in building this dataset. A list of relevant reports describing the experiment and its goals and interpretation may also be included (though perhaps unfortunately BIDS does not require them).

Note
The top-level BIDS *dataset_description.json* file does not contain a text description of the experiment. A full experiment description and a description of the original experiment goals, written in plain language, should instead be contained in the top-level *README* file.

A BIDS file of particular interest for this chapter is the *task-WorkingMemory_events.json* file, which contains metadata about the types of events recorded in the data, including HED annotations that enable machine-actionable event analysis. This file will be discussed in more detail later.

2.5.3 EEG Data Formats

EEG datasets include continuous recordings of multi-channel EEG. Associated with these continuous recordings are event markers (in common practice, event *onset* markers), typically first recorded in log files produced by the experiment control software. BIDS allows EEG data themselves to be stored in any of four widely used formats: EEGLAB (*.set*, or *.set* plus *.fdt*), European Data Format (*.edf*), Biosemi (*.bdf*), and BrainVision (*.vhdr* + *.vmrk* + *.eeg*). EEGLAB has tools for converting each of these formats to its internal *.set* file format.

Note
European and Biosemi data formats store event and channel information within their respective *.edf* or *.bdf* files, while BrainVision data format stores event markers separately (in an *.vmrk* file).

An EEGLAB-based BIDS dataset can store data for a participant session either in a single, continuous *.set* file or in a combination of an *.fdt* file (containing the binary data) and a *.set* file (containing metadata). The latter format allows metadata for a large number of *.set* files to be gathered and stored in computer memory into an EEGLAB *Study* (equivalent to a BIDS *dataset*) and then manipulated, without needing to actually load the much larger *.fdt* data files.

Once loaded, either by calling the EEGLAB *pop_loadset* function or using the EEGLAB GUI, the data is stored in a MATLAB (EEG) data structure. Dataset event onset markers are stored in the *EEG.event* field of this structure, while channel electrode location and other information are stored in the *EEG.chanlocs* field. The EEG data itself are stored in *EEG.data* as a two-dimensional array, scalp channels by time samples.

Note

BIDS requires that event and channel information be stored separately from the data, in *events.tsv* and *channels.tsv* files, respectively. In any of the four supported EEG data formats, data event and channel information are also stored in the EEG data itself. There is, however, no guarantee (and BIDS does not check) that this internally stored information is consistent with the BIDS *.tsv* files. Hopefully, the BIDS formatted *.tsv* file data will be at least as complete and hopefully more complete than that stored in the raw datafiles. EEGLAB allows users to choose which information to import from the *.tsv* files and which to use from the stored data record itself.

2.5.4 HED Quickstart

As mentioned in the introduction, the HED system consists of a standardized term vocabulary of terms organized hierarchically to in the HED standard schema), as well as an extensive tool base supporting HED annotation, validation, and analysis. Figure 4 gives an overview of the standard HED term schema.

The HED schema is organized around six top-level HED tag subcategories (*Event*, *Action*, *Agent*, *Item*, *Property*, and *Relation*). Users create "HED string" event marker annotations as comma-separated lists of ("HED tag") terms from the schema. Because each term can appear in only one place in the schema, during the annotation process users can specify only the end tags (leaves) in the typically shallow tag hierarchies (subtrees) [*see* Appendix], leaving HED machine tools to fill out their full schema tag paths.

A second important point about the HED schema design is that terms lower in the hierarchy are subcategories of supervening terms (i.e., child nodes that satisfy an *is-a* relationship). This is important for search generalization. For example, a search for annotations containing the term *Event* will also return annotations containing *Sensory-event* or *Agent-action*.

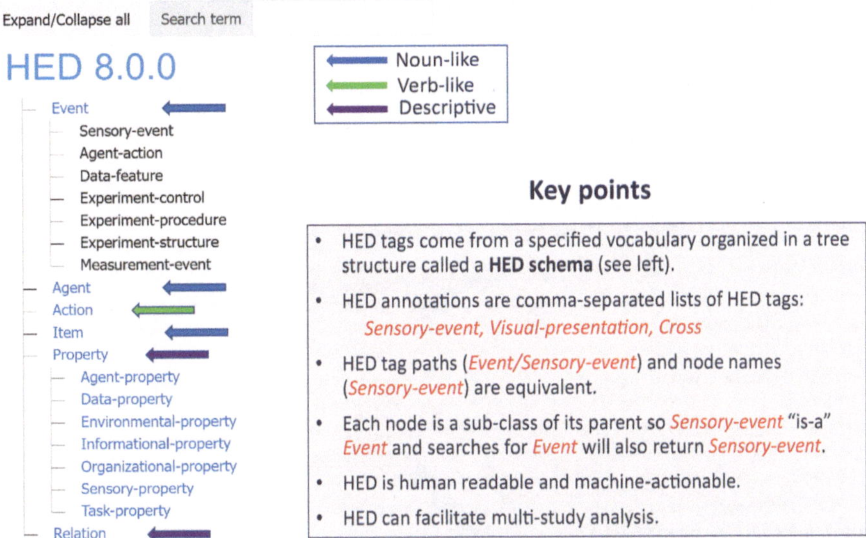

Fig. 4 The standard HED (Hierarchical Event Descriptor) term schema. Left: Partial view of the top-level HED standard schema. See https:/www.hedtags.org/display_hed.html to explore the full schema using an expandable accordion view. Arrows point to top-level tag sub-categories. Right: Some key points are noted

The HED official vocabulary is stored on GitHub in the *hed-schemas* repository of the *hed-standard* organization (https://github.com/hed-standard/hed-schemas). The official HED specification document is available at https://hed-specification.readthedocs.io/en/latest/. The vocabulary is versioned using semantic versioning; HED tools and tools that use HED tools (including the BIDS validator, https://github.com/bids-standard/bids-validator) retrieve a copy of the schema (in XML format [*see* Glossary]) during processing.

A HED schema is needed to perform validation because users usually give a HED annotation as a list of single terms and the software validator must verify that these terms are in the allowed vocabulary and that the term usage is consistent with its properties. For example, terms that take values must have consistent value type and units. Fetching of the schema and processing are done behind the scene by HED tools. These tools (and pertinent tutorials) are described below.

2.5.5 HED in BIDS

As mentioned in the introduction, the HED system is integrated into BIDS [*see* Chapter 4] and is the only BIDS-endorsed mechanism for documenting events using dataset-independent, machine-actionable metadata. However, HED is new to many BIDS users. Thus, (e.g.) a large number of time series datasets on OpenNeuro do not yet have HED tags.

Excerpt from top-level task-WorkingMemory_events.json file

```
{
   "event_type": {
      "HED": {
         "show_cross": "Sensory-event, Visual-presentation, (Cross, (Center-of, Computer-screen))",
         "sound_beep": "Sensory-event, Auditory-presentation, Beep",
                         . . .
      }
   },
   "task_role": {
      "HED": {
         "bad_trial": "(Invalid, Experimental-trial)",
         "feedback_correct": "(Feedback, Correct-action)",
         "fixate": "(Task, Fixate)",
                         . . .
      }
   },
   "trial": {
      "HED": "Experimental-trial/#"
   },
   "letter": {
      "HED": "(Character, Label/#)"
   }
}
```

onset	duration	event_type	task_role	letter	trial	memory_cond
5.716	1257.05	show_cross	fixate	+	1	3
10.96222	305.95	show_letter	to_ignore	D	1	3
12.40393	305.725	show_letter	to_ignore	M	1	3
13.84473	305.95	show_letter	to_ignore	B	1	3
15.28654	305.95	show_letter	to_remember	Y	1	3
16.72834	305.95	show_letter	to_ignore	H	1	3
18.17015	305.7	show_letter	to_remember	Q	1	3
19.61096	305.95	show_letter	to_remember	C	1	3
21.05296	305.9	show_letter	to_ignore	G	1	3
22.49447	842.125	show_dash	work_memory	-	1	3
26.08098	305.95	show_letter	probe_target	C	1	3
27.43899	0	right_click	remembered_correct	n/a	1	3
27.83909	0	sound_beep	feedback_correct	n/a	1	3
30.4181	0	right_click	indicate_ready	n/a	1	3

HED annotation for event at onset 5.716 s.

```
"Sensory-event, Visual-presentation, (Cross, (Center-of, Computer-screen)),
(Task, Fixate), (Character, Label/+), Experimental-trial/1"
```

Fig. 5 The assembly of HED annotations for a BIDS *events.tsv* file. Each column value in the events file (here, "show_letter", "to_ignore", etc.) is defined using a list of HED tags keyed to its name in the JSON *events.json* dictionary. The HED tags for each column value in a row of the .tsv events table are assembled to form the HED annotation for the event

As illustrated schematically in Fig. 3, HED annotation for most BIDS datasets consists of providing a single JSON file containing a dictionary associating values in the columns of the dataset event files with HED annotations. Once this dictionary is provided, tools can automatically take advantage of this metadata during event processing. This means that users can contribute HED annotations post hoc, and that re-annotation focused on enabling further analysis only involves modifying a single text file.

Figure 5 shows the mechanism by which this single top-level *events.json* file is used in conjunction with the dataset events files to produce HED annotations for each event. The top left of Fig. 5 shows an excerpt from the *task-WorkingMemory_events.json* file located in the dataset root directory. This JSON file is a text file containing a dictionary whose keys are the column names of the *events.tsv* files (as in the excerpt on the right in Fig. 5). For example, the *event_type* column has a small number of discrete code values (*show_cross*, *show_letter*, *right_click*, etc.). The JSON file associates a HED annotation with each value in the *event_type* column. Here, for example, the term *show_cross* is associated with the HED annotation "*Sensory-event, Visual-presentation, (Cross, Center-of, Computer-screen)*". The code-value names are arbitrary, but should best be informative.

Additional HED tags might be added to these event type definitions—here, for example, to record the color and/or size of the displayed cross. HED tags for such details might be added later by researchers to the archived and shared data to allow them to investigate questions about brain activity different from those of the original data authors—if these experiment details were preserved in some other way, for example in stimulus images, screenshots or recordings, or were specified in the preserved experiment control script. In general, data authors cannot be expected to anticipate all possible research interests of future data users—particularly when stimuli and/or tasks involved are more complex than in this experiment. Thus, there may thus often be some tension between the aim of data authors to document the exact nature of experiment events and the level of effort and imagination required of them to fully accomplish this.

The *events.tsv* files are columnar text tables with tab characters separating columns. Each row in an *events.tsv* file represents an event-time marker—currently nearly always a marker of its onset. BIDS requires that event files have an *onset* column giving the time in seconds of the event marker relative to the start of the associated data recording. HED also supports event *offset* tags and *inset* tags marking any intervening event phase transitions (for example, the moment of maximum amplitude of a musical sound or the maximum velocity of a movement arc).

BIDS has strict naming conventions for determining which JSON files are associated with each event file [*see* Chapter 4]. A discussion of these conventions is beyond the scope of this chapter, but suffice it to say that the JSON file named *task-XXX_events.json* applies to all *events.tsv* files that have *task-XXX* in their file names. So a single JSON file can be (and typically is) used to hold HED annotations of all the event types in an entire dataset. (See https://bids-specification.readthedocs.io/en/stable/02-common-principles.html#the-inheritance-principle in the BIDS specification document for details.)

Some columns in *events.tsv* such as, here, *letter* and *trial* can have many different numeric or text values. Rather than provide an HED annotation for each value separately, HED allows a single annotation with placeholder (#) to apply to every value in the column. The specific column value for each event marker is substituted for the # when the annotation is assembled.

The bottom of Fig. 5 shows the result of final assembled HED annotation. For each row (event marker) in the *events.tsv* file, HED annotations for the individual row column values are assembled from the JSON dictionary. If an annotation exists, it is included; if no annotation is present, the column value is omitted from the final assembled annotation.

Note

During event annotation the annotator should carefully check that all terms used in the *events.tsv* table are either values that will replace a # placeholder or are separately defined in the corresponding *events.json* file. Otherwise, the table term will not contribute to the assembled HED annotations.

HED has a full suite of online tools (https://hedtools.org/hed) to support HED annotation and processing. We encourage readers to download the Sternberg Working Memory demo dataset (available at https://osf.io/8brgv/) to explore these options. Figure 6 shows a screenshot of the operation to assemble HED annotations for the event file corresponding to the first recording (*run 1*) of participant (subject) *sub-001*. Fill in the options as indicated for the downloaded demo data and then press *Process* to see the output event file containing the assembled HED string event descriptions.

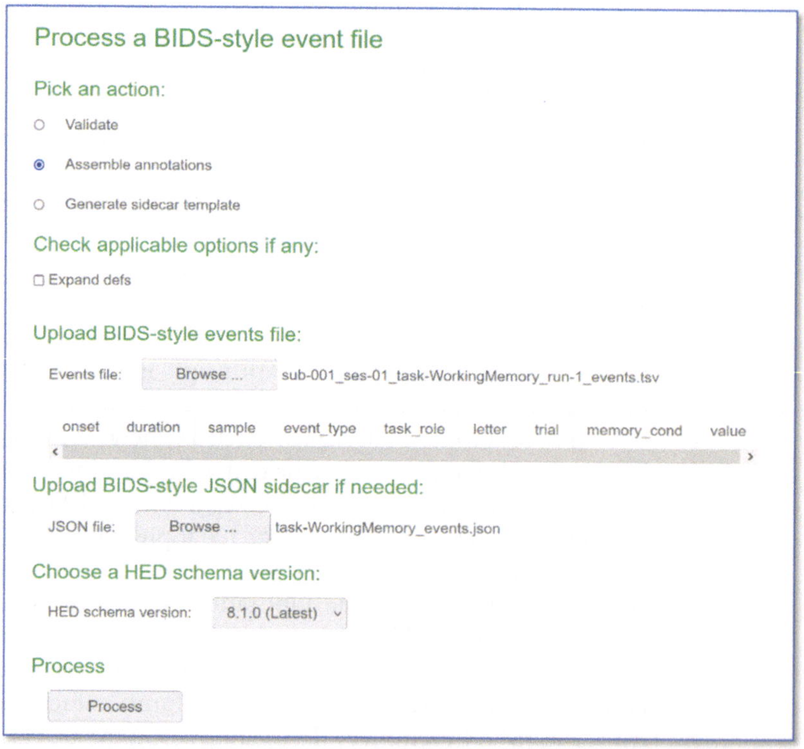

Fig. 6 An example of using the online HED tools to assemble HED annotations for a BIDS events file

The standard process for creating HED annotations in BIDS is to create a JSON template file from a representative *events.tsv* file using the online tool *Generate sidecar template* and then edit the resulting JSON text file directly or using the *Ctagger* tool GUI for assistance in selecting the appropriate HED tags to annotate the data.

A step-by-step process for annotation in BIDS is provided in the following quickstart tutorial:

https://www.hed-resources.org/en/latest/BidsAnnotationQuickstart.html.

CTagger, the HED annotation tool, can be run standalone or EEGLAB plug-in. A step-by-step guide to installing and using CTagger is available in the *Tagging with CTagger* tutorial:

https://www.hed-resources.org/en/latest/CTaggerGuiTaggingTool.html.

All of the tools provided by the HED Online Tools are also available as web REST services. A tutorial on how to use these services within MATLAB is available at:

https://www.hed-resources.org/en/latest/HedMatlabTools.html.

Sample code for calling these services from MATLAB can be found at:

https://github.com/hed-standard/hed-matlab/tree/main/hedmat/web_services_demos.

The most challenging part of learning to perform HED annotation is getting started making annotations. The HED annotation quickstart tutorial:

https://www.hed-resources.org/en/latest/HedAnnotationQuickstart.html

provides a step-by-step recipe for performing basic HED annotations.

A number of annotated sample datasets are available in GitHub *hed-examples* repository of *hed-standard*: https://github.com/hed-standard/hed-examples/tree/main/datasets.

Note

HED annotations vary in their complexity and completeness, and many datasets currently on OpenNeuro do not yet include HED annotations. However, basic HED annotations can now be added quite easily to a BIDS dataset using the methods we describe. As suggested earlier in this chapter, HED annotations (as well as event logs and files themselves) can be enhanced and/or modified at any later time to support analysis goals. This may become particularly important for research using novel objectives, or investigations across studies.

3 Methods

3.1 Setup

3.1.1 Software Installation

The remainder of this tutorial assumes that you are working with EEGLAB in MATLAB. If you do not already have EEGLAB loaded, download and install it by following the instructions available at https://eeglab.org/tutorials/01_Install/Install.html. You should also install the *bids-matlab-tools* and the *HEDTools* EEGLAB plug-ins. To download and install these plug-ins, go to the *Manage EEGLAB extensions* submenu of the *File* menu on the main EEGLAB, then search for and install these plug-ins directly through this menu.

3.1.2 Downloading Data

The demos in this chapter use the Sternberg Working Memory demo dataset, which is available for download at https://osf.io/8brgv/. This single-participant dataset is part of a 24-participant dataset from an experiment performed by Onton et al. (2005) [12] that is available on OpenNeuro (https://openneuro.org/datasets/ds004117).

3.1.3 Importing the BIDS Dataset

The analysis in this chapter assumes that the dataset is in EEGLAB Study format. The EEGLAB *bids-matlab-tools* [14] allow easy import and conversion through the *BIDS tools* submenu item of the *File* menu item in the main EEGLAB GUI. Figure 7 shows the BIDS import tool menu when the demo dataset is imported.

The EEGLAB BIDS import tool does not check the consistency of the BIDS files, but it does allow users to select which external information related to channels and events will overwrite the internal data. MATLAB scripts for checking the consistency between the BIDS external representation of events and of channels are available at:

https://github.com/hed-standard/hed-matlab/tree/main/hedmat/utilities.

3.2 Preprocessing and ICA Data Decomposition

For this demonstration, our preprocessing of the single-participant data began with high pass filtering the data with a cutoff at 1.5 Hz, as this is useful for subsequent ICA decomposition. The data were then re-referenced to common average and excessively noisy portions of the data were removed by running the *pop_clean_rawdata* EEGLAB plug-in [15, 16] *without* using data interpolation. Here, no excessively noisy ("bad") channels were detected or removed in this process. The data were then decomposed using Adaptive Mixture ICA (AMICA) [17] with its automated data rejection option engaged using default parameters. AMICA uses data rejection only during training, to learn an 'unmixing' matrix linearly transforming the data from the input channels to a set of independent component (IC) processes that, when back-projected through the

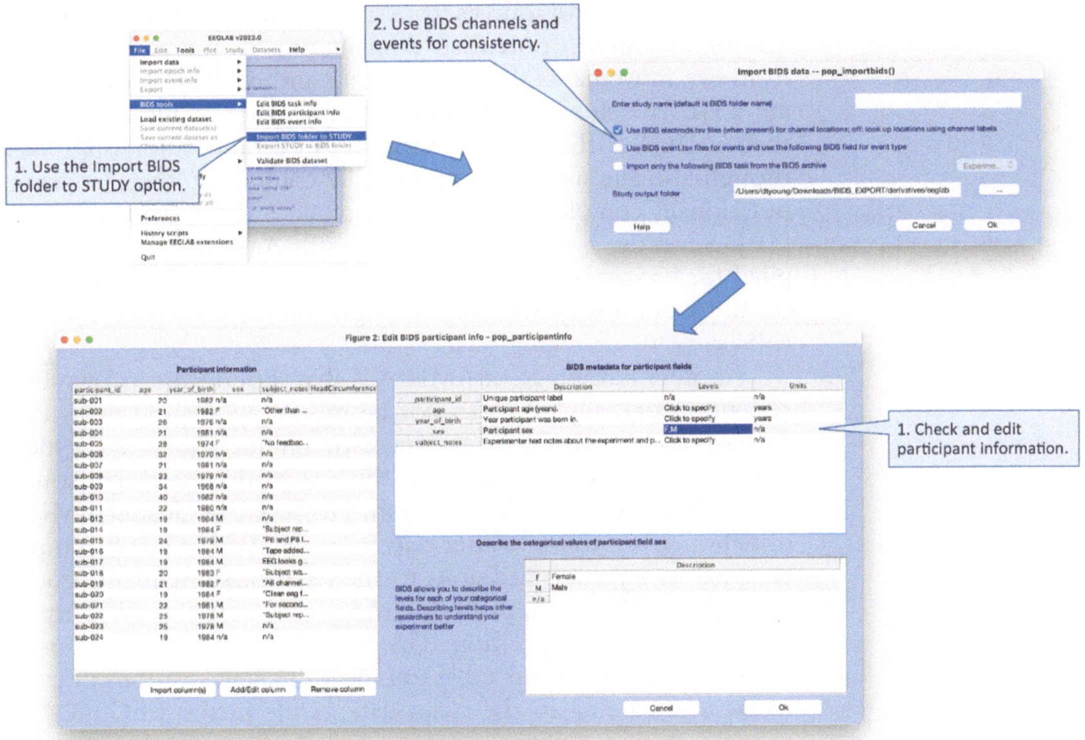

Fig. 7 The BIDS tools import tool for EEGLAB for the demo dataset

'mixing' matrix (the inverse of the unmixing matrix), will reconstitute the original data. ICA decomposition of EEG data finds ICs whose time courses are maximally temporally independent.

ICA decomposition is a powerful source separation method used to isolate both brain and non-brain source contributions in EEG data and data measures including ERPs [18]. The brain generated ICs that ICA discovers in the data may be better called "effective brain sources", as considered at either the neuronal or larger spatial scales the whole cortex is always active. However, local-scale activity in cortex, projected through the brain tissues, skull, and scalp and then summed at the scalp electrodes, is very largely canceled out by phase cancellation (positive voltage projections canceling negative projections). What dominates the "far field" scalp EEG signals are projections from larger areas of local field potential coherence that arise spontaneously by mechanisms that have not yet been well studied or understood. These projected potentials sum to contribute appreciable voltages (positive or negative) at the scalp channels, thereby constituting "effective brain sources" of EEG signals. ICA decomposition separates out the appreciable effective sources from spatially and functionally distinct non-brain sources (potentials contributing to the scalp channels from eye movements, line noise, scalp and neck muscle activities, etc.). We have shown, using mutual information reduction criteria, that AMICA is the most effective ICA decomposition approach, though also the most computationally complex. Other, less computationally complex algorithms approach AMICA in effectiveness [19]. Recently, a very efficient version of AMICA has been compiled for use on high-performance computer resources supported by and made freely available through the Neuroscience Gateway (NSG).

Another advantage of AMICA is that it performs its own data rejection internally, so as to not be misled by atypical noisy data patterns that appear in the data with unique (non-stereotyped) spatial nature incompatible with the assumption of spatial source stationarity used in ICA derivation. However, by default AMICA does not return the data after rejection, and work is just beginning to apply its rejection decisions to the data for subsequent processing. Thus, here we applied *pop_clean_raw_data* for this purpose prior to ICA decomposition. This cleaning appropriately adjusted the set file's EEG.events table to reflect the new locations of event markers remaining in the clean data, while in the EEG.urevents subfield storing pointers to the retained events in their original sequence and timing.

The EEGLAB plug-in *zapline-plus* [20] was then used to remove line noise from the IC timecourses ('activations'). This is a recently introduced approach to line noise removal; another is the *cleanline* plug-in [21]. Removing line noise contamination without losing contributions of other brain sources at the line frequency can be difficult since head movements, moment-to-moment changes in the electrical environment, or the possible presence of electrical line frequency sources with phase differences, together create spatial instability in the line noise contamination pattern throughout the data. Thus, although (as here) ICA decomposition typically gathers much of the line noise contamination in one or a few (here two) ICs, substantial line noise contamination was still present in other, by scalp map and power spectrum clearly effective brain source ICs.

This was supported by applying *ICLabel* [22], a neural network classifier trained on a large body of expert-labeled IC data, to automatically classify components as representing brain sources or as any of several classes of non-brain sources (see discussion below). Applying *ICLabel before* removing line noise correctly identified two strong line noise ICs but also caused *ICLabel* to classify effective brain sources still containing substantial noise as most likely representing line noise rather than brain source activity. *After* running *zapline-plus* on the scalp data and then applying the AMICA weights to the cleaned data to obtain the IC time courses (activations), *ICLabel*, now applied to the IC scalp maps and line-noise

```
% high-pass filter the data and remove (60-Hz) line noise
EEG = pop_eegfiltnew(EEG, 'locutoff',1.5); % High-pass filter

% re-reference the data channels to common average
EEG = pop_reref(EEG, []);

% Remove excessively noisy ('bad') data portions
EEG = pop_clean_rawdata(EEG, 'FlatlineCriterion', 'off', 'ChannelCriterion', 'off',
'LineNoiseCriterion', 'off', 'Highpass', 'off', 'BurstCriterion', 'off',
'WindowCriterion', 0.25, 'BurstRejection', 'off', 'Distance', 'Euclidian',
'WindowCriterionTolerances', [-Inf 20] ); ;

% Perform AMICA decomposition of the data
amicaout = [pwd '/amicaout'];
EEG = pop_runamica(EEG, 'outdir', amicaout, 'numprocs', 1, 'do_reject', 1, 'rejstart',
3, 'numrej', 3, 'rejint', 3, 'pcakeep', EEG.nbchan);

% Remove line noise from data using Zapline_plus plug-in
EEG = clean_data_with_zapline_plus_eeglab_wrapper(EEG,struct('noisefreqs','line'));

% Remove line noise from IC activations using the line noise-removed data
EEG.icaact = EEG.icaweights*EEG.icasphere*EEG.data;

% Apply ICLabel to categorize component types
EEG = pop_iclabel(EEG, 'default');
```

Fig. 8 Preprocessing pipeline script for the examples used in this chapter. The data are first high pass filtered above 1.5 Hz (using the *pop_eegfiltnew* function). Then the data (recorded to a common reference electrode placed behind the right ear) are transformed to average reference. Next we use the *clean_rawdata* plug-in of EEGLAB to remove artifacts, here using conservative parameters (no channel rejection, data-portion rejection with default parameters, no channel interpolation). AMICA decomposition of the data is performed, then 60-Hz line noise is removed, here by *zapline-plus*. The independent component (IC) activations are recomputed following line noise removal, and the ICs are again categorized by *ICLabel*

cleaned activations, confidently classified the ICs in question as representing brain sources, while classifying the two near-wholly line noise ICs only as "Other" (e.g., non-brain).

Figure 8 shows a MATLAB script containing the entire preprocessing pipeline used here. The script can be run on a single participant (as in Fig. 8) or can be applied in a loop or as part of an automated pipeline executed on remote compute resources, as discussed later in this chapter.

Figure 9 shows results of the AMICA decomposition of the 71-channel data in the form of the scalp maps of the largest 35 (of 71) IC processes. All 71 IC scalp maps and activations were then input to the *ICLabel* EEGLAB plug-in for component identification. IC7 and IC15 accounted for much, but not all, of the line noise contamination in the data; the associated scalp maps are quite incompatible with local field activity projecting from a single (or even strongly-connected dual) cortical source area.

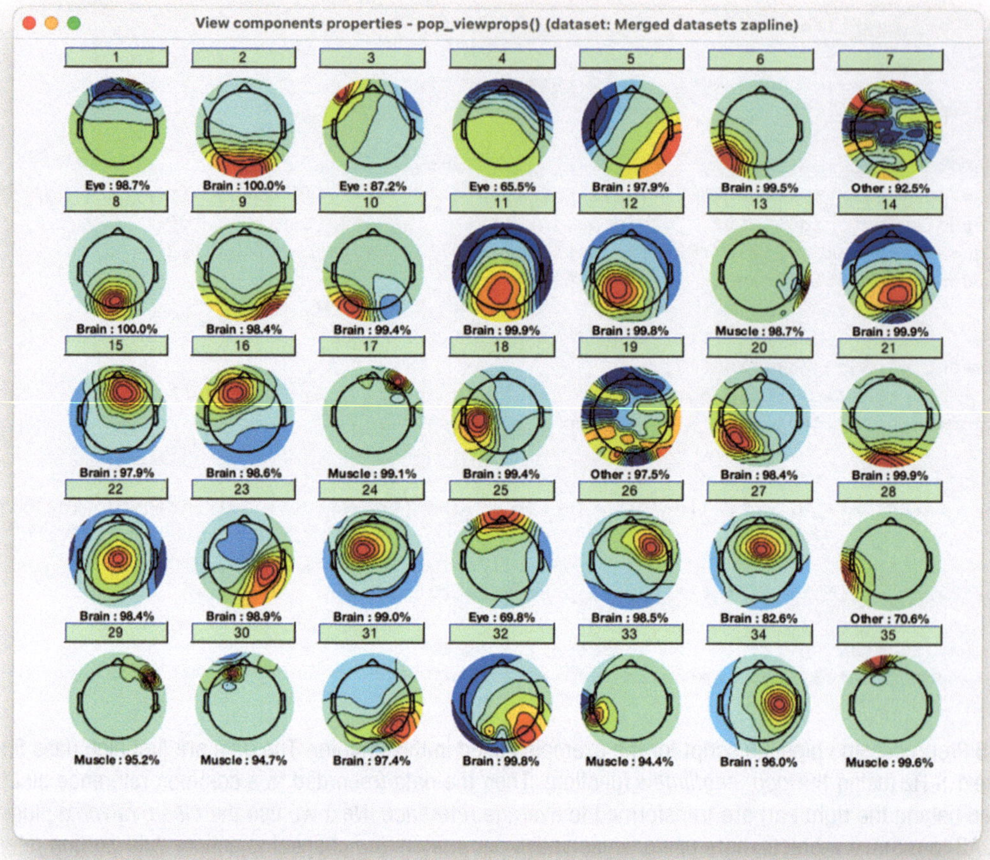

Fig. 9 Output of ICLabel after removing (60-Hz) line noise using *zapline-plus*. Here, ICA decomposition was performed before removing the line noise. The returned ICs accounting for most line noise in the data (here, ICs 7 and 19) are thus here labelled by *ICLabel as 'Other' rather than as 'Line Noise'*

Before removing line noise, *ICLabel* mis-classified brain source ICs such as IC8 as "Line noise" because of remaining line noise contamination. After removing line noise from the IC activations, *ICLabel* here correctly classified ICs such as IC8 as effective brain sources (with high likelihood), while the two major line noise sources (ICs 7 and 19) are now classified simply as *Other*. Several ICs classified as *Muscle* (ICs 13, 17, 29, 30, 33, 35) are compatible with effective sources of surface-recorded electromyographic (EMG) activity, which projects most strongly to the skin from the ends of individual scalp/neck muscles (i.e., from the muscle/tendon interface).

In recent years, several EEG data preprocessing pipelines have been developed and published by different laboratories. To our knowledge, there has been no systematic review of these, nor is it quite clear what measures should best be used for fair comparison. It is best that EEG researchers acquaint themselves with the problems involved in adequate data preprocessing and test for themselves the particular pipeline they use or construct for this purpose. Because there is no agreed-upon standard [23], we hesitate to promote a particular approach (see https://osf.io/8brgv/ for an example of an automated pipeline). We prefer to involve ICA decomposition in this process as it is shown to perform well in identifying and separating out several classes of non-brain source signals typically mixed in the scalp data (eye movements, scalp muscle activities), as well as identifying major, spatially localizable effective brain sources that together account for much of the brain's (largely cortical) contribution to the scalp data.

3.3 Epoching with Event Codes and HED

Onton et al. (2005) analyzed the dynamics of frontal midline theta in this modified Sternberg dataset during presentations of three different letter types. The mean trends in this study replicated previous findings (i.e., stronger frontal midline theta during letter presentations corresponded to more letters being held in memory). Yet these results accounted for relatively little of the trial-to-trial variation in theta power in the frontal sources. Therefore, theta dynamics were compared across working memory loads and then further decomposed into event-related spectral perturbations (ERSPs) across single trials to obtain additional insight. For researchers who are new to the dataset and interested in replicating or extending the analysis results, an integral step after running a preprocessing pipeline is to identify experimental events to use for epoch extraction.

"Epoching" of EEG data refers to extracting sections (epochs) of the data of equal duration time locked to particular classes of experiment events of interest, in order to compute data measures and perform statistical comparison on these and/or the epoched data. The standard method of data epoch extraction in EEGLAB is to use the *pop_epoch* function, specifying specific values in a particular column of the EEGLAB *EEG.event* structure, typically using the *EEG.event.type* field to select the desired class(es) of events. This requires studying the event type terms used by the original investigators, which may likely be idiosyncratic and are often opaque (e.g., "type 17")—and often fail to record distinctions that might in future prove fruitful to analyze.

EEGLAB now also supports epoching using (more detailed and informative) HED tag information. HED tags are stored in the *EEG* structure under EEG.etc. HED, as search terms using the *pop_epochhed* function. Using HED tags allows researchers to query events of interest in a more semantically meaningful way, such as "*Green*, Letter" and "*Press*, Push-*button*" instead of having to work with the originally-assigned cryptic alphanumeric codes (e.g., event types "17" and "256").

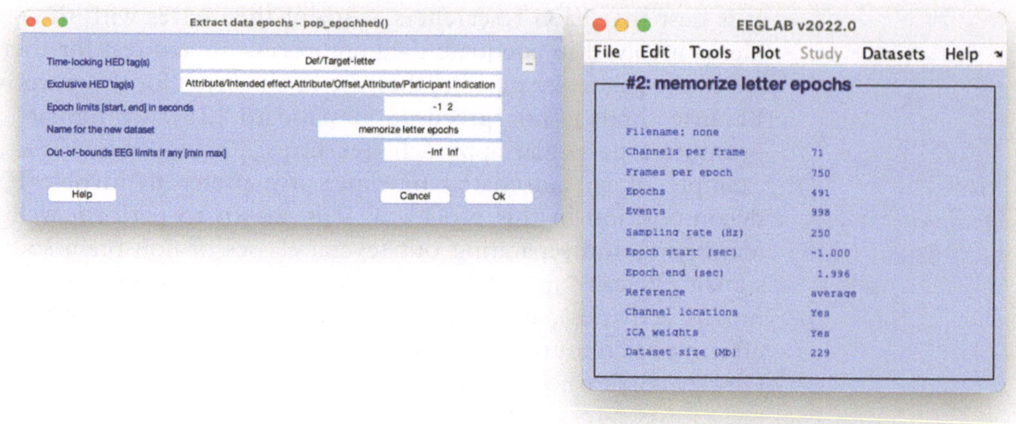

Fig. 10 The *pop_epochhed* function accessed through the Tools menu in EEGLAB

Figure 10 shows the graphic interface for the *pop_epochhed*. Tags can be specified in various ways including in combinations including AND and OR. More complicated search options are also available. The example shows looking for events that have the *Def/Target-letter* tag. Here *Def* indicates a user-defined term *Target-letter*. The definition of this user-defined term, as well as all HED annotations of the dataset can be found in the *events.json* file that accompanies the BIDS dataset (see the excerpt in Fig. 11) and is imported into the EEG.etc. HED field when the BIDS dataset is imported. HED tools look up the tags based on the annotations provided in the JSON sidecar and associate these annotations with the information in *EEG.event*.

Figure 10 shows a search for events containing the user-defined term *Target-letter*. The *Def/* prefix indicates that this term was user-defined using HED tags rather than being drawn directly from the base schema. The epoch start and end times are indicated relative to the position of the specified events.

Figure 11 shows that *Target-letter* is defined, using terms in the base schema, as:

"(Condition-variable/Letter-type, (Target, Memorize), (Letter, Black)))"

This definition includes the *Condition-variable/Letter-type* tag, as do the definitions for *Non-target-letter* and *Probe-letter*. By associating these three terms (*Target-letter*, *Non-target-letter*, and *Probe-letter*) with a common condition variable, the annotator indicates that these terms are three aspects of the same concept.

Most commonly, *Condition-variable* tags are used to group levels of an experiment stimulus or action condition associated with an experimental design. This mechanism allows data authors

```
"task_role": {
    "HED": {
        "bad_trial": "(Invalid, Experimental-trial)",
        "feedback_correct": "(Feedback, Correct-action)",
        "feedback_incorrect": "(Feedback, Incorrect-action)",
        "fixate": "(Task, Fixate)",
        "ignored_correct": "((Recall, Non-target), Correct-action)",
        "ignored_incorrect": "((Recall, Non-target), Incorrect-action)",
        "indicate_ready": "(Appropriate-action, Label/Indicate-ready)",
        "probe_not_shown": "Def/Probe-letter, (Cue, Non-target)",
        "probe_target": "Def/Probe-letter, (Cue, Target)",
        "remembered_correct": "((Recall, Target), Correct-action)",
        "remembered_incorrect": "((Recall, Target), Incorrect-action)",
        "to_ignore": "Def/Non-target-letter",
        "to_remember": "Def/Target-letter",
        "work_memory": "(Cue, Recall)"
    }
},
"curr_memory_load": {
    "Description": "The number of target letters shown up to this point",
    "HED": "(Condition-variable/Working-memory-load, ((Letter, Memorize),
Item-count/#))"
},
"condition_def": {
    "HED": {
        "memorize_cond_def": "(Definition/Target-letter, (Condition-variable/Letter-type,
(Target, Memorize), (Letter, Black)))",
        "ignore_cond_def": "(Definition/Non-target-letter,
(Condition-variable/Letter-type, (Non-target, Ignore), (Letter, Green)))",
        "probe_cond_def": "(Definition/Probe-letter, (Condition-variable/Letter-type,
(Cue, (Press, Mouse-button)), (Letter, Red), Recall))"
    }
}
```

Fig. 11 An excerpt of the top-level events.json sidecar for the Sternberg dataset

to encode the *experiment design matrix* in a standardized format that is consistent with the event structure. HED tools can automatically extract design matrices and also factor vectors associated with the *Condition-variable tags*. For additional information see the tutorial on HED Conditions and Design Matrices at:

https://www.hed-resources.org/en/latest/HedConditionsAndDesignMatrices.html.

Note also that the *Target-letter* definition above might easily be extended to note that the letters presented were in *upper_case, non_serif* font in the *Roman alphabet*, and their displayed width (in approximate degrees of viewing angle). These details might be of little or no interest for the analyses planned by the data authors themselves, but after the data were shared publicly might prove to be of real interest to some future studies, for example a study of reading of different character sets. However, when that interest arose, information about the size and case of the letters presented in this study might no longer be available.

Such considerations might provide data author annotators some incentive to fill in such additional details during initial annotation. In fact, however, the current base HED schema does not contain terms for letter case, font, and alphabet. Therefore, the annotator wishing to record these details would need to either introduce off-schema terms into the annotation or else wait for them to be introduced into HED, most likely in the newly released HED *LANG* (language) library schema for linguistic terms (now in its first release). However, less extensive HED descriptions (such as those in Fig. 11) do indeed represent a real advance beyond the long practice of recording event onset markers only as "*type, 17*" and the like.

3.4 ERP Analysis Using HED-Based Epoch Extraction

The ICA decomposition by AMICA described in Fig. 9 was used to identify the effective brain source components in the data for source-resolved analysis. Using the HED-based epoch extraction demonstrated above, we here extracted epochs time locked to letter presentation onsets of the three types of stimulus conditions used in the experiment (*Def/Target-letter*, *Def/Non-target-letter*, and *Def/Probe-letter*). ERP trial averaging was then performed on the resulting epochs, with the results shown in Fig. 12 as visualized by EEGLAB function *envtopo*. These plots use a pair of thick black traces to plot the outer *envelope* of the bundle of 71 individual scalp channel ERP traces (i.e., the respective maximum and minimum channel values at each ERP trial latency).

The six independent component (IC) scalp maps show the scalp projection patterns learned by ICA decomposition for the six brain-source ICs making the largest contributions to the ERPs (across the 71 scalp channels) within the ERP trial latency window (50–400 ms) indicated by the dotted vertical lines. The top and bottom edges of the blue shaded areas show the envelope (again, the max and min channel values at each latency) of the joint (i.e., summed) projections of these six IC processes. Colored trace pairs show the envelope of the respective IC projections in the ERP data (again, the max- and min-value IC channel projections at each latency).

Plotting the data and channel envelopes in this way allows inspection of the time courses of multiple ICs in the ERP data period in comparison to the whole scalp ERP data. Note that most IC brain sources contribute to more than one ERP peak. Note also that the seeming much larger negative ERP peak near 120 ms (N1) following probe letter presentation onsets (bottom panel) is predominantly accounted for by larger peaks in the projections of the lateral occipital IC6 (with left posterior scalp projection, red traces) and IC9 (projecting predominantly to right posterior scalp, blue traces).

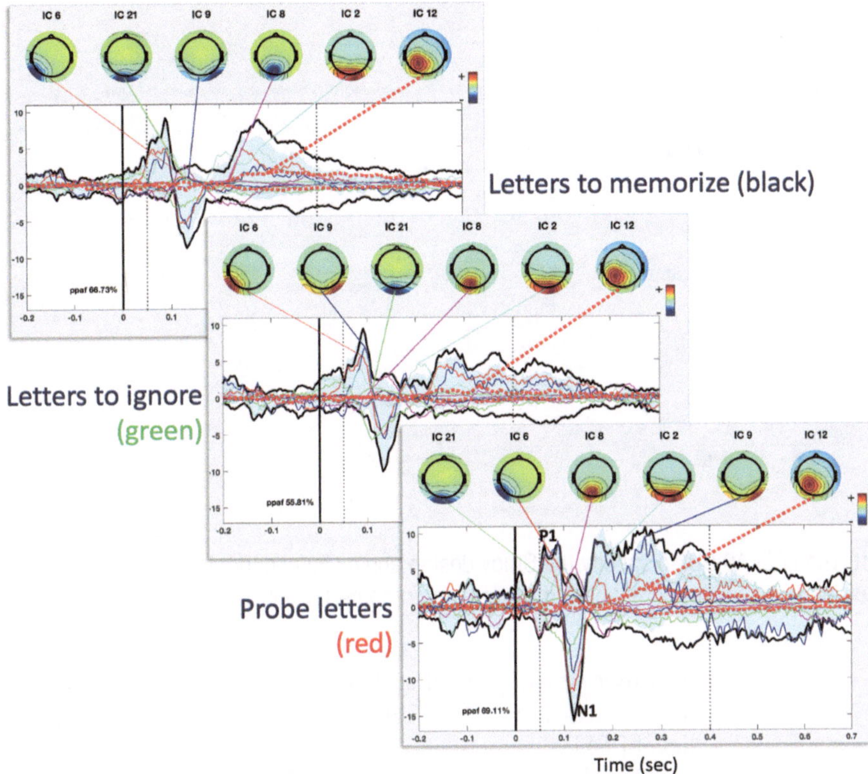

Fig. 12 Brain source resolved ERP trial averages, after removing large IC projections from eye movements, scalp muscle noise, plus small contributions to the data from other non-brain source ICs. The first two ERP peaks are conventionally referred to as P1 and N1. The three averages are time locked to onsets letter presentations of three types: (black) to be memorized letters ($n = 491$), (green) to be ignored letters ($n = 296$), and (red) probe query letters ($n = 99$). See text for further details

3.5 Computing Statistics for an EEGLAB Study

EEGLAB *Studies* (the EEGLAB equivalent of BIDS *datasets*) and EEGLAB *Study designs* provide a convenient way to compute single-participant and group level statistics for various EEG measures. For each design, users can define its independent variables by selecting the column of the event structure as the variable and the column's values as the variable's levels. Once the *Study design* is specified, EEG measures, including ERP, ERSP, power spectrum, and Intertrial coherence can be computed automatically. These measures apply to both the channel space and the independent component source space. Figure 13 shows sample menus for selection of statistics within an EEGLAB Study.

Once EEG measures are computed, users can choose to perform statistical analysis to compare measure values across levels of the specified independent variables. The LIMO-EEG plug-in extends EEGLAB statistical capability by allowing for an arbitrary

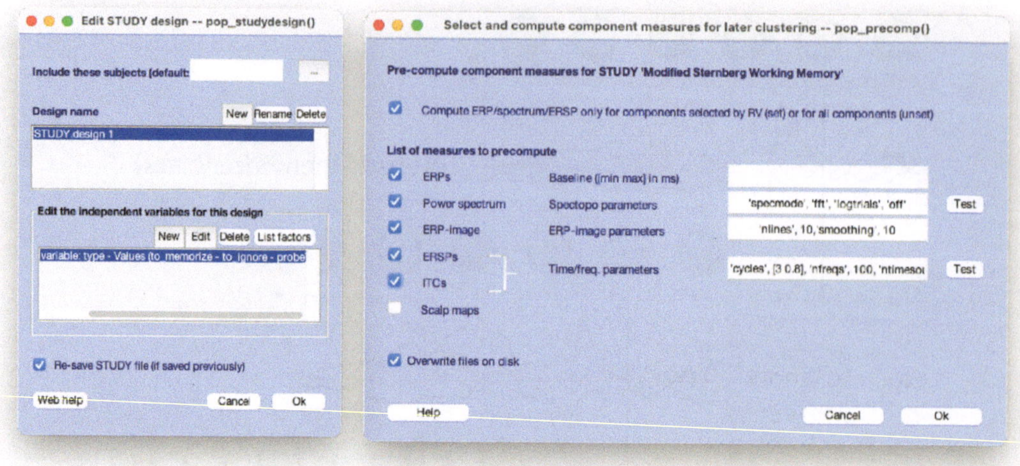

Fig. 13 (Left) The EEGLAB GUI to specify the Study design and its independent variable(s). (Right) The EEGLAB GUI to select the independent component measures to compute for subsequent statistical comparison

number of categorical and continuous variables in trial averages to be contrasted statistically by comparisons at the single-trial level [24].

Work is now underway to integrate *HEDTools* facilities for automated extraction of the experiment design matrix more tightly into the EEGLAB Study infrastructure. These facilities will allow users who import a HED-annotated BIDS dataset to run a chosen preprocessing pipeline and automatically compute all measures needed for statistical comparisons based on the *Study design* extracted from HED condition variable annotations themselves. Thus for a well-formatted HED-annotated BIDS dataset, the steps from importing those datasets into EEGLAB to reproducing statistical results as intended by the original data authors can be made automated. Of course, the primary uses for HED annotation lie in making possible more flexible (as well as larger scale) computing on neuroimaging data. It will create and fit new, more informative models of event-related brain dynamics occurring within a wide array of contexts and purposes (e.g., for basic understanding of brain function, neurological and psychiatric diagnosis, cognitive monitoring).

Many other functions are available within the EEGLAB platform and its associated plug-ins. While executing an analysis through the GUI is useful for exploratory work, it becomes tedious for large-scale analysis. After executing commands using the EEGLAB GUI during exploratory analysis, researchers can create a script for automating the process by editing the *ALLCOM* structure that accumulates the underlying EEGLAB commands that

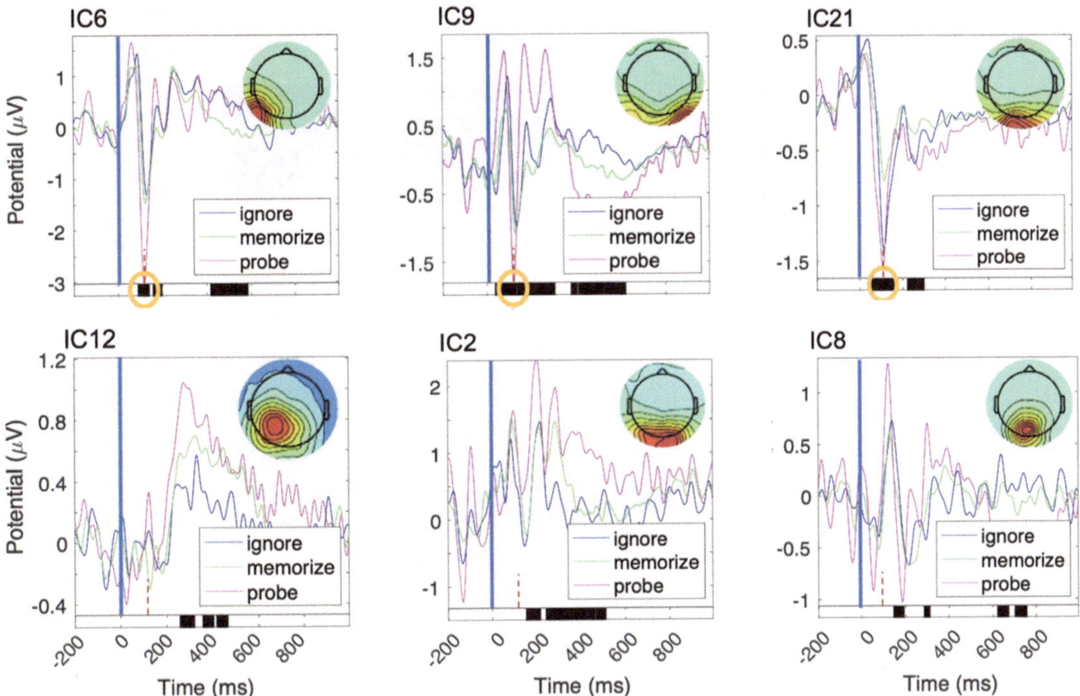

Fig. 14 Independent component (IC) ERPs in the three conditions, with regions of statistical significance for condition differences shown below ($p < 0.01$). Three IC ERPs in the top row (ICs 6, 9, and 21) exhibit pronounced N1 peaks. However, their condition differences differ: for medial occipital IC21 (top right), the N1 peak for letters to be memorized (green traces) is smaller than for the other two letter types, while for the lateral occipital ICs 6 and 9 (top left), the N1 peak in the ERP time locked to probe letters (magenta trace) is stronger than in responses to either the memorize or ignore letters

were executed at each step. This script can then be used for automated analysis exploiting cloud resources as described in the next section.

Figure 14 plots the projections of the six ICs contributing most strongly to the ERPs in the three analysis conditions. Here, each trace represents the trial-averaged IC activations in units of rms microvolts per scalp channel across all channels. (The projection of each IC to each channel ERP is the product of the averaged IC activation time course with the 71 IC scalp map values.)

To perform statistical testing of these results, we made a single-participant EEGLAB Study including the three sets of epoched trial data. We then tested for statistical difference between the three conditions at each ERP latency using methods from the LIMO toolbox [25]. Figure 14 highlights the three IC sources that exhibit a significant condition difference at the "N1" ERP peak (near 120 ms). For two of these ICs (6 and 9), the N1 peak projection is stronger in response to probe letter onsets (magenta). For IC21, the response to target (memorize) letter onsets is weaker than in

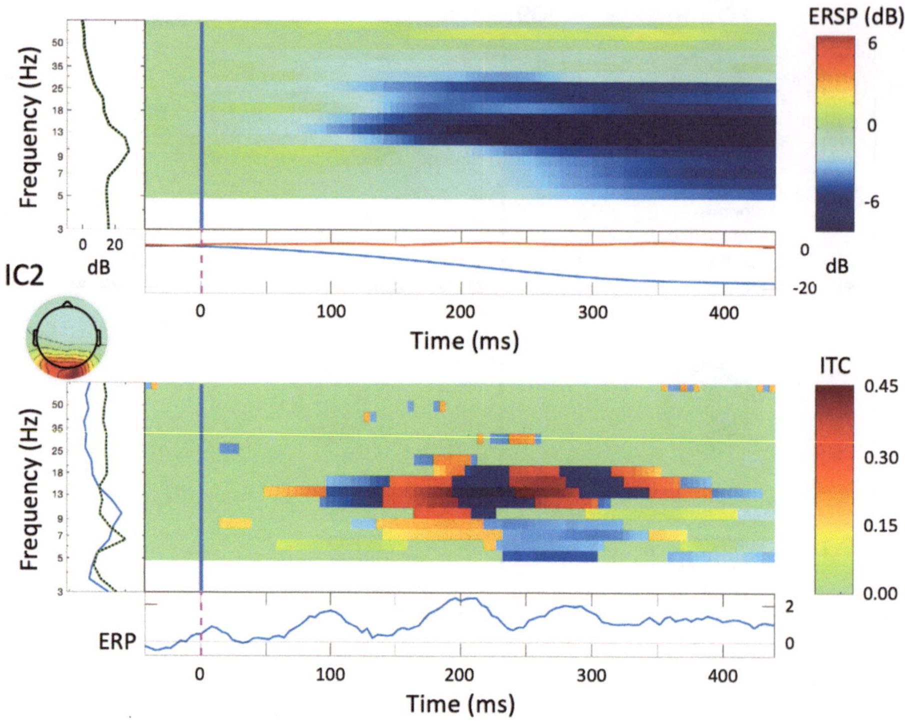

Fig. 15 (Upper panel) Event-related spectral perturbation (ERSP), and (lower panel) Inter-trial coherence (ITC) plots for IC2 epochs time locked to probe letter onsets (at time 0). Figure produced by EEGLAB function *pop_timef*, with later text enhancement for publication. The trial ERP is shown below the ITC plot. See text for further details

the other two conditions. The other three ICs contributing most strongly to the ERPs do not exhibit a peak projection at the N1 peak latency, nor a projection condition effect at that latency.

3.6 Source-Resolved Time/Frequency Analysis

Although ERP measures of event-related brain dynamics dominated EEG research in cognitive psychology and psychiatry for more than 40 years, ERPs alone do not reveal and cannot be used to model every important aspect of event-related EEG brain dynamics. Another set of measures based on time/frequency analysis—measuring mean changes in EEG power and phase spectra time locked to events of interest—reveal information complementary to that revealed by ERPs. Figure 15 shows the trial-mean event-related spectral perturbation (ERSP) [26] and inter-trial coherence (ITC) for IC2 (compatible with a bilateral occipital pole brain source) time locked to probe letter presentation onsets.

The upper panel (ERSP) shows that the upper part (~13 Hz) of the (~10 Hz) alpha peak in the baseline log power spectrum (the end of the memory maintenance period; top panel, left box) is profoundly suppressed (top panel, lower box) during probe stimulus presentation—a nearly 20 dB decrease from the pre-probe baseline period.

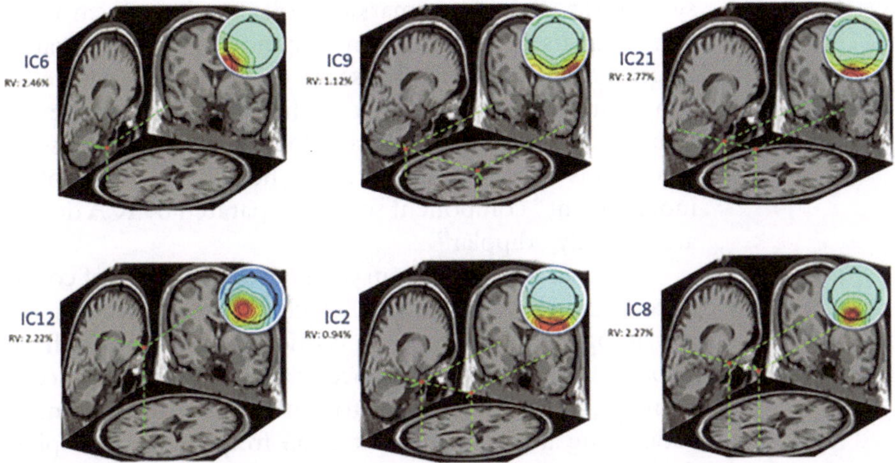

Fig. 16 Equivalent dipole models of source locations for the six sources making largest contributions to the letter presentation ERPs. Here "RV" refers to residual variance in the IC scalp map not explained by the indicated single or bilaterally symmetric dual equivalent dipole model projection to the scalp (Fig. 14). See text for details

The lower panel Fig. 15 shows that while in the pre-stimulus baseline period trial phase with respect to the upcoming stimulus onset is random (green background), phase of remaining activity near 13 Hz (and less strongly so at 7–9 Hz) becomes regularized (blue/red line segments representing negative and positive phase) with respect to stimulus onset. This causes partial failure of phase cancellation across trials, thus producing the trial-mean ERP (bottom panel, lower box) that resembles a ~13-Hz oscillation in the trial-average ERP for this effective brain source component.

The strong low-beta band activity suppression in lateral occipital cortices (cf. Fig. 16) might be tentatively interpreted as ending during and after Probe letter presentations, the suppression of activity in object recognition centers that was encouraged by "alpha flooding" of these areas during task Memory maintenance periods—possibly to enhance object memory retrieval involving these same areas.

3.7 Localizing Sources

The many effective brain sources separated from EEG data by ICA decomposition have a common feature not anticipated nor promoted by the decomposition algorithm itself. They have highly "dipolar" scalp maps, meaning that their scalp maps, learned from the data during decomposition almost exactly resemble the projection of single oriented equivalent dipoles—an imaginary infinitesimal battery located somewhere in the brain volume [19]. In some instances, these "brain IC" scalps may require a dual equivalent dipole model, most often the locations of the two equivalent dipoles (though not their orientations) being nearly bilaterally symmetric. This is a remarkable observation, given that the ICA algorithm itself is given *no* information about the locations of the electrode channels on the scalp or about the nature of current propagation from the cortex to the scalp through the intervening brain, skull, and skin media. One might put it this way: "ICA algorithms don't know there is a head". Yet many of the "most independent" component sources separated by ICA decomposition are strongly "dipolar".

This can only arise from the neurophysiology of cortical coherence in local field potentials, which can only spread through direct physiological connections in cortical neuropile, which in turn are strongly weighted toward local (<0.1 mm) connections between neurons, in particular inhibitory neurons that play a major role in supporting field dynamics at EEG frequencies. The physiological fact means that synchrony (or near synchrony) in cortical tissue must typically spread out from an origination point (Walter Freeman likened these to "pond ripples"). When field synchrony spreads out from a central field oscillation, its phase near-uniformity across a cortical area gives it far stronger contribution to the electrically distant scalp electrodes than the same area when it is out of synchrony. This is most likely the reason that the projection patterns of effective brain source ICs so often closely resemble the projection patterns of a single equivalent model dipole, e.g., one located at or near the origin of the cortical spreading field pattern and oriented near perpendicularly to the local cortical surface. The origin of the dual-symmetrical dipole models required to fit the scalp maps of other effective brain ICs then likely arises from synchronous field activity patterns arising in two cortical patches that are bidirectionally coupled, either directly by white matter tracts such as *corpus callosum* or possibly indirectly by being driven by strong common input.

Residual variance (RV) in the IC scalp maps learned by ICA decomposition from the data is computed by regressing out the projection of the (best-fitting) equivalent dipole in a (here, template) electrical head model, across all 71 scalp channels and measuring the variance of the residual map values. The ratio of the residual map to the orignal IC scalp map variance is the residual variance (RV). The RVs in Fig. 16 range from 0.94% to 2.27%, values likely no larger than those expected based only on the expected rough fit of the template head model to the participant's actual head (i.e., their exact head geometry and tissue conductance values).

Note that the model equivalent dipole location for IC12 (lower left) is in cortical white matter rather than gray matter. The foremost error in EEG source localization is that created by using a uniform template value for skull conductance, which in fact varies widely across individuals, even across adult individuals. Using an inaccurate value for skull conductance in the electrical head model used in for source localization will typically drive the implied source location deeper into the head (or less deep) than actual. The SCALE algorithm for using ICA decomposition to estimate the largest source of error in EEG source localization, namely skull conductivity [27], requires an individual MR head image, so could not be used here. The equivalent dipoles for IC9 (top middle) are in the cerebellum; this is likely not their actual locations, as effective sources in the cerebellum of sufficient strength to be captured as an (larger) effective source IC have not so far been demonstrated. Rather, the mislocalization more likely arose here through the projection of the basal posterior bilateral effective source being at the lower edge of the scalp electrode montage and therefore not well enough represented in the IC scalp map, and/or through the use of an incorrect (here) template value in the electrical head model used to compute the equivalent dipole locations.

Note that the *equivalent* dipole (or dual-symmetric dipole) modeling used here does not rest on an assumption that the true sources are infinitesimally small patches of cortex—this would clearly be physiologically impossible. Rather, the attraction of a single (or symmetry-constrained dual) equivalent dipole model is the oft demonstrated fact that the current projecting from an equivalent dipole located in the brain AND of one (or two) cortical patches surrounding the equivalent dipole(s) will be nearly identical. That is, an equivalent dipole is the model dipole whose scalp projection should nearly equal the projection of a cortical patch covering it. EEGLAB now includes a plug-in, the "Neuroelectromagnetic Forward problem head modeling Toolbox" (NFT) [28] to estimate the mean location and cortical surface extent of cortical effective sources learned from the data by ICA decomposition, though these tools could not be applied to these data as they also require availability of an individual MR head image for the participant.

3.8 Processing EEG Data Using High-Performance Computing

We have demonstrated an end-to-end EEGLAB workflow for a BIDS dataset with HED annotation from importing to statistical analysis. Such workflows can be performed locally on a user's own workstation. However, as public sharing of datasets becomes more prevalent and analysis dataset sizes grow, and/or as computational processing applied in EEG brain imaging becomes more intensive, researchers may find their local compute resources too limited for the desired data analyses. Fortunately, online data portals and cloud computing resources are being made publicly available for researchers. The "NeuroElectroMagnetic Archive and compute Resource" (NEMAR) and the Neuroscience Gateway (NSG) are two such facilities being developed and made to function cooperatively to support the analysis and meta-analysis of human electrophysiology data using publicly available data, tools, and compute resources [*see* Resources].

The joint OpenNeuro/NEMAR/NSG resource recently brought online by a collaboration between the developers of NEMAR, EEGLAB, the Neuroscience Gateway (NSG, http://www.nsgportal.org) [29], and the OpenNeuro neuroimaging data archive (https://openneuro.org) represents, we believe, a new category of open data science facility, a publicly available ***integrated data, tools, and compute resource*** (i.e., ***datacor***). The NSG compute resource is openly and freely available to researchers working on not-for-profit projects, providing neuroscience community access to multiple software environments that are widely used by neuroscience researchers such as analysis environments (e.g., MATLAB, python, R), and toolsets (e.g., EEGLAB, Open Brain, Freesurfer, TensorFlow). NSG enables users to submit custom analysis scripts running in any of these environments, and making use of users' own – or any NEMAR-hosted toolsets – to direct processing on NSF-supported high-performance computing (aka supercomputer) resources.

Furthermore, NSG provides an easy-to-use web portal-based user environment as well as programmatic access via a REST software interface. Users of the EEGLAB software environment, in particular, may use a set of REST-based EEGLAB tools (*nsgportal*) to launch, monitor, and examine results of NSG jobs directly from the EEGLAB menu window. Using NSG, neuroscientists can thus process and model data, run simulations, and train AI/ML networks on internationally supported high-performance supercomputer networks.

The NEMAR/NSG/OpenNeuro collaboration allows users to bypass the slow and computationally and energywise costly processes of downloading NEMAR datasets of interest, then re-uploading to NSG or elsewhere for analysis. Instead, NSG analysis scripts can directly access data hosted on the NEMAR server, encouraging intensive exploration of NEM datasets that have been made public by their authors on OpenNeuro. NEM data in OpenNeuro are copied to NEMAR where they are further curated and their quality is assessed. These datasets are then made available for user inspection, and visualization within NEMAR and for open-ended analysis via NSG.

The process for users to analyze an NEMAR dataset using NSG is simple: (1) Use the NEMAR.org website data search and visualization tools to identify one or more BIDS-formatted NEM datasets of interest. Then, (2) include the identified dataset control number (s) (example, *ds000123*) in an NSG data processing script. The NSG processing script may use any environment and software tools that NSG supports. (3) When processing is complete, the user will be

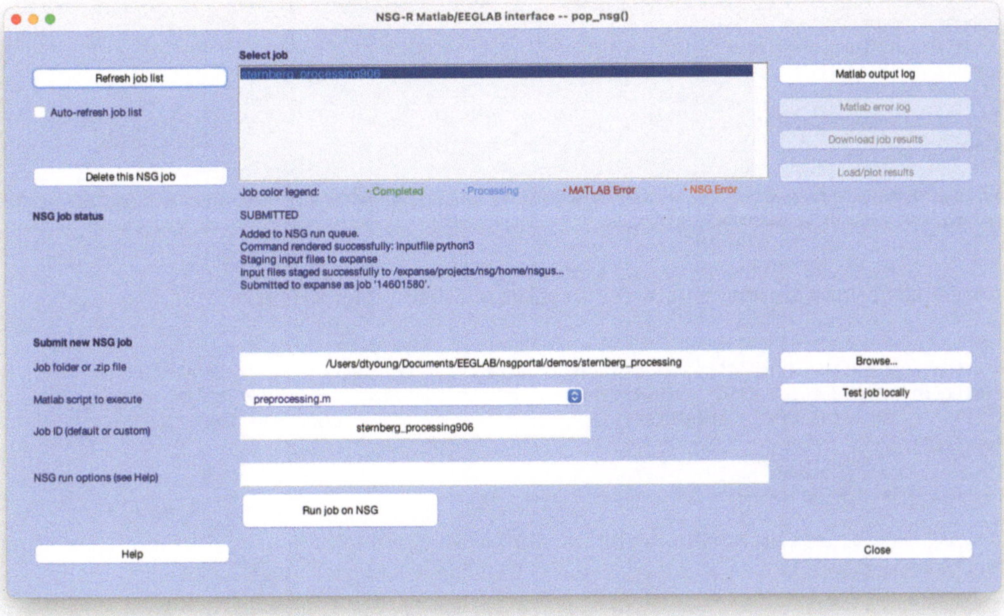

Fig. 17 The nsgportal plug-in GUI in EEGLAB

informed by email that the results specified in the processing script are available by NSG for download. More detailed information can be found at https://nemar.org/nsg_for_nemar.

EEGLAB users can use the *nsgportal* plug-in to directly submit NSG jobs processing NEMAR datasets from their local MATLAB environment [30]. Users place their analysis script in a directory then zip the directory for upload. They can then use the *pop_nsg* GUI shown in Fig. 17 to submit the job zip file, monitor job status in the processing queue, and when completed download results to display or further process locally.

To use the portal, first create a "job directory" containing at least one MATLAB script (the job) and the EEGLAB Study on which the script should be run. In the *Submit new NSG job* section of the GUI browse to that directory. A pull-down menu with the list of potential MATLAB scripts to be run then becomes visible (*preprocessing.m* is shown in Fig. 17). After selecting the script to run and filling in a job name and any desired options, push *Run job on NSG* to submit the job. EEGLAB zips the directory (which can be quite big) if not already zipped and uploads to NSG for execution. The status of the job on NSG is displayed in the top window.

Clearly, uploading an entire BIDS dataset as an EEGLAB Study to run a job is not ideal. The integration of NEMAR with NSG responds to this problem by allowing users, via NSG, to process data originally stored on OpenNeuro and staged on NEMAR, and

```
% Set the NEMAR/OpenNeuro ID of the 24-participant Modified Sternberg dataset
bidsName = 'ds004117';
% Clear the MATLAB workspace
clear

% Open EEGLAB within the NSG session
eeglab;

% Build the NEMAR address of the dataset using the global NEMARPATH variable from NSG MATLAB
filePath = [ getenv('NEMARPATH') bidsName ];   % NEMARPATH points to where NEMAR stores the data

% Open and convert the BIDS dataset to an EEGLAB Study
[STUDY, ALLEEG] = pop_importbids(filePath, 'outputdir', [pwd '/SternbergSTUDY']);

% Output total number of EEG files in the imported Study into an ASCII file
fid = fopen('results.txt', 'w'); % Open a text file for returning dataset information
fprintf(fid,'BIDS dataset %s\n', bidsName);   % Output the OpenNeuro/NEMAR dataset ID
fprintf(fid,'%d datasets\n', length(EEG));    % Output the total number of EEG files
fclose(fid);                                  % Close the file

% ... [Perform some analysis on this Study as desired] ...

% Remove the imported Study directory and its associated files
rmdir('SternbergSTUDY');                      % To avoid NSG also returning these
```

Fig. 18 Example script for submission to NSG from the EEGLAB nsgportal plug-in to process a BIDS dataset available in NEMAR

to receive and work with results of the processing using local computer resources. When working in MATLAB with a dataset made available on NEMAR, users only need to provide the dataset accession number (an 8-character string beginning with "ds" returned by NEMAR data search tools) and call *getenv("NEMAR-PATH")* to run their NSG job. The job directory needs only to contain the MATLAB script(s). Figure 18 shows an excerpt of an example of a MATLAB script based on this approach:

Note

The temptation for beginning users of remote resources is to write a processing script and launch a job using it without testing it first on a small dataset. Running scripts by remote execution makes script debugging more difficult than running the same script on the desktop. The MATLAB desktop has excellent facilities for single-stepping through a script or function and observing the data as execution proceeds. It is important to test a script thoroughly on less data before consuming NSG resources to apply it to more data.

4 Conclusions

This chapter covered end-to-end processing of EEG data using standardized BIDS and HED format to organize and describe information about the dataset. We believe that HED will have a crucial role in the use of public data and will be an essential enabler of cross-study and large-scale analysis. It should be emphasized that although BIDS enforces strict formatting standards, which enable tools to run automatically, it currently does not enforce standards for data quality and curation, and also allows some metadata fields to be left empty or with insufficient information. Thus, the researcher using the data must carefully check for inconsistencies and missing data. In the future, we suspect that more investment in careful curation of data to be shared publicly will be required, to better enable automated discovery.

This chapter is supported by extensive online tutorials and documentation. See https://osf.io/8brgv/ for links to the supporting materials and downloadable demo datasets.

Acknowledgments

We would like to acknowledge the seminal vision and contributions of Nima Bigdely-Shamlo, who initially conceived the HED system concept and demonstrated its potential for practical use and importance for data search and analysis within and across studies. We also thank Jonathan Touryan of the Army Research Laboratory and Tony Johnson of DCS Corporation for their work in supporting the development of HED for EEG data sharing. Ian Callanan and Alexander Jones are primary tool developers of the HED supporting infrastructure. This project received support from the Army Research Laboratory under Cooperative Agreement Number W911NF-10-2-0022 (KR) and from NIH projects R01 EB023297-03, R01 NS047293-14, and R24 MH120037-01 (SM). The Swartz Center for Computational Neuroscience is supported in part by a generous gift from The Swartz Foundation (Old Field, NY).

Appendix: List of Terms Used in This Chapter

AMICA	*Adaptive Mixture Independent Component Analysis*, a powerful ICA algorithm
BIDS	*Brain Imaging Data Structure*—a set of formatting specifications for storing and sharing of neuroimaging data (*bids.neuroimaging.io*)

(continued)

EEGLAB	A software environment for analysis of electrophysiological data running on MATLAB (The Mathworks, Inc.) (eeglab.org)
ERP	*Event-Related Potential*, mean of electrophysiological data trials time locked to events of a specified type
ERSP	*Event-Related Spectral Perturbation*, mean data trial spectrograms, trials time locked to events of a specified type
Event	A process unfolding through time during neuroimaging time series recording, especially any occurring process that may affect the experience and/or behavior of a participant. Also called an *event process*
Event context	The set of ongoing *event processes* at the time point of any *event marker*
Event marker	A pointer from a specified *event process* to a time point on the experiment timeline that marks a critical point, phase transition, or time point of interest within the event process, e.g., marking the event onset or offset. Also called an *event phase marker*
HED	*Hierarchical Event Descriptors*—a system for specifying the nature of events occurring during neuroimaging time series recordings (hedtags.org)
HED Schema	A dictionary of terms for use in *HED tags*, also indicating their allowed syntax
HED Base Schema	The common root *HED schema* of terms in common use across HED annotations (https://www.hedtags.org/display_hed.html)
HED Library Schema	A *HED schema* supplementing the *HED base schema,* adding terms needed to specify the nature of events in some neuroimaging research subfield (for example, language, movement, or clinical diagnosis)
HED String	A comma-separated list of *HED tags* specifying the nature of an event occurring during a neuroimaging experiment
HED Tag	A formatted list of *HED schema terms* giving some fact about the tagged event
IC	*Independent Component* process, identified in data by ICA decomposition
ICA	*Independent Component Analysis*
ITC	*Inter-Trial Coherence*—a measure of the phase angle coherence of a set of trials
NEMAR	*NeuroElectroMagnetic data Archive and compute Resource* (nemar.org)
NSG	*The Neuroscience Gateway*, enabling free research use of a high-performance computing network in neuroimaging research (nsgportal.org)
OpenNeuro	An open neuroimaging data archive using BIDS data formatting (openneuro.org)
RV	*Residual Variance* (percent) variance remaining in data following the removal of some other data
Trial	A task *event process* recurring during a recording, typically encompassing a unit of task performance during which a participant performs some action following one or more sensory and/or action events

References

1. Marek S, Tervo-Clemmens B, Calabro FJ, Montez DF, Kay BP, Hatoum AS, Donohue MR, Foran W, Miller RL, Hendrickson TJ, Malone SM, Kandala S, Feczko E, Miranda-Dominguez O, Graham AM, Earl EA, Perrone AJ, Cordova M, Doyle O, Moore LA, Conan GM, Uriarte J, Snider K, Lynch BJ, Wilgenbusch JC, Pengo T, Tam A, Chen J, Newbold DJ, Zheng A, Seider NA, Van AN, Metoki A, Chauvin RJ, Laumann TO, Greene DJ, Petersen SE, Garavan H, Thompson WK, Nichols TE, Yeo BTT, Barch DM, Luna B, Fair DA, Dosenbach NUF (2022) Reproducible brain-wide association studies require thousands of individuals. Nature 603:654–660. https://doi.org/10.1038/s41586-022-04492-9
2. Gorgolewski KJ, Poldrack RA (2016) A practical guide for improving transparency and reproducibility in neuroimaging research. PLoS Biol 14:e1002506. https://doi.org/10.1371/journal.pbio.1002506
3. Pernet CR, Appelhoff S, Gorgolewski KJ, Flandin G, Phillips C, Delorme A, Oostenveld R (2019) EEG-BIDS, an extension to the brain imaging data structure for electroencephalography. Sci Data 6:103. https://doi.org/10.1038/s41597-019-0104-8
4. Niso G, Gorgolewski KJ, Bock E, Brooks TL, Flandin G, Gramfort A, Henson RN, Jas M, Litvak VT, Moreau J, Oostenveld R, Schoffelen J-M, Tadel F, Wexler J, Baillet S (2018) MEG-BIDS, the brain imaging data structure extended to magnetoencephalography. Sci Data 5:180110. https://doi.org/10.1038/sdata.2018.110
5. Holdgraf C, Appelhoff S, Bickel S, Bouchard K, D'Ambrosio S, David O, Devinsky O, Dichter B, Flinker A, Foster BL, Gorgolewski KJ, Groen I, Groppe D, Gunduz A, Hamilton L, Honey CJ, Jas M, Knight R, Lachaux J-P, Lau JC, Lee-Messer C, Lundstrom BN, Miller KJ, Ojemann JG, Oostenveld R, Petridou N, Piantoni G, Pigorini A, Pouratian N, Ramsey NF, Stolk A, Swann NC, Tadel F, Voytek B, Wandell BA, Winawer J, Whitaker K, Zehl L, Hermes D (2019) iEEG-BIDS, extending the brain imaging data structure specification to human intracranial electrophysiology. Sci Data 6:102. https://doi.org/10.1038/s41597-019-0105-7
6. Bigdely-Shamlo N, Kreutz-Delgado K, Robbins K, Miyakoshi M, Westerfield M, Bel-Bahar T, Kothe C, Hsi J, Makeig S (2013) Hierarchical event descriptor (HED) tags for analysis of event-related EEG studies. In: 2013 IEEE global conference on signal and information processing, pp 1–4
7. Robbins K, Truong D, Appelhoff S, Delorme A, Makeig S (2021) Capturing the nature of events and event context using hierarchical event descriptors (HED). NeuroImage 245:118766. https://doi.org/10.1016/j.neuroimage.2021.118766
8. Robbins K, Truong D, Jones A, Callanan I, Makeig S (2022) Building FAIR functionality: annotating events in time series data using hierarchical event descriptors (HED). Neuroinformatics 20:463–481. https://doi.org/10.1007/s12021-021-09537-4
9. Beniczky S, Aurlien H, Brøgger JC, Hirsch LJ, Schomer DL, Trinka E, Pressler RM, Wennberg R, Visser GH, Eisermann M, Diehl B, Lesser RP, Kaplan PW, Nguyen The Tich S, Lee JW, Martins-da-Silva A, Stefan H, Neufeld M, Rubboli G, Fabricius M, Gardella E, Terney D, Meritam P, Eichele T, Asano E, Cox F, van Emde BW, Mameniskiene R, Marusic P, Zárubová J, Schmitt FC, Rosén I, Fuglsang-Frederiksen A, Ikeda A, MacDonald DB, Terada K, Ugawa Y, Zhou D, Herman ST (2017) Standardized computer-based organized reporting of EEG: SCORE - second version. Clin Neurophysiol 128:2334–2346. https://doi.org/10.1016/j.clinph.2017.07.418
10. Delorme A, Makeig S (2004) EEGLAB: an open source toolbox for analysis of single-trial EEG dynamics including independent component analysis. J Neurosci Methods 134:9–21. https://doi.org/10.1016/j.jneumeth.2003.10.009
11. Makeig S, Bell A, Jung T-P, Sejnowski TJ (1995) Independent component analysis of electroencephalographic data. In: Advances in neural information processing systems. MIT Press
12. Onton J, Delorme A, Makeig S (2005) Frontal midline EEG dynamics during working memory. NeuroImage 27:341–356. https://doi.org/10.1016/j.neuroimage.2005.04.014
13. Pavlov YG, Adamian N, Appelhoff S, Arvaneh M, Benwell CSY, Beste C, Bland AR, Bradford DE, Bublatzky F, Busch NA, Clayson PE, Cruse D, Czeszumski A, Dreber A, Dumas G, Ehinger B, Ganis G, He X, Hinojosa JA, Huber-Huber C, Inzlicht M, Jack BN, Johannesson M, Jones R, Kalenkovich E, Kaltwasser L, Karimi-Rouzbahani H, Keil A, König P, Kouara L, Kulke L, Ladouceur CD, Langer N, Liesefeld HR, Luque D, MacNamara A, Mudrik L, Muthuraman M,

Neal LB, Nilsonne G, Niso G, Ocklenburg S, Oostenveld R, Pernet CR, Pourtois G, Ruzzoli M, Sass SM, Schaefer A, Senderecka M, Snyder JS, Tamnes CK, Tognoli E, van Vugt MK, Verona E, Vloeberghs R, Welke D, Wessel JR, Zakharov I, Mushtaq F (2021) #EEGManyLabs: investigating the replicability of influential EEG experiments. Cortex J Devoted Study Nerv Syst Behav 144:213–229. https://doi.org/10.1016/j.cortex.2021.03.013

14. Delorme A, Truong D, Martinez-Cancino R, Pernet C, Sivagnanam S, Yoshimoto K, Poldrack R, Majumdar A, Makeig S (2021) Tools for importing and evaluating BIDS-EEG formatted data. In: 2021 10th international IEEE/EMBS conference on neural engineering (NER), pp 210–213

15. Kothe CAE, Jung T-P (2015) Artifact removal techniques with signal reconstruction

16. Chang C-Y, Hsu S-H, Pion-Tonachini L, Jung T-P (2018) Evaluation of artifact subspace reconstruction for automatic EEG artifact removal. Annu Int Conf IEEE Eng Med Biol Soc IEEE Eng Med Biol Soc Annu Int Conf 2018:1242–1245. https://doi.org/10.1109/EMBC.2018.8512547

17. Palmer JA, Makeig S, Kreutz-Delgado K, Rao BD (2008) Newton method for the ICA mixture model. In: 2008 IEEE international conference on acoustics, speech and signal processing, pp 1805–1808

18. Makeig S, Westerfield M, Jung T-P, Enghoff S, Townsend J, Courchesne E, Sejnowski TJ (2002) Dynamic brain sources of visual evoked responses. Science 295:690–694. https://doi.org/10.1126/science.1066168

19. Delorme A, Palmer J, Onton J, Oostenveld R, Makeig S (2012) Independent EEG sources are dipolar. PLoS One 7:e30135. https://doi.org/10.1371/journal.pone.0030135

20. Klug M, Kloosterman NA (2022) Zapline-plus: a Zapline extension for automatic and adaptive removal of frequency-specific noise artifacts in M/EEG. Hum Brain Mapp 43:2743–2758. https://doi.org/10.1002/hbm.25832

21. Mullen T (2012) CleanLine EEGLAB plugin. San Diego CA Neuroimaging Inform Toolsand Resour Clgh NITRC

22. Pion-Tonachini L, Kreutz-Delgado K, Makeig S (2019) ICLabel: an automated electroencephalographic independent component classifier, dataset, and website. NeuroImage 198:181–197. https://doi.org/10.1016/j.neuroimage.2019.05.026

23. Robbins KA, Touryan J, Mullen T, Kothe C, Bigdely-Shamlo N (2020) How sensitive are EEG results to preprocessing methods: a benchmarking study. IEEE Trans Neural Syst Rehabil Eng Publ IEEE Eng Med Biol Soc 28:1081–1090. https://doi.org/10.1109/TNSRE.2020.2980223

24. Pernet CR, Martinez-Cancino R, Truong D, Makeig S, Delorme A (2020) From BIDS-formatted EEG data to sensor-space group results: a fully reproducible workflow with EEGLAB and LIMO EEG. Front Neurosci 14:610388. https://doi.org/10.3389/fnins.2020.610388

25. Pernet CR, Chauveau N, Gaspar C, Rousselet GA (2011) LIMO EEG: a toolbox for hierarchical LInear MOdeling of ElectroEncephaloGraphic data. Comput Intell Neurosci 2011:831409. https://doi.org/10.1155/2011/831409

26. Makeig S, Inlow M (1993) Lapse in alertness: coherence of fluctuations in performance and EEG spectrum. Electroencephalogr Clin Neurophysiol 86:23–35. https://doi.org/10.1016/0013-4694(93)90064-3

27. Akalin Acar Z, Acar CE, Makeig S (2016) Simultaneous head tissue conductivity and EEG source location estimation. NeuroImage 124:168–180. https://doi.org/10.1016/j.neuroimage.2015.08.032

28. Acar ZA, Makeig S (2010) Neuroelectromagnetic forward head modeling toolbox. J Neurosci Methods 190:258–270. https://doi.org/10.1016/j.jneumeth.2010.04.031

29. Sivagnanam S, Yoshimoto K, Carnevale NT, Majumdar A (2018) The neuroscience gateway: enabling large scale modeling and data processing in neuroscience. In: Proceedings of the practice and experience on advanced research computing. Association for Computing Machinery, New York, pp 1–7

30. Martínez-Cancino R, Delorme A, Truong D, Artoni F, Kreutz-Delgado K, Sivagnanam S, Yoshimoto K, Majumdar A, Makeig S (2021) The open EEGLAB portal Interface: high-performance computing with EEGLAB. NeuroImage 224:116778. https://doi.org/10.1016/j.neuroimage.2020.116778

Chapter 7

Actionable Event Annotation and Analysis in fMRI: A Practical Guide to Event Handling

Monique J. M. Denissen, Fabio Richlan, Jürgen Birklbauer, Mateusz Pawlik, Anna N. Ravenschlag, Nicole A. Himmelstoß, Florian Hutzler, and Kay Robbins

Abstract

Many common analysis methods for task-based functional MRI rely on detailed information about experiment design and events. Event recording and representation during cognitive experiments deserves more attention, as it forms an essential link between neuroimaging data and the cognition we wish to understand. The use of standardized data structures enables tools to directly use event-based metadata for preprocessing and analysis, allowing for more efficient processing and more standardized results. However, the complex paradigms utilized by cognitive neuroscience often have different requirements for event representation. The process of generating event files from experimental logs and to iteratively restructuring these event files is a time-intensive process. Careful planning and effective tools can reduce the burden on the researcher and create better documented and more shareable datasets. This chapter discusses event representation within the BIDS (Brain Imaging Data Structure) framework. We discuss some of the common pitfalls in event representation and introduce tools to easily transform event files to meet specific analysis requirements. We demonstrate these tools and the corresponding analysis by comparing two BIDS datasets in which participants performed a stop-signal task. We work through the required event restructuring, and use *Fitlins* to calculate several comparable contrasts across the two datasets.

Key words fMRI, BIDS, Events, Event annotation, HED

1 Introduction

In task-based fMRI, the most common type of analysis models BOLD signal response under different conditions to identify brain areas associated with those conditions. In such analyses, conditions are usually defined by aspects of experimental events that are relevant to human experience, cognition and behavior; the neural correlates of which we wish to understand. Thus, experimental events provide an essential link between the neural activity captured in the BOLD signal and the human experience and/or behavior we are trying to understand. Experimental events are direct drivers of

Robert Whelan and Hervé Lemaître (eds.), *Methods for Analyzing Large Neuroimaging Datasets*, Neuromethods, vol. 218, https://doi.org/10.1007/978-1-0716-4260-3_7,

this experience. Because so much of the primary analysis for fMRI is directly based on comparison of conditions—encoded in the dataset events—the accuracy, completeness, and structure of the reported events can dramatically affect the usability of the data and the correctness of the final results. Unfortunately, there is no generally accepted standard for how events should be encoded, reported, and documented.

When an experiment is conducted, information about the sensory presentations, participant responses, and other control information is usually orchestrated by stimulus presentation software such as Presentation® (Neurobehavioral Systems, Inc., Berkeley, CA, www.neurobs.com), E-Prime [1], or PsychoPy [2]. Logged information is often user-controlled, and different software packages generate log files containing different structures and data types. The experimental logs, along with the fMRI recordings and auxiliary data such additional anatomical MRI recordings comprise the raw data of an experiment, which then must be converted into a standardized format for uploading to repositories or for input to analysis tools.

The Brain Imaging Data Structure (BIDS) [*see* Chapter 4] provides a standard for organizing and storing neuroimaging data along with event information, the focus of this chapter [3]. Event files in BIDS are tab-separated value (*.tsv*) files. Each row in an event file represents an event, and each column represents a different aspect of the event. BIDS only requires *onset* and *duration* columns in event files. The *onset* gives the time of the event marker in seconds relative to the start of the file's associated data recording, and the *duration* gives the duration of the event in seconds. Researchers may describe other properties of events by including additional columns in the event files. These additional columns and the values contained in them may be described in accompanying JSON files [*see* Glossary] (sidecars), but these sidecars are not required.

BIDS also supports Hierarchical Event Descriptors (HED) [*see* Chapter 6] for creating machine-actionable annotation [4, 5]. HED is a vocabulary designed to describe experiment events in a structured human-readable and machine-actionable way. This means events annotated with HED can be searched across datasets and understood outside of the context of any specific study. A HED annotation is a string of HED terms that is associated with an event in a BIDS event file. Terms are organized hierarchically to allow for search on broad categories as well as specific terms. However, the addition of HED is optional in BIDS, and so a BIDS-compliant dataset does not necessarily contain any documentation of events.

BIDS is now a well-established standard for storing neuroimaging data. Neuroimaging repositories such as OpenNeuro [6] [*see* Resources] expect datasets to be in BIDS. *BIDSApps* provide data management and analysis tools for BIDS datasets, leading to more standardized data processing and more reproducible neuroscience [7]. These apps are powerful because they can utilize standardized APIs (application program interfaces) to automatically access metadata such as scanning parameters or event data as necessary. By having a single software tool format, analyses stay flexible while also becoming more accessible, open, and reproducible. For the use of these tools, however, the *quality* of the data stored in BIDS is essential. This book chapter focuses on how to manage events and prepare event files for analysis in the context of requirements for downstream automated analysis.

To illustrate the process of generating extensive and analysis-ready event files, we demonstrate a simple analysis of two datasets: AOMIC-PIOP2 available on OpenNeuro (ds002790) [*see* Chapter 2] and SOCCER [https://osf.io/93km8/]. Both of these employed a stop signal task. In order to process both datasets and make them comparable, we carefully consider the event files, their structure, and the potential pitfalls that occur during setup.

Figure 1 illustrates a typical lifecycle of data from a neuroimaging experiment. To start, researchers use a combination of the data recordings and the experimental log files to convert their datasets to BIDS. The log files from the presentation software packages are

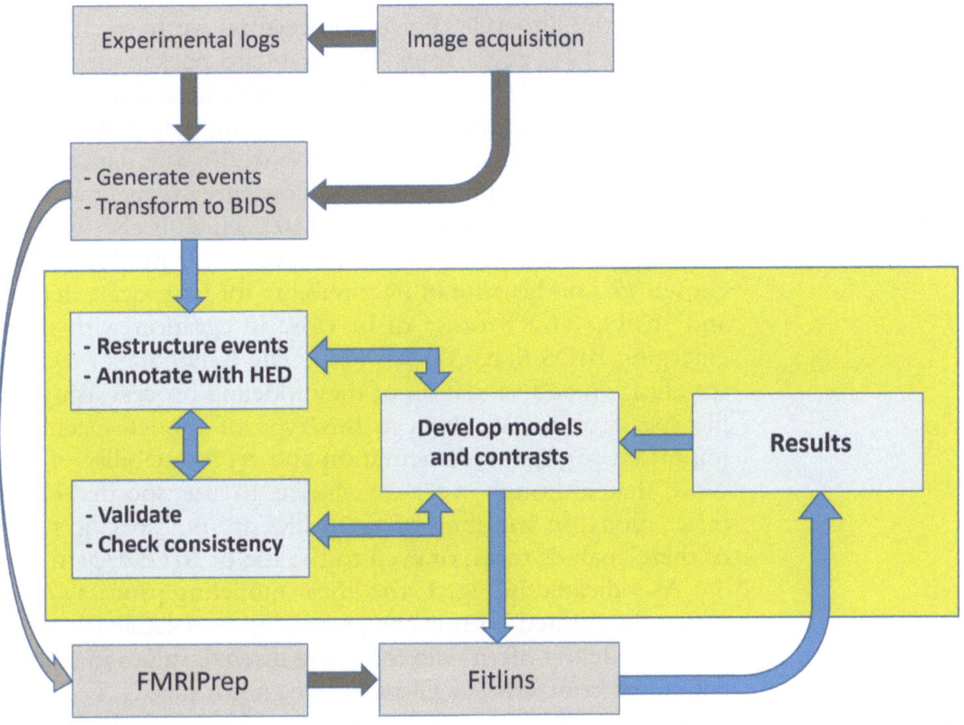

Fig. 1 The lifecycle of data from a neuroimaging experiment. The yellow box is the focus of this chapter. *FMRIPrep* and *FitLins*, which are the widely used open-source preprocessing and analysis tools used in this chapter, are placeholders for preprocessing and analysis, respectively

used to generate the experimental event files. When obtaining a BIDS dataset from an external source this process will have already been completed. The two example datasets demonstrate these respective use cases.

The AOMIC-PIOP2 dataset is part of a large set of data collected at the University of Amsterdam and made available on OpenNeuro. For this dataset, the analysis must rely only on the event data available on OpenNeuro plus any published descriptions of the dataset available in related publications [8]. This is the typical situation for researchers using open repositories to obtain data for analysis.

The second dataset, SOCCER, contains data collected at the University of Salzburg and provides a more typical use case for researchers working with their own data. Here we start directly from the experimental log files, allowing revision and clarification of the event encodings taken from experiment logs as would be the situation for the original experimenter who is working on initial publications and release of the data. After generating a fully BIDS compliant dataset that has been adequately preprocessed, we can start the linear modeling using *FitLins*. However, as we will show, effective linear modeling often requires event files to be restructured while executing the desired models and checking the results.

Preprocessing is by far the most computationally intensive aspect of the life-cycle. For preprocessing, we have chosen to use *FMRIPrep* [9] [*see* Chapter 8], a standard package for fMRI preprocessing. *FMRIPrep*, which is available as a *BIDSApp* [7] [*see* Glossary], relies only on the BIDS imaging data and does not use the event files. Thus, it can be run once for the dataset independently of the event processing and downstream analysis. For linear modeling we have chosen *FitLins* [10], which is also available as a *BIDSApp* and performs basic multilevel linear modeling. We have chosen *FitLins* because of its suitability for large-scale deployment and analysis and because of its close integration with the newly emerging BIDS Stats Models [11]. The BIDS Stats Model allows specification and validation of the modeling process using a JSON file specification. We believe this type of model specification is important for the documentation and reproducibility of analyses. Note that although we have chosen to use specific *BIDSApps*, instructions on handling of event files are not specific to the use of these analysis tools, or even to the use of *BIDSApps* in general.

As indicated in Fig. 1, the linear modeling process (*FitLins*) is closely integrated with event processing in a feedback loop. The fMRI modeling often relies on using discrete values in an event file column as contrasts [*see* Glossary] in a linear model. For a modeler wanting to ask the exact question envisioned by the person who created the dataset, using these discrete values might be possible, but in most cases it is unlikely to be sufficient. Hence, modelers will need to rework their event files, either through their own coding or through the emerging pybids-transforms [*see* Resources] mechanism [12], in order to perform significant analysis.

Thus, a feedback cycle emerges in which a model is developed, results are reviewed, and potentially additional reorganization of event files is done to enable additional analysis. While preprocessing of the imaging data is computationally intensive, it can be automated. In contrast, the event restructuring and model development is researcher-time intensive and challenging; however, it can be made more effective by better community guidelines and more tools for user-friendly changes and updates to event files. Here we present some guidelines on what to consider when building event files. We also discuss some open-source tools that we have developed to help to fix issues when they are encountered.

It is possible to follow along with the demo data. In the following section we will first go over the requirements for working along with the tutorial. We will describe the datasets used in the tutorial and highlight the differences that we will later compare in the example analysis. Next, we will discuss the layout of the initial event files for both datasets and issues we encountered as a consequence of this layout, along with general recommendations for the event file structure. We then briefly discuss the BIDS Stats Model and explain how it bridges between BIDS data and linear analysis of this data. Based on the issues we encountered in the event files, we present a series of tools that can be used to restructure event files for analysis. We illustrate the use of these tools, the BIDS Stats Model, and *FitLins* on our example data. The demo data and supporting material for this chapter are available at [https://osf.io/93km8/]. This chapter emphasizes event processing from the perspective of fMRI analysis, but the general concepts also apply to other imaging modalities such as EEG and MEG.

2 Methods

2.1 Starting Point for Data

The analysis requires a dataset in BIDS format as would be downloaded from OpenNeuro [https://openneuro.org] [*see* Chapter 2], or transformed directly from collected data. At least two subjects are required, since multilevel analysis is discussed. Clearly, using a larger number of subjects is essential to obtain adequate statistical power for reproducible group level results, but statistical power is not the focus of this chapter. One should also be aware—as we shall demonstrate—that getting processing to work with a small number of subjects does not guarantee that this processing will work with many subjects due to variances in the runs and missing or bad data.

We assume that the data is preprocessed using *fMRIPrep*, which produces files in a BIDS-derivative compliant format that can be automatically ingested by *FitLins* without intervention (see the supplementary material for additional information on usage [https://osf.io/93km8/]). Issues such as inconsistency or missing values in the events are part of the event restructuring process addressed in this chapter.

2.2 Data Storage and Computing

fMRIPrep and *FitLins* are—in theory—platform-independent. Because our demos involve small datasets, the computations can be done on a desktop computer (although *FMRIprep* may take several hours per subject depending on the processing power of the desktop computer being used). *FitLins* for a small number of subjects usually takes less than an hour on a moderately powered desktop computer.

The three-subject AOMIC-PIOP2 dataset that we have provided as a demo [https://osf.io/93km8/] has raw data of approximately 220 MB and processed data derived from *FMRIPrep* of approximately 1.12 GB. The processed results from *FitLins* for this dataset are about 50 MB for each model. The computing time and storage required for event processing and restructuring is negligible.

2.3 Software and Coding

Some coding experience is necessary. Running of *fMRIprep* can be done by typing a single command, which does not require coding skills but does require an understanding of command-line arguments and their meaning. *FMRIprep* is available as a Docker image, but it can also be installed as a python package to run without Docker.

In theory, *FitLins* only requires knowledge of command-line arguments and their meaning as well as access to a Docker-enabled system. *FitLins* also requires a model file in JSON format, which can be created using an ordinary text editor. The construction of the JSON file requires knowledge of how the models are encoded using the BIDS Stats Model. The interaction between these model files and the event files is the focus of this book chapter.

The tools that we have developed for event structuring and re-coding are written in Python and assume at least Python version 3.7. These tools are available for support of event processing with minimal programming. These tools are being integrated in the Hierarchical Event Descriptors (HED) tool suite [https://www.hed-resources.org/] for support of event handling and annotation [*see* Chapter 6]. These tools will soon be available via a command-line interface in a Docker container and will also be available as web services and through online tools with no coding or software installation [https://hedtools.ucsd.edu/hed].

2.4 Framing the Problem

This chapter uses the problem of comparing results from two datasets (AOMIC-PIOP2 and SOCCER) on a particular task (stop-signal) as a motivation and a focus for discussing event restructuring with an end goal of comparing results across these experiments.

2.4.1 The Task

The stop-signal task is a well-studied test for understanding response initiation and inhibition [13, 14]. In a series of trials [*see* Glossary], participants are asked to respond to frequent "Go" stimuli but to cancel their response when a particular stop signal is presented at an interval (the "stop signal delay") after the Go stimulus is presented. The selection of the Go stimulus is the primary discrimination task, with the stop-task as a secondary detection task. Although the task structure and instructions adhere to a strict format, there can be considerable variation possible in stimuli used and the selection criteria for the primary task as illustrated by the two datasets (AOMIC-PIOP2 and SOCCER) discussed in this chapter.

AOMIC-PIOP2

In AOMIC-PIOP2, participants performed four different tasks during each of four functional runs in a single session. Participants also completed several demographic and psychometric questionnaires, either before or after scanning, on the same day. Scanning lasted about 60 min, and the entire stop-signal run lasted around 7 min.

The primary discrimination task was a gender decision task using face images. The stop signal was an auditory stimulus, presented with a delay starting from 250 ms but shortened stepwise if the average number of failed stop trials was higher than 50% and lengthened if the average number of failed trials was lower than 50%. Participants performed a total of 100 trials, as well as approximately 10 additional null trials in which no stimuli were presented for an average trial duration of 4 s.

SOCCER

In SOCCER, the stop-signal task was the only task participants performed in the MRI scanner. Scanning lasted around 45 min during which the stop-signal task was performed during a single 12-min run. This run was organized into six experimental blocks of 64 trials each, with 16 s of rest between each block. At the end of the six blocks, participants completed a survey in which they described the strategy they used during the task by indicating agreement on a 1–100 scale to several questions. The same controller buttons were used to record responses during the survey portion as during the task.

The primary selection task consisted of selecting between left and right pointing arrows, and pressing a left or right button accordingly. The stop signal was a color change in the presented arrow. The delay was updated constantly, based on a thresholding procedure designed to keep stop-signal performance around 50%. Note that although this is similar to the aim in the AOMIC-PIOP2, the exact rules to get to a 50% average were different. Also in contrast to the AOMIC-PIOP2 study, which had a broader focus, participants in SOCCER were familiarized with the task extensively by practicing at home as well as one additional time inside the scanner. The practice block in the scanner had trial-by-trial feedback.

2.4.2 Dataset Acquisition

The two datasets represent the two most common use cases in analyses of neuroimaging—downloaded versus locally acquired data. The AOMIC-PIOP2 dataset is a public dataset, and the information about the events is restricted to that available from the dataset itself and the referenced papers. There is limited opportunity to ask for clarification or correction. SOCCER is an ongoing experiment at the time of this writing, so the experimental logs and control scripts are available, as is access to the experimenter.

AOMIC-PIOP2

AOMIC-PIOP2 is part of a large open collection from the University of Amsterdam which has been deposited on OpenNeuro (ds002790) [*see* Chapter 2]. The collection includes diffusion-weighted images and fMRI runs from four different tasks, but this chapter only considers fMRI data from the stop-signal task. Extensive details on scanning parameters, demographic and other participant characteristics are provided in Snoek et al. (2021) [8], and some information at the participant level has been distributed with the dataset. All participants of the AOMIC-PIOP2 were students, aged between 18 and 26 years old.

SOCCER

SOCCER is part of an ongoing experiment from the University of Salzburg. In this experiment male adolescent (16–17 years old) amateur soccer players are invited to participate in a study on the link between performance, development, and low-level response and inhibition processes. Participants are scanned during a single session. Functional and structural neuroimaging data were collected with a Siemens Magnetom Trio 3 T Scanner (Siemens AG, Erlangen, Germany) using a 64-channel head-coil. Functional images consisted of a T2*-weighted gradient echo EPI sequence (TR 1050 ms, TE 32 ms, matrix 80 × 80, FOV 192 mm, flip angle 45°). Within the TR 56 slices with a slice thickness of 2.4 mm were acquired. In addition to the functional images, a gradient echo field map (TR 623 ms, TE 1 = 4.92 ms, TE 2 = 7.38 ms, flip angle 60°) and a high resolution (0.8 × 0.8 × 0.8 mm) structural scan with a T1-weighted MPRAGE sequence were acquired from each participant. Additional structural files were collected but have not been shared for the purposes of this demo. Participants spent around 60 min in the MRI scanner for the entire session. Of this time, the functional run containing the stop-signal task lasted around 12 min.

2.4.3 A Side-by-Side Comparison

Although both datasets use the stop-signal task paradigm, the experiments differ significantly. Table 1 shows a side-by-side comparison of various implementation aspects of the two experiments. The similarities and dissimilarities between the tasks determine the

Table 1
An overview comparison of the two datasets used for this chapter

Aspect	AOMIC-PIOP2	SOCCER
Participant pool	University students	Amateur soccer players
Number of participants (Demo/Total)	**Demo:** 3 **Total:** 226	5
Participant age (Demo/Total)	**Demo:** 20–24, mean: 22.67 **Total:** 18–25, mean: 21.96	16–17, mean 16.4
Participant gender (Demo/Total)	**Demo:** 66% female, 33% male **Total:** 57% female, 42% male	100% male
Choice images	Female (right) and male (left) faces	Left and right white arrows
Choice image duration	0.5 s	Up to 1.25 s depending on button press
Discrimination task	Gender	Direction
No go indicator	Auditory tone 450 Hz for 0.5 s	Color change (white to red)
Initial No-go delay	0.25 s	0.2 s
Go response time average	1.0243 s	0.371 s
Button press configuration	Index fingers of left and right hands	Index and ring finger of right hand
Trials per run	100 trials	384 in 6 blocks of 64 trials each
Approx trial time	2.5 s	0.57 s
Total trials	15,119 go trials 4370 successful stop and 3111 unsuccessful stop trials	1452 go trials 233 successful stops and 232 unsuccessful stop trials
% of stop trials	33%	25%
% successful versus unsuccessful stop	~58% successful stops with large variation Three subjects had no successful stops, and one subject had no unsuccessful stops. In all, 61 subjects had a low number of successful or unsuccessful stops	~50% successful stops Evenly distributed across participants
Spacing	Has 10% null trials	Rest blocks

questions we can ask about either dataset or the types of comparisons we can make to contrast the two datasets.

As Table 1 states, AOMIC-PIOP2 is a large published study with a mix of participant genders, while SOCCER is an ongoing study of all male participants tightly grouped by age. SOCCER will ultimately collect data for 55 participants.

The task implementation also varied in several respects. While the AOMIC-PIOP2 study used face discrimination as a primary decision task, SOCCER uses a simpler direction discrimination task, which is expected to exert a lower cognitive load. Face discrimination has significant social correlates, and we expect to see activations related to face processing. In contrast, the identification of left versus right arrows is unlikely to be related to specific brain areas. The stop signal modalities also differed for the two datasets: AOMIC-PIOP2 used an auditory signal, while SOCCER used a visual color cue.

Reaction times vary strongly between datasets. This can be explained by differences in other aspects of the implementation: the complexity of the discrimination task, the relative athletic capabilities of the participant pool, and the amount of practice on the stop-signal task (see also below). The response hand usage was also different in the two experiments, with the AOMIC-PIOP2 participants using index fingers on right and left hands for right and left responses, possibly inducing strong hemispheric differentials. On the other hand, participants in the SOCCER experiment use two fingers on the right hand so little hemispheric differential is expected. The distribution of trial types within the files also differed, with AOMIC-PIOP2 having several participants with an imbalance between stop and go trials, while SOCCER aggressively maintained a balance between stop and go trials in each run.

2.4.4 The Impact of Event Encoding

Within a BIDS dataset, the event files fulfill multiple functions including documenting the experiment and providing direct input to the analysis. Different analyses can put their own requirements on the organization of the event file and careful consideration of the organization of the event files is necessary. *FitLins*, for example, extracts the onsets and durations for modeling BOLD responses directly from the event files.

Foundational to building models for functional MRI data analysis is the convolution of event-based boxcar functions with the hemodynamic response function (HRF) [*see* Glossary]. The relationship between neural activity and brain hemodynamics is a well-studied topic [15]. *FitLins* uses the canonical HRF based on the one used by SPM [*see* Resources] as a standard, although there are other options available [16]. For short events in event-related experimental designs [*see* Glossary] the HRF is convolved with an impulse function. For longer events that are analyzed in a block design [*see* Glossary] the HRF is convolved with a boxcar function. For this, as well as for analysis of individual HRFs, modeling event durations precisely is important.

For the purpose of creating optimal event files, we focus on the accurate and precise reflection of event onsets and durations. In the following sections, differences in how onsets and durations are represented in event files across datasets are emphasized, and we discuss different options for representation.

While the events in the AOMIC-PIOP2 were already encoded in a BIDS event file, the starting point for the SOCCER events is a log file generated by NeuroObs presentation software. The first step here is to read out this log file and transform it into BIDS event format. Creating a minimally valid BIDS file from a log file is easy, but in order to prepare for analysis, additional details and context information must be added.

With well curated, finalized event files, there can still be many hurdles for a researcher who wants to reanalyze the data, necessitating additional time be spent on restructuring the event file and recategorizing the events.

2.4.5 Trial-Level Versus Event-Level Encoding

Trial-level encoding represents a trial by a single line in the event file with the *onset* column value marking the time of a particular anchor event (often the primary stimulus presentation) within the trial. Other trial events are either omitted or represented in other columns as offsets (or a combination of offsets) from the event onset time. Correctly analyzing the timing of any event except the anchor event can be quite complicated and require a careful reading of the documentation and accompanying literature. Automated processing can easily fail or incorrectly interpret these offsets. The practice of trial-level encoding also encourages the skipping of the extra events that occur within the trial, which can limit the scope of the data use.

The alternative is to split the trial into multiple events. While this event-level encoding results in additional rows in the event files, it allows more flexibility in downstream analysis for contrasts anchored on different internal events in the trial. Further, durations of the events within a trial can be correctly distinguished for convolution with the hemodynamic response function (HRF), an essential step in building a model for fMRI data. However, context that relates to an entire trial can be lost.

In a hybrid approach, each individual event is represented as a row as with event-level encoding. Additional events are added before the first event in each trial with a duration representing the extent of the entire trial. This trial event can be associated with information that should be modeled on a trial level. Additional events representing the start of experimental blocks may also be included.

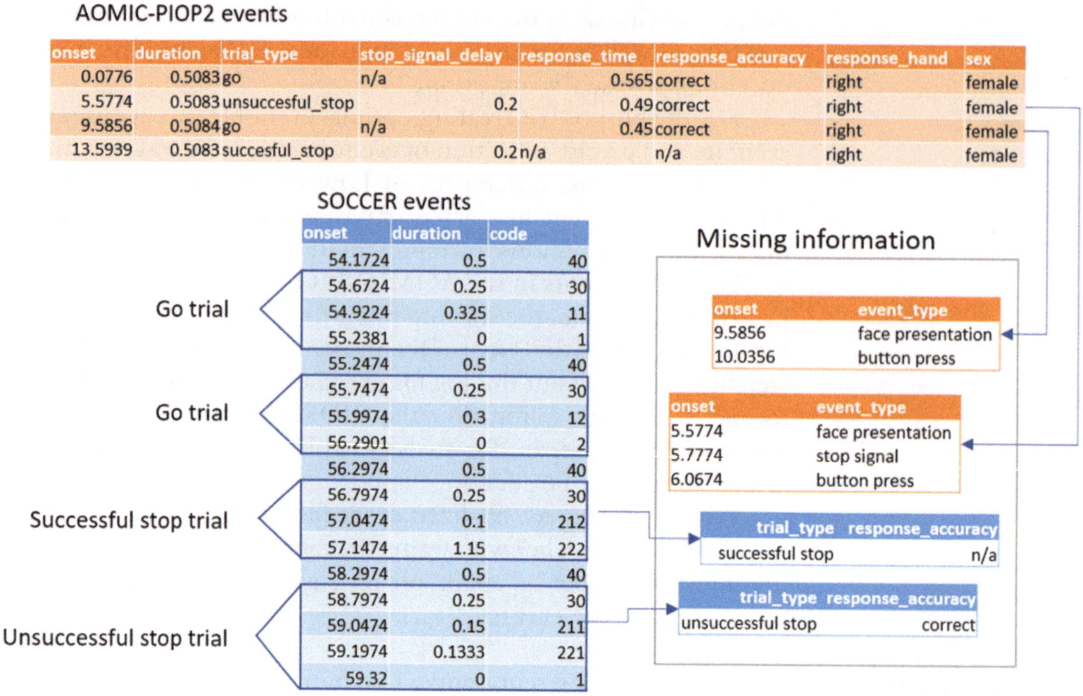

AOMIC-PIOP2 events

onset	duration	trial_type	stop_signal_delay	response_time	response_accuracy	response_hand	sex
0.0776	0.5083	go	n/a	0.565	correct	right	female
5.5774	0.5083	unsuccesful_stop	0.2	0.49	correct	right	female
9.5856	0.5084	go	n/a	0.45	correct	right	female
13.5939	0.5083	succesful_stop	0.2	n/a	n/a	right	female

SOCCER events

	onset	duration	code
	54.1724	0.5	40
Go trial	54.6724	0.25	30
	54.9224	0.325	11
	55.2381	0	1
	55.2474	0.5	40
Go trial	55.7474	0.25	30
	55.9974	0.3	12
	56.2901	0	2
	56.2974	0.5	40
Successful stop trial	56.7974	0.25	30
	57.0474	0.1	212
	57.1474	1.15	222
	58.2974	0.5	40
Unsuccessful stop trial	58.7974	0.25	30
	59.0474	0.15	211
	59.1974	0.1333	221
	59.32	0	1

Missing information

onset	event_type
9.5856	face presentation
10.0356	button press

onset	event_type
5.5774	face presentation
5.7774	stop signal
6.0674	button press

trial_type	response_accuracy
successful stop	n/a

trial_type	response_accuracy
unsuccessful stop	correct

Fig. 2 Comparison of AOMIC-PIOP2 and SOCCER event file structure. Event rows in AOMIC-PIOP2 (in orange) represent entire trials with additional columns reflecting offsets of various trial events relative to trial onset. As a result, the event file onset information for stop signals and responses is implicit, as reflected in the missing information box (right box). Event rows in SOCCER (in blue) represent individual events, but information on context of the overall trial is implicit

The two datasets considered in this chapter initially used different encoding strategies, and both had to be restructured in order to perform comparative analyses. Figure 2 compares an excerpt from an event file for each dataset. The two event files represent the same number of trials, but are shaped differently because of their different encoding. The trial-level encoding used in the AOMIC-PIOP2 dataset contains information about the events in additional columns, but onsets anchored to neural data recording are missing for the stop signal and the response. In the SOCCER data, individual events are represented with onsets anchored to the neural data recording, but context information about trials is missing. In either case, restructuring the events so all information is adequately captured in the event files normally requires a programmer to manually code transformations of the event files. Later in this chapter we introduce tools that allow researchers to list the transformations needed in a JSON text file and run remodeling tools to automatically perform the transformations without additional programming.

AOMIC-PIOP2

As shown in Fig. 2, AOMIC-PIOP2 uses trial-level encoding, representing an entire trial by a single row with eight event file columns: *onset*, *duration*, *trial_type*, *stop_signal_delay*, *response_time*, *response_accuracy*, *response_hand*, and *sex*. The *onset* column corresponds to the presentation of the face for 0.5083 s and also marks the start of the trial. Other key events in a trial: the stop signal presentation and the participant response, are encoded implicitly using the *stop_signal_delay* and *response_time* as times offset from the row's *onset* column value. If there was no stop signal or no participant response in a trial, these columns are filled with *n/a*, following BIDS convention.

Trial-level encodings present two difficulties for downstream analysis. The first is that unusual events such as extra button presses cannot be represented although they elicit motor responses. The second difficulty is that these relative onsets may need to be unfolded into event markers before certain analyses (e.g., analysis linked to participant response onsets) because this information is hidden in additional columns. In some datasets, multiple offsets must be combined to get the correct time of a relative event. Without a very careful reading of the available documentation for each experiment, it may not be possible to correctly compute the position of these markers.

For example, the documentation for AOMIC-PIOP2 states that the *stop_signal_delay* and the *response_time* are both given in seconds relative to the go image presentation onset. However, the original experiment on which AOMIC-PIOP2 is based (Fig. 1a [17]) shows a *stop_signal_onset* (also of 250 ms) relative to the end of the face image presentation rather than the start of the presentation. Figure 1 of the Jahfari paper further indicates that participants were not allowed to respond until after the go image disappeared (500 ms). If the *stop_signal_onset* were actually at 750 ms rather than at 250 ms relative to the start of the trial, then the large differential in participant response times between the AOMIC-PIOP2 and SOCCER would be essentially eliminated, resulting in little potential effects of participant skill and discrimination task complexity on response time. Another possibility is that stop signals occurred 250 ms after the face image onset, but that participants were instructed not to respond until after the face image offset, since almost all response times were greater than 500 ms. While the most likely explanation for the difference in response times is the complexity of the discrimination task, the other possibilities cannot be completely eliminated since the original experimental logs are not available.

SOCCER

The SOCCER events are read out directly from the log files and distinguished by basic codes as illustrated in Table 2. Common events, such as the presentation of left and right arrows, the fixation

Table 2
Overview of common event codes used in the original SOCCER log

Code	Event description	Context string for the log
40	Blank screen	Experiment block
30	Fixation dot	
11	Left go arrow	Go trial
12	Right go arrow	Go trial
211	Left go arrow	Stop trial
221	Left stop arrow	Stop trial
212	Right go arrow	Stop trial
222	Right stop arrow	Stop trial
1	Left button	
2	Right button	
70	Block feedback	
80	Blank screen	Rest block

dot or a blank screen, are associated with unique numerical codes. Other codes, particularly those associated with participant responses, have different meanings depending on the phase of the experiment (i.e., the experimental context) during which the corresponding event occurred. Unambiguous coding, so important for downstream analysis, can be achieved either by reassigning ambiguous codes or by providing context marker events in the data.

In general, the log files generated by experiment control software contain markers for individual events, not for entire trials. If trial-level event encoding (as in AOMIC-PIOP2) is desired, code must be written to identify sequences of event markers in the experimental log and generate the appropriate trial events. Figure 3 shows some examples of the sequences that must be identified and collapsed.

For example, the first boxed sequence of events on the left in Fig. 3 contains the event marker sequence 30, 11, 1 starting at *onset* 340.9269 s. This sequence contains a fixation dot, followed by a left go arrow, followed by a left button press and indicates a *go* trial. The corresponding box on the right of Fig. 3 shows the insertion of an event marker representing the start of this *go* trial. The *onset* of this structure marker is the same as the *onset* of the earliest event marker in the sequence. The *duration* is the difference between its onset and the end of the last event marker in the sequence.

Once these trial event markers are inserted and the trial type is identified, it is much easier to transform to trial-level coding or to

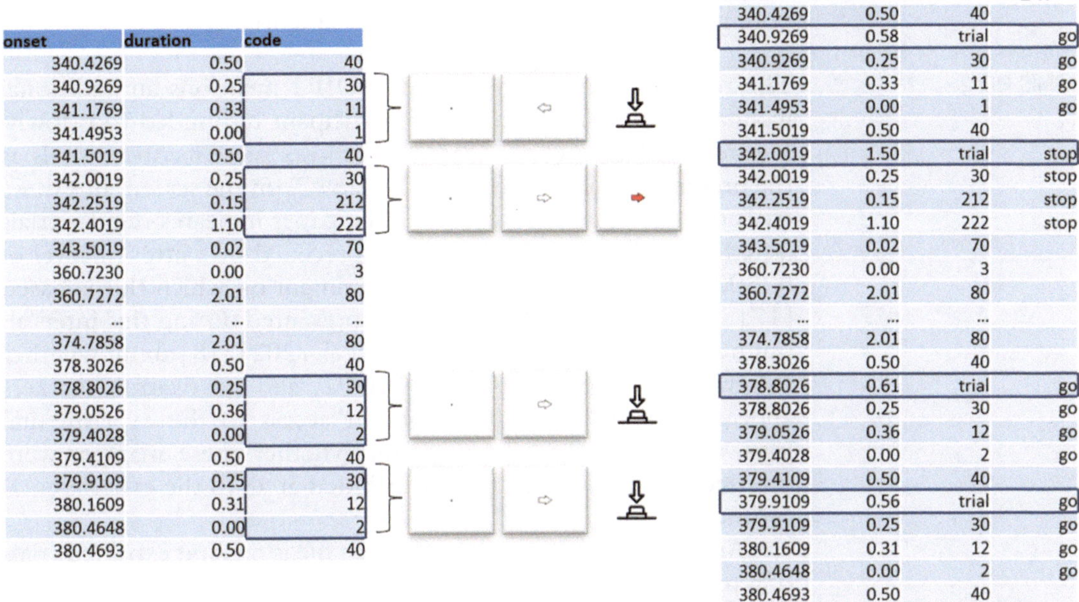

onset	duration	code
340.4269	0.50	40
340.9269	0.25	30
341.1769	0.33	11
341.4953	0.00	1
341.5019	0.50	40
342.0019	0.25	30
342.2519	0.15	212
342.4019	1.10	222
343.5019	0.02	70
360.7230	0.00	3
360.7272	2.01	80
...	...	...
374.7858	2.01	80
378.3026	0.50	40
378.8026	0.25	30
379.0526	0.36	12
379.4028	0.00	2
379.4109	0.50	40
379.9109	0.25	30
380.1609	0.31	12
380.4648	0.00	2
380.4693	0.50	40

onset	duration	code	trial_type
340.4269	0.50	40	
340.9269	0.58	trial	go
340.9269	0.25	30	go
341.1769	0.33	11	go
341.4953	0.00	1	go
341.5019	0.50	40	
342.0019	1.50	trial	stop
342.0019	0.25	30	stop
342.2519	0.15	212	stop
342.4019	1.10	222	stop
343.5019	0.02	70	
360.7230	0.00	3	
360.7272	2.01	80	
...	...	...	
374.7858	2.01	80	
378.3026	0.50	40	
378.8026	0.61	trial	go
378.8026	0.25	30	go
379.0526	0.36	12	go
379.4028	0.00	2	go
379.4109	0.50	40	
379.9109	0.56	trial	go
379.9109	0.25	30	go
380.1609	0.31	12	go
380.4648	0.00	2	go
380.4693	0.50	40	

Fig. 3 Events in original SOCCER events. Events are represented with simple code, but specific sequences represent structural elements of the experiments such as trials. Based on the sequence of events within a trial, we can determine what the trial type is. To appropriately associate context with entire trials, instead of only the events within, one can add trial events. (This Figure has been designed using resources from Flaticon.com)

disambiguate event marker codes. The tools introduced later in this chapter allow users to specify in a JSON text file how to insert various structure markers into the event files and to subsequently disambiguate codes or to transform between event-level and trial-level encoding with no or little additional programming.

2.4.6 Missing Events and Event Values

Events that are not encoded will only contribute to implicit baseline for the signal but cannot be modeled explicitly. Visual or auditory cues indicating that participants should get ready for the trial start, as well as feedback events and other sensory cues that occur during a trial or at the beginning and ending of a block, are often omitted. Often these omitted events are seen as trivial, or they are simply not the object of inquiry for the curating researchers, yet their inclusion could result in a better fitting model.

Another problem with missing information is that values representing expected conditions can be missing entirely from a dataset because they were not recorded—or more frequently because the experimental participant failed to follow the protocol. This can cause serious problems for downstream analysis. Sometimes these types of omissions make it difficult to determine exactly where a trial or block ends when translating individual log entries to a single event marker representing the whole trial. For event-level encodings including a trial number column can be very helpful for downstream analyses, so that specific code to check for all the potential omissions does not have to be written.

AOMIC-PIOP2

The reference paper for AOMIC-POIP2 mentions an additional feedback trial of 2000 ms if the participant responded too slowly. Figure 2 of Snoek et al. (2021) [8] shows these feedback trials as the presentation of the words "Too slow!" on the screen, but these trials did not appear in the data. The paper indicates regular trials lasted 4000 ms and that trials were preceded by a jitter interval of 0, 500, 1000 or 1500 ms. The experiment on which this is based [17] stated that a fixation cross was presented during this interval, but no mention was made in Snoek et al. (2021) [8] about whether this was the case for AOMIC-PIOP2, and no fixation markers appeared in the event file. Without access to the experimental logs, users of the data cannot tell whether these markers were omitted as unimportant or not included at all in the experimental protocol.

Another issue arises when the data file is missing expected trials of a particular type or has an unexpected distribution of trials in a particular condition. For example, some event files in AOMIC-PIOP2 contained no successful stops, and some event files contained no unsuccessful stops, although the experimental goal was to adjust the *stop_signal_delay* to achieve a 50% balance between successful and unsuccessful stops. This missing data caused crashes downstream in the *FitLins* processing when the *trial_type* column was converted to factors for modeling and a factor value previously encountered during processing other files (e.g., a successful stop or an unsuccessful stop) was found to be missing from an event file. While a large number of runs in this dataset achieved a reasonable balance of successful versus unsuccessful stop trials, the burden is on the analyst to check the event file contents and only use runs and trials appropriate for the analysis.

SOCCER

The coding of events in the SOCCER presentation log was not complete. Because the user controls which events created are pushed to the log file, it is possible to present stimuli or acquire responses without logging any information. In this case, there was a short "get ready" message before the start of each experimental block that was not pushed into the log file. Often, events are unreported because they are not the object of inquiry in the study. In this case the event may be viewed as trivial, but the event is the first after a 16-s period of rest and meant to prepare the participants for a new period of activity. Based on the experiment coding we can make a reasonable estimate of these event onsets and add them to the event file.

Another situation in the SOCCER study requiring special handling were rest blocks of 16 s between the experimental blocks. During these rest blocks an empty white screen was presented continuously. In the presentation software logs, this presentation was coded as 8 repeat events, each lasting 2 s. If these rest events were factored for the model, there would be 8 onsets and durations. In case of an event-related analysis all these onsets would be modeled with impulse function individually. Note that HRFs interact nonlinearly when events quickly succeed each other [18]. Modeling multiple quick successive events where there are none creates an inappropriate model for the BOLD response and will likely negatively affect the results. If there are no new onsets there should be no new events, even if the experiment software internally refreshes the presentation of such an event. To analyze resting blocks a researcher would have to fold in these events with appropriate duration before analyzing based on the event file.

2.4.7 Ambiguous Encoding

Event meanings are encoded in event files using custom labels. Many downstream analyses are predicated on using these labels to define factors or contrasts for analysis. If labels (e.g., values in a *trial_type* or *code* column) are ambiguously encoded, downstream analysts will need to disambiguate them before starting analysis, usually by writing special-purpose code. Events can be ambiguous in multiple ways. Often, ambiguous encoding comes down to a lack of contextual information. The event itself could be a simple button press, but the events occurring before generally determine whether this button press was appropriate or a correct solution to the task given to the participant. Another common case is multiple button presses due to participant errors. Here, we go over some of the ambiguous codes in the example datasets.

AOMIC-PIOP2

The three *trial_type* values were: *go*, *succesful_stop* (sic), and *unsuccesful_stop* (sic). However, this did not completely encode all possible trial types, because in some cases the participant missed pressing a button entirely during a *go* trial. Because this case wasn't encoded as a separate *trial_type*, analysts downstream must write code to exclude these trials from comparisons of the go condition with stop condition. In addition to missed *go* trials, there were trials in which the participant pressed the wrong button during discrimination.

SOCCER

The SOCCER study consisted of multiple blocks, including one survey block at the end of the experiment in which the participants responded to statements on their task strategies. Throughout the entire run participants used a single button box. During experimental blocks participants used the left and right button to indicate whether there was a left or right arrow. During the survey blocks these same buttons were used to move a slider left and right. To analyze the neural correlates of button presses during experimental trials, button presses during experimental trials must be made distinct from button presses during the survey block.

Notes

1. *Ideally a trial type (or event code) should have unique codes for each possible type of trial (or event). Unambiguous encoding enables correct downstream splitting of type into factors for modeling without specialized programming.*
2. *Careful summarization of values and combinations of values in event files should be done before setting up statistical models. Appropriate rows should be dropped and missing or unusual column values handled.*
3. *All the events in a trial should be reported, ideally using event-level rather than trial-level encoding to facilitate correct automated downstream processing.*
4. *Context should be actively associated with all events to allow differentiate between identical events that have different implications for participant cognition.*

As shown in Fig. 1 on the lifecycle of data and further illustrated by the event encoding issues discussed in this section, analysis and modeling are iterative processes. No data curator, no matter how conscientious or proficient, can anticipate all of the possible questions that might be pursued by downstream analysts. Further, even when the representation of events in a data set is relatively complete, these encodings are likely to be incompatible when organized for larger analyses across multiple datasets. In the next section we discuss the modeling infrastructure represented by BIDS Stats Models and introduce tools for restructuring events without re-coding.

3 Modeling and Event Structure

In this section we give an overview of the BIDS Stats Model format and discuss the relationship of models to the event structure. We then introduce event file restructuring tools designed to help researchers restructure their event files to address particular questions.

3.1 The BIDS Stats Model Framework

Our example analyses use the *FitLins* linear modeling package to illustrate the interaction between modeling and event organization. *FitLins* is designed for large-scale automated processing using containers. Installation and examples of running *FitLins* for the examples discussed in this paper are contained in the supplementary materials at https://osf.io/93km8/. An important aspect of

BIDS Stats Model JSON

```
{
   "Name":"leftvsrighthands",
   "BIDSModelVersion":"1.0.0",
   "Description":"Test go left versus go right hand",
   "Input":{"task":"stopsignal"},
   "Nodes":[
      {"Level": "run", "Name":"Subject-level", "GroupBy":["subject"], ... }
      {"Level": "dataset", "Name": "T-test-contrasts", "GroupBy": ["contrast"], ...}
      {"Level": "dataset", "Name": "F-test-left-right", "GroupBy": [],... }
   ],
   "Edges":[
      {"Source": "Subject-level", "Destination": "T-test-contrasts", ... }
      {"Source": "Subject-level", "Destination": "F-test-left-right", ...}
   ]
}
```

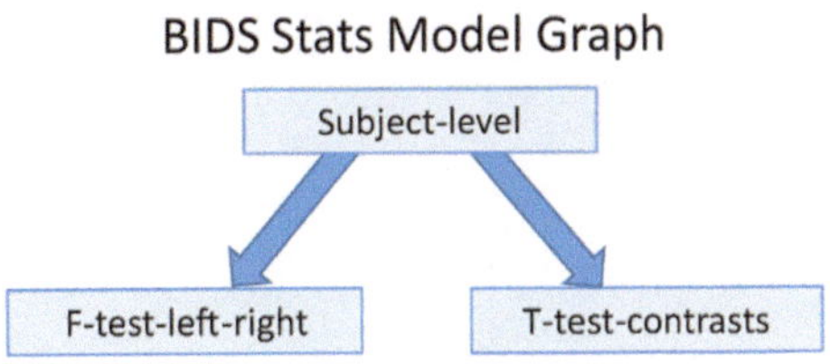

BIDS Stats Model Graph

Subject-level

F-test-left-right

T-test-contrasts

Fig. 4 An example of a BIDS Stats Model for comparing responses between left and right hands. On the left: the JSON model file used as input for *FitLins*. On the right: the execution graph for the computational model

FitLins for reproducibility is its use of model specifications, which are then included with the output to fully document the computation and the results. The *FitLins* model specifications use the newly-standardized BIDS Stats Model format, an example is shown in Fig. 4.

A BIDS Stats Model encodes a multilevel hierarchical statistical model, where statistics for individual runs are combined at higher levels to obtain group statistics representative of the entire dataset or across multiple datasets. The purpose of these models is to provide a complete, re-executable record of the computations.

BIDS Stats Models have only recently been incorporated into BIDS, and while currently focused mainly on fMRI and hierarchical modeling, will likely be extended in the future to support other imaging modalities and analysis techniques. A nice introduction to BIDS Stats Models is available at [https://bids-standard.github.io/stats-models/index.html].

3.2 BIDS Stats Model Graphs

A BIDS Stats Model is represented by a JSON file (Fig. 4, left) specifying the nodes (computations) and edges (input/output relationships between nodes) of a computational graph. The graph in Fig. 4 indicates that after the computational block *Subject-level* is executed, its output is fed into two additional computational blocks, *F-test-left-right* and *T-test-contrasts*. These latter blocks pool results from individual runs across the entire dataset, as indicated by the *Level* parameters of the respective blocks. The input data to the run level computational block depends on the tool, but in this case it is the fMRI imaging files. BIDS Stats Models currently focus on general linear models (GLMs), and the event files associated with the individual imaging files provide critical input for the models as shown by the expansion of the *Subject-level* node of Fig. 4 shown in Fig. 5.

The model indicates that the BOLD signal of each imaging file should be convolved at specific time markers in the data, as specified

BIDS Stats Model Node at Run Level

```
{
  "Level":"run",
  "Name":"Subject-level",
  "GroupBy":["subject"],
  "Transformations":{
     "Transformer": "pybids-transforms-v1",
     "Instructions": [
        {"Name": "Factor", "Input": ["trial_type", "response_hand"]},
        {"Name": "And", "Input": ["trial_type.go", "response_hand.left"], "Output": "go_left"},
        {"Name": "And", "Input": ["trial_type.go", "response_hand.right"], "Output": "go_right"},
        {"Name": "Convolve", "Input":["go_right", "go_left"], "Model":"spm"}
     ]
  },
  "Model":{"X":["go_right", "go_left", "trans_*", "rot_*", 1], "Type": "glm"},
  "DummyContrasts": {"Conditions": ["go_left", "go_right"], "Test":"t"},
  "Contrasts": [
     {"Name": "left_bigger_than_right", "ConditionList": ["go_left", "go_right"], "Weights":[1, -1], "Test":"t"},
     {"Name": "right_bigger_than_left", "ConditionList": ["go_left", "go_right"], "Weights":[-1, 1], "Test":"t"},
     {"Name": "go_trial_vs_baseline", "ConditionList":["go_left", "go_right"], "Weights":[0.5, 0.5], "Test":"t"},
     {"Name": "Movement-related effects", "ConditionList":,"Weights": [...],"Test": "F"}
  ]
}
```

Fig. 5 An expansion of the Subject-level node of Fig. 4

by *go_right* and *go_left*, using an HRF model based on the canonical HRF as provided in *spm* [16]. Here *go_right* and *go_left* are factor vectors of the same length as the *events.tsv* file associated with the imaging file being processed. These factor vectors have 1's in positions for events that satisfy the *go_right* and *go_left* conditions, respectively. The remaining positions are 0.

3.3 BIDS Transformations

If *go_right* and *go_left* were values in a column of the *events.tsv* files, the computation could proceed directly, but since they were not, the events file must undergo some transformations prior to the application of the statistical model. BIDS Transformations, which are under development for incorporation into the BIDS specification, perform logical, selection, and other operations on columns of an events file to generate factor vectors that can be used as input to the models.

The BIDS Transformations can be specified directly in the BIDS Stats model and Fitlins runs these internally without a requirement for additional coding. BIDS Transformations consist of a list of dictionaries, each specifying an operation. The operations are performed in succession. For the transformations in the example of Fig. 5, the input consists of column names from the internal representation of the events file, and the outputs are also names of derived columns computed from the same internal event file.

Figure 6 shows how this process works for the transformations specified in Fig. 5. To obtain the factor columns *go_right* and *go_left*, the *Factor Transformation* creates new factor columns from the unique values in the input columns *trial_type* and *response_hand*. These newly created columns follow the naming

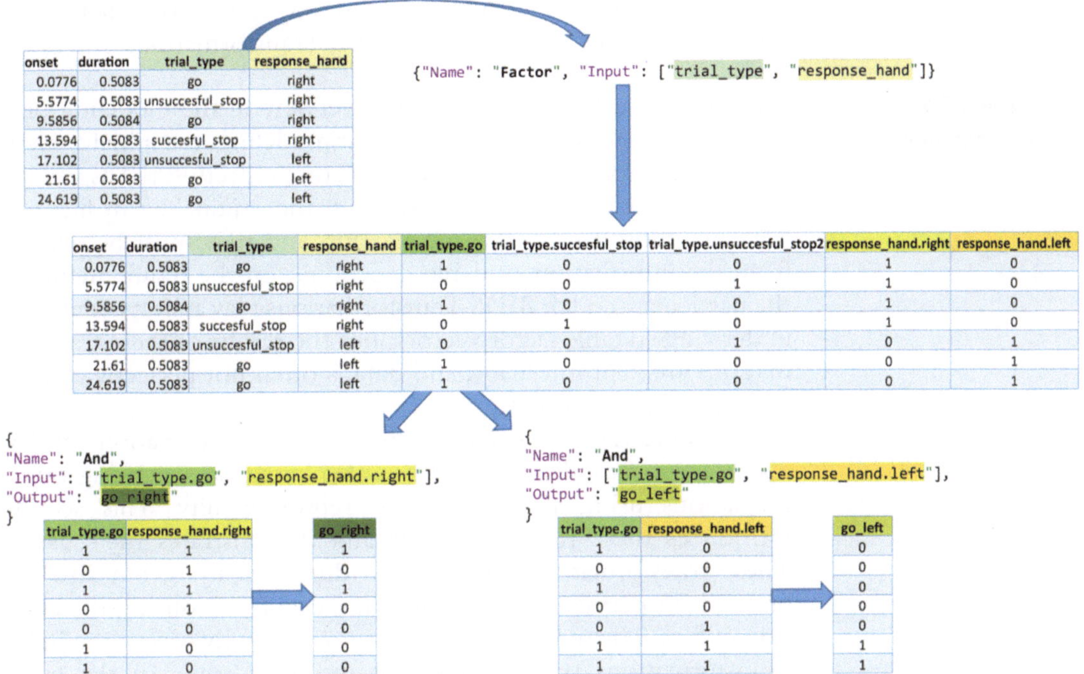

Fig. 6 The process of using BIDS Transformations to create factor vectors out of event file columns

convention *column_name.column_value*, so the factor column representing *go* trials is called *trial_type.go*. Since we only want to analyze *go* trials, we apply the **And** transformation to the factor vectors for *trial_type.go* and *response_hand.right* to create a new column, we have chosen to call *go_right*. A similar operation is specified for the left hand responses. These factors, which never actually appear in the events file itself, are used as input for subsequent computations.

The BIDS Transformations are necessary because the event file columns often do not directly correspond to the needed factors. They promote reproducibility because they are incorporated as part of the model itself. However, there are a limited number of transformations available, and the results are sometimes hard to debug since the actual event file that goes into the computation is not directly visible (though it is possible to include something to dump the internal event files as part of the computation). For further information BIDS Stats Models (BIDS Extension Proposal 2) see:

https://docs.google.com/document/d/1bq5eNDHTb6Nkx3WUiOBgKvLNnaa5OMcGtD0AZ9yms2M/edit and on the BIDS Transformations specification:

https://docs.google.com/document/d/1uxN6vPWbC7ciAx2XWtT5Y-lBrdckZKpPdNUNpwRxHoU/edit#heading=h.kuzdziksbkpm).

The next section introduces an alternative approach that can be used instead of or in addition to BIDS Transformations.

3.4 Event File Transformations

Previous subsections presented an overview of the modeling process and showed that, at least for linear modeling, the initial analysis relies on the specification of factor vectors reflecting the aspects of the data to be modeled. In most cases, the original event files will not have those factors directly present, and they must be derived from the information available depending on the requirements of the particular model. BIDS Transformations play an important role in deriving suitable factor vectors, but the results of these transformations are kept internally, and data is often not perfectly aligned with the required structure.

We have developed an external event transformation mechanism (event remodeling), which is also based on specifying transformations using JSON files for reproducibility. This section introduces these transformations and demonstrates their usage. These transformations can be used during analysis, but also can be used for permanently restructuring event files during the curation of files. The transformations are not currently part of the BIDS Transformations specification, but we hope some of the event remodeling operations will eventually become part of this specification.

The event transformation strategy relies on creating backup event files as shown in Fig. 7.

The process begins by creating a backup copy of the original events files. This backup remains the same throughout the process. When a transformation is to be done, remodeling always makes a clean copy of the events file from the backup. Thus, the remodeling should always assume that it is starting with the original events file. This assures a consistent state of the data throughout the process. Remodeling takes the newly copied *events.tsv* file and the JSON specification file, performs the specified transformations, and then rewrites the *events.tsv* file.

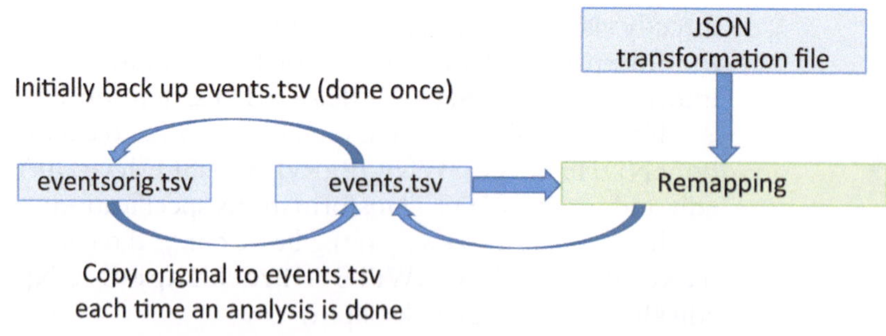

Fig. 7 The process of event file remodeling

Note
After transformation, users can review their files and use the summary tools to make sure that the transformations have the expected results. This review is particularly important in practice as most data has some unexpected quirks such as missing data, missing events as described above. Once satisfied with the results, researchers can perform analyses using tools such as *FitLins* as though the event files were the originals.

3.5 Tools for Restructuring Events

We have developed some event file restructuring tools patterned after the BIDS Transformations. Like BIDS Transformations and BIDS Stats Models, these event remapping tools use a JSON specification file to list the transformations to be performed in succession on the rows and/or columns of every *events.tsv* file in a BIDS dataset. Unlike the BIDS Transformations, however, these operations save the transformed files, so that they can be verified manually and by summarization tools.

The top-level structure of the remodeling file is a list of transformations rather than a dictionary, because a transformation of a particular type may occur at multiple stages during the remapping process. The transformations are performed in order.

Each transformation is represented as a dictionary with keys: *operation*, *description*, and *parameters*. Table 3 summarizes the transformations that are currently available. The toolbox is part of a larger toolset that supports event handling and annotation using the HED (Hierarchical Event Descriptors) framework. A more detailed listing of the event file remodeling operations and their parameters along with tutorials can be found at the File Remodeling Quickstart tutorial [https://www.hed-resources.org/en/latest/HedRemodelingQuickstart.html] and the File Remodeling Tools documentation [https://www.hed-resources.org/en/latest/HedRemodelingTools.html].

Figure 8 shows a simple example of an event restructuring specification that deletes the *sample* and *value* columns and then reorders the columns so the *onset*, *duration*, *event_type*, and *task_role* columns are the first four columns. Since *keep_others* is true, other columns are placed at the end.

When restructuring is performed using the HED remodeling tools with this file and the BIDS root directory as input, these operations will be performed on every *events.tsv* file in the BIDS dataset.

Table 3
Some standard event transformations available in the HED remodeling tools

Operation	Purpose
factor_column	Produce factor columns based on presence or absence of specified values in a column.
factor_hed_tags	Produce factor columns based on a HED tag search string.
factor_hed_type	Produce factor columns from HED type (e.g., *condition-variable* creates factor columns based on the annotated experimental design).
merge_consecutive	Merges several consecutive events of the same type into one event with duration being the span of the merged events.
remap_columns	Map the values of *n* columns into new values in *m* columns using a dictionary lookup.
remove_columns	Remove the specified columns if present.
remove_rows	Remove rows where specified columns take particular values.
rename_columns	Rename columns by providing old names and new names.
reorder_columns	Reorder the columns in the specified order. Columns not included are discarded or placed at the end.
split_rows	Split specified rows into multiple rows in the event file and adjust the meanings of the columns—Usually for unfolding trials into individual events.

```
[
  {
    "operation": "remove_columns",
    "description": "Get rid of the sample and the value columns.",
    "parameters": {
      "remove_names": ["sample", "value"],
      "ignore_missing": true
    }
  },
  {
    "operation": "reorder_columns",
    "description": "Want event_type and task_role columns after onset and duration.",
    "parameters": {
      "column_order": ["onset", "duration", "event_type", "task_role"],
      "ignore_missing": false,
       "keep_others": true
    }
  }
]
```

Fig. 8 A JSON specification for event file restructuring using the HED remodeling tools

4 Example Data Analysis

In this section we present the results of two contrasts on the AOMIC-PIOP2 and SOCCER data: *successful stop* (sic) versus *unsuccessful stop* (sic) and *left* versus *right*. Before we are able to run these contrasts we must resolve some of the problems we described earlier, mainly this considers the issue of trial-level encoding versus event-level encoding. We consider simple *t*-statistic maps [*see* Glossary] as outcomes.

A few notes should be given on the interpretation of these maps, which are created by applying statistical tests to individual voxels in the fMRI images. The process of deriving whole brain fMRI results is strongly influenced by thresholding procedures. Many strategies have been developed to deal with the massive multiple comparison problem, without applying the most conservative Bonferroni approaches, such as calculating the Family Wise Error, or the False Discovery Rate [19]. Often these approaches are combined with some cluster thresholding, meaning we require a specified number of voxels to pass the threshold before we accept the result as significant. When looking at the basic *t*-statistic maps we cannot say definitively whether higher values are a reflection of true differences between conditions. Here we broadly view patterns across datasets to showcase which differences could be points of interest.

The Nilearn python library [https://nistats.github.io/] [*see* Resources] has several functions for applying statistical thresholds to data, including FDR and cluster thresholds. We have included a simple niistats script that can be used to load in the unthresholded statistical maps created by *FitLins* and apply some basic thresholds in the supplementary material. More information on this process can be found here: [https://nilearn.github.io/stable/auto_examples/05_glm_second_level/plot_thresholding.html].

All files necessary to run the *FitLins* analysis for both datasets, as well as all *FitLins* results can be found in the supplementary materials [https://osf.io/93km8/]. In the case of the AOMIC-PIOP2 dataset, the remapping and model files can be applied to the entire dataset as found on OpenNeuro, to obtain results for the full 200 participants if desired.

4.1 Successful Versus Unsuccessful Stops

The primary focus of stop-signal tasks is to study the interaction of response and inhibition. Because the task load and task type were very different in the two datasets, one would expect some differences in the interplay of response and inhibition in the two cases. However, there might be some overlap in regions related to inhibition processes themselves. As an example comparison, we choose successful stop versus unsuccessful stop trials for the two datasets. We use the onset of the go signal as the marker for these events in both datasets. We use the encoding of successful and unsuccessful stop trials from the AOMIC dataset without event remodeling. The models for this example are available in the supplemental materials [https://osf.io/93km8/].

Figure 9 shows results for the contrast of successful versus unsuccessful stops in the two datasets. As expected the results from this smaller sample (top two graphs) show limited convergence of regional activation across participants. Interpretation here is difficult because of the small sample size. However, the larger AOMIC-PIOP2 dataset shows distinctive patterns which require additional thresholding.

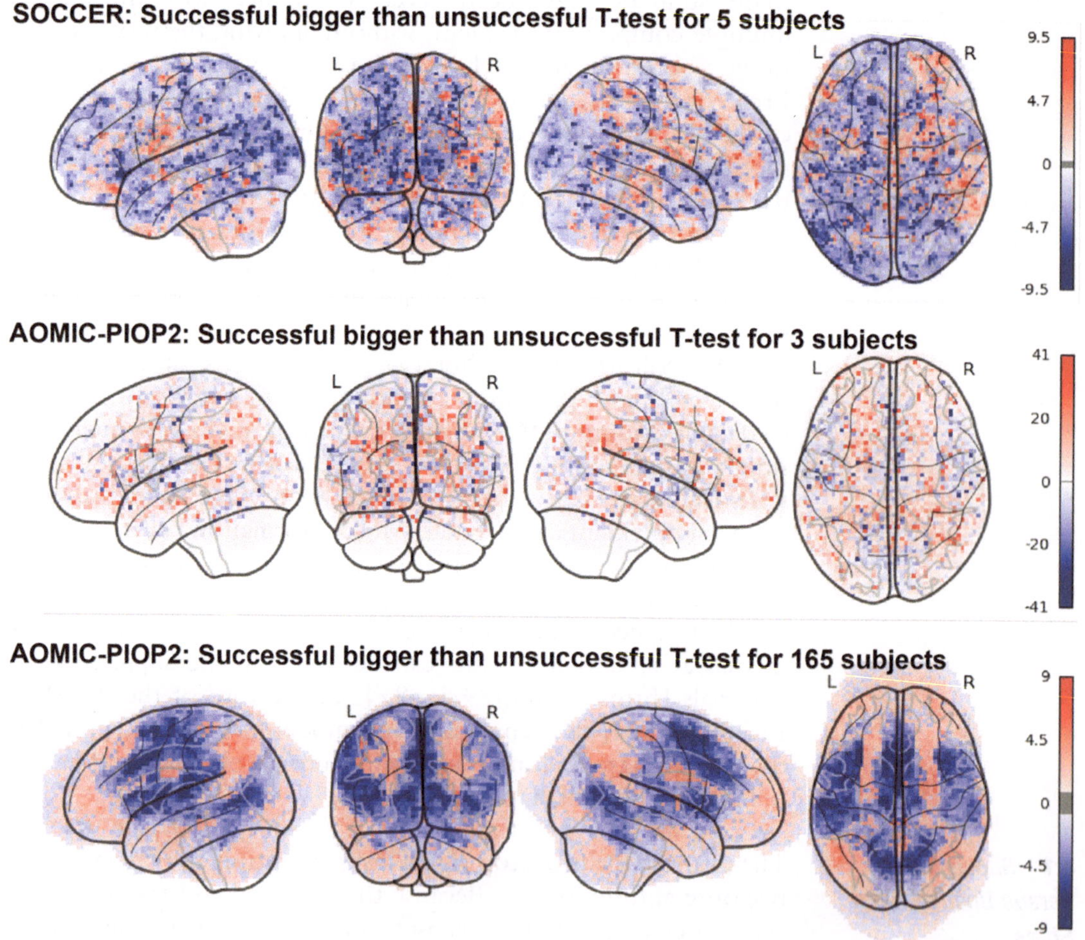

Fig. 9 *T*-test comparison of regions where response for successful stop trials was greater than for unsuccessful stop trials. Top graph: SOCCER for the five-subject demo data. Middle graph: AOMIC-PIOP2 for the three-subject demo data. Bottom graph: AOMIC-PIOP2 for 165 subjects. Note: three subjects in AOMIC-PIOP2 had no successful stops and one subject had no unsuccessful stops. In all, 61 subjects had fewer than ten trials of one of the types and were removed. Color scales represent *t*-values

Based on the t-statistic maps of the all participants we see a larger difference to the successful versus unsuccessful trials in frontal cortex, cingulate gyrus, as well as inferior parietal gyrus and regions around the supramarginal gyrus. Some of these regions have been found previously in studies related to response inhibition. A meta-analysis into different inhibition tasks found peak activations across several stop signal tasks in the right middle cingulate gyrus, bilateral supramarginal gyrus, and right inferior parietal gyrus, as well as several other regions that do not show a clear difference here [20]. By comparing results with other types of inhibition tasks they found response inhibition was most likely controlled by a fronto-parietal network and ventral network, which is congruent with the pattern we see.

4.2 Left Versus Right Responses

Both datasets had participants press left versus right buttons to indicate their responses in the experiment's discrimination task. While AOMIC-PIOP2 experimental setup used separate fiber-optic response pads with four buttons for each hand, SOCCER only used a single response pad, located at the participant's right hand. Participants in the SOCCER dataset were instructed to use their right index finger and ring finger for responses, while participants in the AOMIC-PIOP2 study used their left and right hand index fingers to press buttons on the left and right response pads. Based on this configuration, we expect a large difference in neural activation between left and right motor cortex for left and right button presses in the AOMIC-PIOP2 results, while we expect no such differences for the SOCCER data.

Because SOCCER uses event-level encoding, the button presses occupy their own rows in the event tables with onsets corresponding to the times of the button presses. In contrast, AOMIC-PIOP2 uses trial-level encoding, and the onset of the trial is the presentation of the face image, not the time of the button press. One possible model for the AOMIC-PIOP2 left versus right comparison was introduced in Figs. 5 and 6 using the BIDS Transformations. The difficulty with using this model for comparison of left-right responses with SOCCER is that SOCCER encodes the response events individually and the times of these events are recorded as the times of the button presses rather than the times of the image presentations. The AOMIC-PIOP2 response times average 1.02 s, but there is significant variability among subjects and trials. An alternative is to recode the AOMIC-PIOP2 event files to more closely match the response events of SOCCER. In the next section we show how to do this re-coding using the remodeling tools.

4.2.1 AOMIC-PIOP2 Event Preparation

Figure 10 shows the remodeling steps required. The goal is to produce a new event file where the rows represent the times of the button presses in response to the face images. Since the new

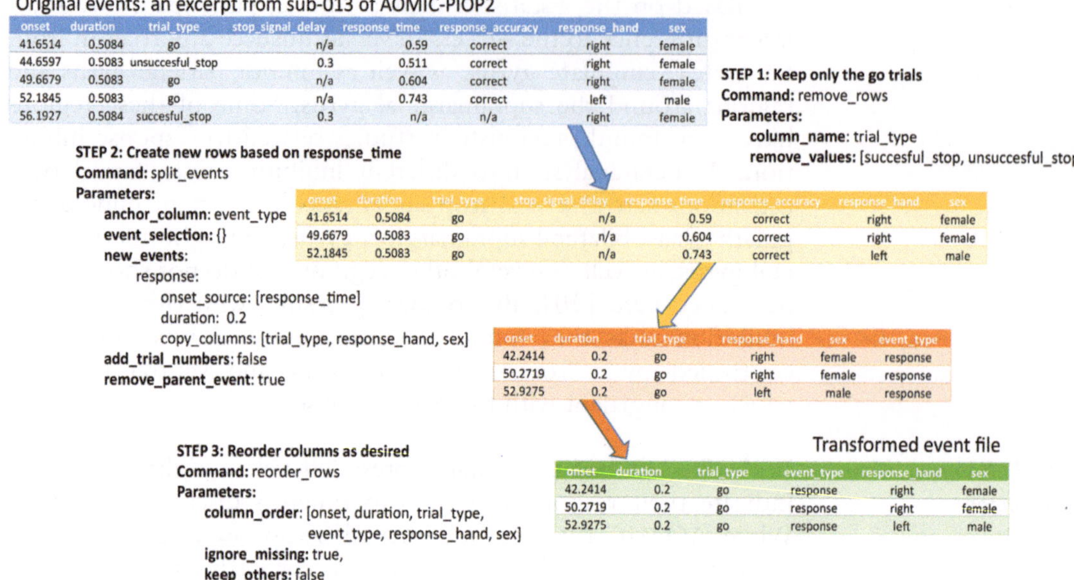

Fig. 10 Remodeling AOMIC-PIOP2 response events. The *response_time* events are transformed into new events rows, with the original events being removed

response times will be the original *onset* plus the *response_time*, we first remove any rows where the *response_time* is not defined.

The next step is to create new events from the original trial event. Since *remove_trial_parent* is true, the original trial event will not appear in the remodeled file. Any number of new events can be created during a *split_event*, but the example only shows creation of *response* events. The new_column parameter designates the column in which the "codes" representing these new events are recorded. Here the only code is *response*. The *onset_source* designates a list of values (and/or column names from which to extract the values for each event) that are added to the onset value to produce the onset of the new event. Some values in the list could be negative. The created events don't have to be created in order of their onsets as the onsets are resorted at the end of the split_event process.

Figure 11 shows the remodeling transformation JSON file that performs these operations. The file consists of a list of four dictionaries corresponding to the three steps in transforming the event file. The first two transformations focus on go trials and make sure that all response times are defined. The next transformation creates a new event for each go trial with onset at the time of button press. The final transformation reorders the columns as required. Additional information and tutorials on using the remodeling facilities can be found in the File Remodeling tools documentation: [https://www.hed-resources.org/en/latest/HedRemodelingTools.html].

```
[
    {
        "command": "remove_rows",
        "description": "Keep only the go trials",
        "parameters": {
            "column_name": "trial_type",
            "remove_values": ["succesful_stop", "unsuccesful_stop"]
        }
    },
    {
        "command": "remove_rows",
        "description": "Remove any other rows where response_time is n/a",
        "parameters": {
            "column_name": "response_time",
            "remove_values": ["n/a"]
        }
    },
    {
        "command": "split_events",
        "description": "Create event files that only have response events",
        "parameters": {
            "anchor_column": "event_type",
            "event_selection": {},
            "new_events": {
                "response": {
                    "onset_source": ["response_time"],
                    "duration": "0.2",
                    "copy_columns": ["trial_type", "response_hand", "sex"]
                }
            },
            "add_trial_numbers": false
            "remove_parent_event": true
        }
    },
    {
        "command": "reorder_columns",
        "description": "Order columns for the output",
        "parameters": {
            "column_order": ["onset", "duration", "trial_type", "event_type",
                             "response_hand", "sex"],
            "ignore_missing": true,
            "keep_others": false,
        }
    }
]
```

STEP 1: Keep only go *trial_type* and make sure all *response_time* values are defined.

STEP 2: Create new event for each trial with onsets corresponding to time of button push.

STEP 3: Reorder the columns as desired.

Fig. 11 JSON file with instructions for restructuring AOMIC-PIOP2 event files for Model 2 above

The script of Fig. 11 includes *remove_rows* operations for both *response_time* equal to *n/a* and for trial_type values of either *succesful_stop* or *unsuccesful_stop* to assure that only successful *go* trials with *response_time* defined are included. For simplicity of explanation, we did not also include a *remove_rows* operation for *response_accuracy* values of *incorrect*, as there were relatively few trials where this occurred. However, it is a good idea to consider all unusual cases in developing a remodeling script.

4.2.2 SOCCER Event Preparation

To run a comparable contrast on the SOCCER data, two issues needed to be addressed: resolution of ambiguity in the button press codes and detection of invalid trials. The first issue illustrates inserting block markers, the second issue illustrates insertion of trial structure.

As described earlier (Subheading "SOCCER"), the SOCCER data contained mainly event level information. Left and right button presses were coded as either code 1 or code 2 in the log file. But these same buttons were used to answer questions during the survey block, as well as to perform the experiment task. Because of this, button presses during experimental trials must be distinguished from button presses in the survey trials. To address this ambiguity, we select all events between code "40" and "80" and define them as part of the experiment block group. Once blocks have been distinguished, the unique codes for the survey blocks may be substituted to distinguish from the experimental blocks.

The invalid trial problem illustrates the use of the second approach based on labeling specific event sequences. We are interested in the left versus right responses during *go* trials. *Go* trials can be identified based on specific sequences within trial groups. Once we have identified the trial groups, we provide individual labels for each sequence in the group.

Table 4 shows an overview of all the sequences found in trial groups labeled as go trials, successful stop trials, or unsuccessful stop trials. These labeled sequences can be used not only for labeling trial types, but also for providing detailed summaries of the trials that occurred in the dataset.

Some sequences in the table are more difficult to label such as trial groups only containing a single participant button press event. We consider these trials invalid. These responses were either corrections from incorrectly solved *go* trials, or double responses. The participant pressed the button before the stop signal was presented in some trials. In some cases, the button presses were so early that the stop signal was never even presented. We classified these latter trials as *go* trials as this is what it would be perceived from the participant perspective.

Table 4
Trial sequences in SOCCER dataset along with appropriate trial labels

Sequence	Trial category
30, 11, 1	Go
30, 12, 2	Go
30, 12, 1	Go
30, 211, 221	Successful stop
30, 212, 222	Successful stop
30, 211, 221, 1	Unsuccessful stop
30, 212, 222, 2	Unsuccessful stop
30, 211, 221, 2	Unsuccessful stop
30, 211, 1	Go
30, 212, 2	Go
30, 211, 2	Go
30, 211, 1, 221	Premature signal response
1	Invalid
2	Invalid

The details of these transformations are complex and are beyond the scope of this chapter. For more information, see [https://www.hed-resources.org/en/latest/HedRemodelingTools.html].

4.2.3 Left Versus Right Results

Figure 12 shows a comparison of the unthresholded *FitLins* statistical maps for the *T*-test for go trials where activation for the left button press is greater than for the right button press. As expected, the SOCCER demo data (top graph) does not show an obvious difference across hemispheres.

On the other hand, even in a dataset as small as the three-subject demo dataset, AOMIC-PIOP2 does show a visual

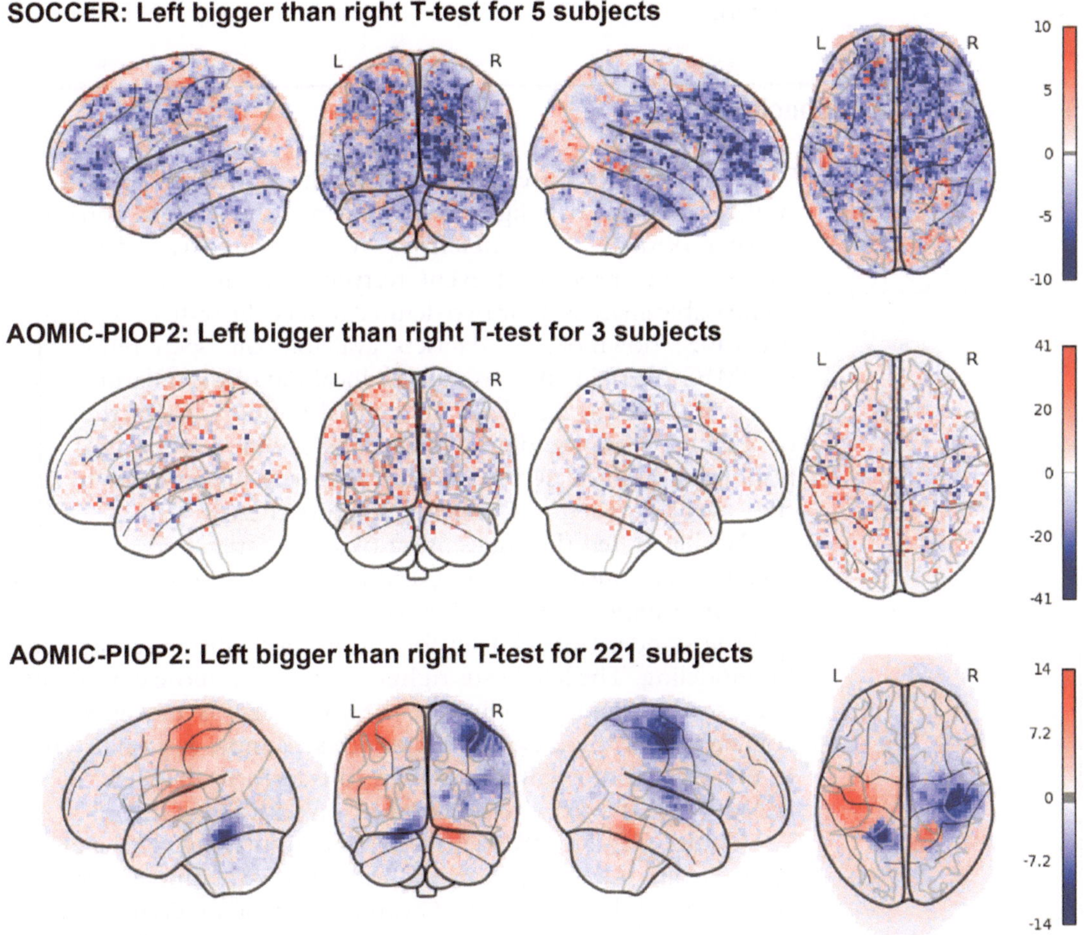

Fig. 12 *T*-test comparison of regions where response for left button press is greater than for right button presses in go trials. Top graph: SOCCER for the five-subject demo data. Middle graph: AOMIC-PIOP2 for the three-subject demo data. Bottom graph: AOMIC-PIOP2 for 221 subjects. Note: three subjects in AOMIC-PIOP2 had no *go* trial left button presses and two had no *go* trial right button presses and were excluded. Color scales represent *t*-values

difference between hemispheres for left larger than right button presses (middle graph). The distinction is much more clearly demonstrated in the 221-subject AOMIC-PIOP2 dataset (bottom graph). Unfortunately, the difference is in the opposite direction of our expectations—with left hand actions associated with higher *t*-values in the left hemisphere. We also see the contralateral (normally ipsilateral) activation in the cerebellum. This discrepancy could be a mistake in the model, a reversal of designations in the event files, or even be a reversal of the hemispheres during image processing. Since the original log files are not available we have not been able to rule out any of these possibilities. However, this result demonstrates the usefulness of running even relatively trivial comparisons to cross-check the consistency of the data and results. All of the model files and results are available in the supplementary materials.

5 Discussion and Conclusion

The stop signal task used as an example throughout this chapter is a well-established paradigm for studying the mental processes *response initiation* and *inhibition*. We have used two datasets to showcase the process of event restructuring and annotation for comparable analysis. Our two demo datasets illustrate two common use cases: downloading datasets available on open repositories (AOMIC-PIOP2) and structuring local data in standardized formats (SOCCER). Based on the limited sample sizes of the demo datasets, we cannot draw conclusions about regions related to response inhibition across studies, rather our purpose was to illustrate event restructuring issues and approaches.

We illustrated the issues using two comparison problems: successful versus unsuccessful stops and left versus right responses. In the first example, there are comparable events in the two datasets, and the standard BIDS Transformation mechanisms are sufficient for modeling. The left versus right comparison is more complicated because the event encoding of the two datasets prevents direct comparison of these events. We use this example as an illustration of the process of event remodeling to transform the event files of AOMIC-PIOP2 for a better comparison with SOCCER.

Events play an essential role in neuroimaging data management, from initial preparation using the experimental logs to advanced task-based analysis. Creating appropriate event representations for a given application often requires considerable time investment, and the problem is compounded when the analysis includes datasets from different experiments and laboratories as demonstrated by the comparison of the AOMIC-PIOP2 and SOCCER datasets presented in this chapter.

Event files also play an important role in reproducible and transparent analyses. Transparency requires thorough documentation of events and use of standardized terms for describing cognitive tasks and their implied role in cognition. Careful documentation of control variables and specific targeting of mental concepts is an important part of the practice of cognitive science. More recent incorporations of naturalistic paradigms such as movie watching also depend on careful structuring and flexible annotation of events occurring during the experiment.

To address these issues, we have developed a framework and event restructuring tools (remodeler) that allow users to modify a dataset's event files by specifying a series of operations in a JSON text file. Using such a file not only allows researchers to avoid writing one-off code for each analysis, but also results in a file that clearly documents the operations performed on the events files so that they can be easily reproduced. These restructuring tools are well-integrated with BIDS and augment the BIDS Stats Model and BIDS Transformations. The tools are also integrated with HED (Hierarchical Event Descriptors) [5] for more advanced analysis and standardized annotation of events. Event restructuring, as well as the role of HED in restructuring, is discussed in [https://www.hed-resources.org/en/latest/HedRemodelingQuickstart.html] and [https://www.hed-resources.org/en/latest/HedRemodelingTools.html].

Aggregating neuroimaging data is an important aspect of verifying neuroimaging results from often small sample sizes, and establishing consistent results across studies with often variable designs and participant groups. Besides questions about the generalizable results, however, there are also questions in neuroscience about the effect of variations in experimental design on neural activation. In order to learn more about this, it is important to compare experiments on an event level. Data sharing has also proven useful for method testing and validation as well as for the development and testing of software [6].

Event representation for documentation and analysis rarely receives the attention it deserves, given its complexity and how essential it is to a valid analysis. We have shown that synchronization of multiple datasets to allow for comparable analysis results requires a significant investment of time and resources, even for relatively straightforward stimulus-response paradigms. A deep understanding of the executed paradigms and the events remains essential. However, with better guidelines and more tools for event handling, event processing can be made less time intensive for users of shared data as well as for those collecting new data.

Author Contributions

KR and MD conceived software and demonstration and drafted the manuscript. FR and JB were responsible for data collection in the SOCCER study. MP worked on data curation for SOCCER study. All authors contributed useful comments to tutorial conception and the finalization of the manuscript.

Funding

This study was funded by NIMH grant R01MH126700-01A1 and by a grant to the first and fifth authors (MJMD, ANR) from the Doctoral College "Imaging the Mind" [FWF W 1233-B].

References

1. (2017) E-Prime® | Psychology Software Tools. In: Psychology Software Tools | Solutions for Research, Assessment, and Education. https://pstnet.com/products/e-prime/. Accessed 5 Mar 2024
2. Peirce J, Gray JR, Simpson S, MacAskill M, Höchenberger R, Sogo H, Kastman E, Lindeløv JK (2019) PsychoPy2: experiments in behavior made easy. Behav Res Methods 51: 195–203. https://doi.org/10.3758/s13428-018-01193-y
3. Gorgolewski KJ, Auer T, Calhoun VD, Craddock RC, Das S, Duff EP, Flandin G, Ghosh SS, Glatard T, Halchenko YO, Handwerker DA, Hanke M, Keator D, Li X, Michael Z, Maumet C, Nichols BN, Nichols TE, Pellman J, Poline J-B, Rokem A, Schaefer G, Sochat V, Triplett W, Turner JA, Varoquaux G, Poldrack RA (2016) The brain imaging data structure, a format for organizing and describing outputs of neuroimaging experiments. Sci Data 3:160044. https://doi.org/10.1038/sdata.2016.44
4. Robbins K, Truong D, Appelhoff S, Delorme A, Makeig S (2021) Capturing the nature of events and event context using hierarchical event descriptors (HED). NeuroImage 245:118766. https://doi.org/10.1016/j.neuroimage.2021.118766
5. Robbins K, Truong D, Jones A, Callanan I, Makeig S (2022) Building FAIR functionality: annotating events in time series data using hierarchical event descriptors (HED). Neuroinformatics 20:463–481. https://doi.org/10.1007/s12021-021-09537-4
6. Markiewicz CJ, Gorgolewski KJ, Feingold F, Blair R, Halchenko YO, Miller E, Hardcastle N, Wexler J, Esteban O, Goncavles M, Jwa A, Poldrack R (2021) The OpenNeuro resource for sharing of neuroscience data. elife 10:e71774. https://doi.org/10.7554/eLife.71774
7. Gorgolewski KJ, Alfaro-Almagro F, Auer T, Bellec P, Capotă M, Chakravarty MM, Churchill NW, Cohen AL, Craddock RC, Devenyi GA, Eklund A, Esteban O, Flandin G, Ghosh SS, Guntupalli JS, Jenkinson M, Keshavan A, Kiar G, Liem F, Raamana PR, Raffelt D, Steele CJ, Quirion P-O, Smith RE, Strother SC, Varoquaux G, Wang Y, Yarkoni T, Poldrack RA (2017) BIDS apps: improving ease of use, accessibility, and reproducibility of neuroimaging data analysis methods. PLoS Comput Biol 13:e1005209. https://doi.org/10.1371/journal.pcbi.1005209
8. Snoek L, van der Miesen MM, Beemsterboer T, van der Leij A, Eigenhuis A, Steven Scholte H (2021) The Amsterdam open MRI collection, a set of multimodal MRI datasets for individual difference analyses. Sci Data 8:85. https://doi.org/10.1038/s41597-021-00870-6
9. Esteban O, Markiewicz CJ, Blair RW, Moodie CA, Isik AI, Erramuzpe A, Kent JD, Goncalves M, DuPre E, Snyder M, Oya H, Ghosh SS, Wright J, Durnez J, Poldrack RA, Gorgolewski KJ (2019) fMRIPrep: a robust preprocessing pipeline for functional MRI. Nat Methods 16:111–116. https://doi.org/10.1038/s41592-018-0235-4
10. Markiewicz CJ, De La Vega A, Wagner A, Halchenko YO, Finc K, Ciric R, Goncalves M, Nielson DM, Kent JD, Lee JA, Bansal S,

Poldrack RA, Gorgolewski KJ (2022) Poldracklab/fitlins: 0.10.1

11. Markiewicz C, Bottenhorn K, Chen G, Vega A de L, Esteban O, Maumet C, Nichols T, Poldrack R, Poline J-B, Yarkoni T (2021) BIDS Statistical Models – An implementation-independent representation of General Linear Models p. 1
12. Yarkoni T, Markiewicz CJ, de la Vega A, Gorgolewski KJ, Salo T, Halchenko YO, McNamara Q, DeStasio K, Poline J-B, Petrov D, Hayot-Sasson V, Nielson DM, Carlin J, Kiar G, Whitaker K, DuPre E, Wagner A, Tirrell LS, Jas M, Hanke M, Poldrack RA, Esteban O, Appelhoff S, Holdgraf C, Staden I, Thirion B, Kleinschmidt DF, Lee JA, di Oleggio V, Castello M, Notter MP, Blair R (2019) PyBIDS: python tools for BIDS datasets. J Open Source Softw 4:1294. https://doi.org/10.21105/joss.01294
13. Logan GD, Cowan WB (1984) On the ability to inhibit thought and action: a theory of an act of control. Psychol Rev 91:295–327. https://doi.org/10.1037/0033-295X.91.3.295
14. Ramautar JR, Slagter HA, Kok A, Ridderinkhof KR (2006) Probability effects in the stop-signal paradigm: the insula and the significance of failed inhibition. Brain Res 1105: 143–154. https://doi.org/10.1016/j.brainres.2006.02.091
15. Logothetis NK, Pauls J, Augath M, Trinath T, Oeltermann A (2001) Neurophysiological investigation of the basis of the fMRI signal. Nature 412:150–157. https://doi.org/10.1038/35084005
16. BIDS Stats Models—BIDS Stats Models Specification. https://bids-standard.github.io/stats-models/. Accessed 5 Mar 2024
17. Jahfari S, Waldorp L, Ridderinkhof KR, Scholte HS (2015) Visual information shapes the dynamics of corticobasal ganglia pathways during response selection and inhibition. J Cogn Neurosci 27:1344–1359. https://doi.org/10.1162/jocn_a_00792
18. Friston KJ, Josephs O, Rees G, Turner R (1998) Nonlinear event-related responses in fMRI. Magn Reson Med 39:41–52. https://doi.org/10.1002/mrm.1910390109
19. Logan BR, Rowe DB (2004) An evaluation of thresholding techniques in fMRI analysis. NeuroImage 22:95–108. https://doi.org/10.1016/j.neuroimage.2003.12.047
20. Zhang R, Geng X, Lee TMC (2017) Large-scale functional neural network correlates of response inhibition: an fMRI meta-analysis. Brain Struct Funct 222:3973–3990. https://doi.org/10.1007/s00429-017-1443-x

Chapter 8

Standardized Preprocessing in Neuroimaging: Enhancing Reliability and Reproducibility

Oscar Esteban

Abstract

This chapter critically examines the standardization of preprocessing in neuroimaging, exploring the field's evolution, the necessity of methodological consistency, and the future directions shaped by artificial intelligence (AI). It begins with an overview of the technical advancements and the emergence of software tools with standardized neuroimaging processes. It also emphasizes the importance of the *Brain Imaging Data Structure* (*BIDS*) and data sharing to improve reproducibility. The chapter then discusses the impact of methodological choices on research reliability, advocating for standardization to mitigate analytical variability.

The multifaceted approach to standardization is explored, including workflow architecture, quality control, and community involvement in open-source projects. Challenges such as method selection, resource optimization, and the integration of AI are addressed, highlighting the role of openly available data and the potential of AI-assisted code writing in enhancing productivity.

In conclusion, the chapter underscores *NiPreps*' contribution to providing reliable and reproducible preprocessing solutions, inviting community engagement to advance neuroimaging research. The chapter envisions a collaborative and robust scientific culture in neuroimaging by promoting standardized practices.

Key words Computational neuroscience, fMRIPrep, MRIQC, Neuroimaging, Reliability, Reproducibility, Python, Open-source

1 Introduction

Neuroimaging has seen remarkable technical developments over the past three decades, reflecting its singular adequacy for probing the brain's structure and its intricate workings in vivo. The continuous innovation in image formation technologies has bolstered the development of domain software tools and applications. Correspondingly, new theories of the brain and new experimental approaches have also stimulated progress with the demand for new hardware and software instruments [1]. Consequently, the field has produced a multiplicity of software instruments over time, developed with high engineering standards and readily

Robert Whelan and Hervé Lemaître (eds.), *Methods for Analyzing Large Neuroimaging Datasets*, Neuromethods, vol. 218, https://doi.org/10.1007/978-1-0716-4260-3_8,

available to neuroimagers. Among many others, *AFNI* [2], *FreeSurfer* [3], *FSL* [4], and *SPM* [5] have achieved remarkable adoption. For transparency, most neuroimaging packages enable researchers to independently scrutinize the implementations they add to their tool belts by making the source code accessible—if not fully open-source.[1]

With the advancement of imaging techniques, these toolboxes have been substantially expanded to support ever-growing spatial, and temporal resolutions, as well as new modalities and acquisition approaches. Moreover, neuroimaging research has also seeded convergent efforts toward multimodal fusion, where features are extracted from several measurement types. These two drivers prompted the establishment of neuroimaging "pipelines" that stage processing steps and encompass data management operations. Traditionally, these steps are drawn from a single toolbox of choice for a particular application or analysis, as compatibility between tools poses substantial problems. Tools such as *Nibabel* [6] and *Nipype* [7] have enabled "mixing-and-matching" from available neuroimaging tools to select the best-in-class implementations across them by standardizing access to data (*Nibabel*) and the user interface to tools (*Nipype*).

While the redundancy of implementations for a given task is positive from a knowledge formalization and accessibility perspective, it has gradually become apparent that methodological variability is an obstacle to obtaining reliable results and interpretations, a problem only exacerbated by inaccurate or insufficient reporting [8]. Indeed, variations in processing methods across different modalities, research groups, studies, and even individual researchers have contributed to inconsistencies and discrepancies in reported findings [9–11]. This problem was more recently surfaced with the *Neuroimaging Analysis Replication and Prediction Study* (*NARPS*; [12]), where 70 teams of functional MRI (fMRI) experts were provided with the same dataset and tasked with testing a closed set of nine hypotheses. The results highlighted an overall poor agreement in conclusions across teams. Considering that no two teams fully coincided in the design of their analysis pipelines, the *NARPS* authors interpreted that methodological variability was at the core of the divergent results. As potential counter-measures, Botvinik-Nezer and colleagues discussed the value of preregistration to avoid methodological variability introduced *post-hoc*, that is, fine-tuning the processing pipeline until the results align with expectations. Additionally, they also envisioned multiverse analyses where many combinations of different implementations of a pro-

[1] A source code may be made accessible (e.g., shared over a private email) while open-source implies a license stating unambiguous terms for reuse and redistribution.

cessing and analysis pipeline are explored, and results are either interpreted as a range of possibilities or aggregated statistically [13], e.g., by means of active learning [14]. Both—preregistration and multiverse analyses—are powerful tools for reproducibility that operate in the domain of methodologies, either limiting the researcher's degrees of freedom and incentives to workaround nonnegative findings (preregistration; [15, 16]) or embracing the exploration of the breadth of methodological alternatives and combinations thereof (multiverse; [17]). Because shallow reporting bears great responsibility for how analytical variability may undermine the reliability, best practices in reporting such as checklists (for instance, the Organization for the Human Brain Mapping's *COBIDAS*; *Committee on Best Practice in Data Analysis and Sharing*; [18]) have been proposed to solve the problem. Nonetheless, it is worth noting that all the teams involved in *NARPS* completed the *COBIDAS* checklist. That alone did not guarantee that the reported methods could be adequately replicated. Taylor and colleagues showed evidence that the *NARPS* results are more convergent than initially interpreted when outputs are examined without standardly applied simplifications such as thresholding of statistical maps [19]. Nonetheless, analytical variability remains a concerning issue that undermines the reliability of neuroimaging research.

Over the last decade, researchers have harnessed their neuroimaging workflows targeting reliability. The *Brain Imaging Data Structure* (*BIDS*; [20, 21]) has proven to be a hallmark example of how standard dataset organization is critical to implement reproducible neuroimaging research (*see* Chapter 4 for a detailed guide to *BIDS*). Not only has *BIDS* deeply transformed the neuroimaging landscape by establishing a consistent agreement on how data and metadata must be organized, maximizing the shareability of datasets and ensuring proper data archiving, it has also spurred a body of research addressing aforementioned challenges to reproducibility. Indeed, data sharing has been recognized as a powerful reproducibility tool, and outstanding resources such as *OpenNeuro* [22] have contributed to solidifying the development of neuroimaging workflows with a clear and standardized interface for input—*BIDS*— and output—*BIDS-Derivatives*—data. Leveraging *BIDS* and following the *BIDS Apps* principles [23], our *fMRIPrep* application [24] has shaped the development of standardized neuroimaging workflows and given rise to the *NeuroImaging PREProcessing toolS* (*NiPreps*; [25]). Using the *NiPreps* development experience as a foundation, this chapter explores the rationale, benefits, and potential trade-offs of standardizing the preprocessing stage as a way to account for analytical variability in a significant stage of every neuroimaging pipeline.

2 Standardizing Preprocessing: What and Why?

Generally, neuroimaging analyses cannot be carried out directly with "unprocessed" data, that is data after reconstruction from the "raw" recordings collected by an imaging device (c.f., Chapter 15). While *BIDS* helps organize unprocessed data and provides a reliable ingress interface into subsequent processing, data needs preprocessing before it can be analyzed [25, 26]. Preprocessing involves a series of essential operations, including data cleaning, spatiotemporal normalization and alignment, artifact removal, and other steps required by statistical modeling [24]. Analytical variability quickly emerges in the design of such pipelines as each processing step with its associated parameters involves methodological choices. These choices will likely undermine the reliability of the outcomes unless the pipeline abides by strict self-consistency and transparent implementation and reporting. The variability introduced by preprocessing compounds with the variability of the data collection and subsequent steps, such as image reconstruction, exacerbating the overall pipeline's unreliability.

When looking through the lens of classical test theory [27–29], the neuroimaging "scores" that are statistically modeled in the final analysis step are indeed "preprocessed" data. For example, a morphometry analysis quantifying T_1-weighted MRI properties such as cortical thickness—that is, a neuroimaging "score",— requires preprocessing steps involving brain extraction or reconstruction of brain surfaces. The classical theory posits two approaches to improve the reliability of scores, such as the cortical thickness in the example: aggregation and standardization. Please note that although "reliability", "reproducibility", "repeatability", and "replication" are different terms understood in many ways across disciplines [30], here we will define reliability as the property that the score or measurement is consistently correlated with the true value of it [29].

The aggregation approach follows the "Spearman-Brown prophecy formula" [31, 32], and supports that random error components cancel out by aggregating items of the same true score, thereby providing more reliable measurements. This aggregation approach is at the core of recent dense and "personalized" imaging data collection efforts [33]. These approaches repeat the same experiment on reduced numbers of individuals to analyze them independently and focus only on within-subject variability, thereby improving the within-individual reliability. Within the example of cortical thickness analysis, aggregation could be implemented by collecting several images for every subject and extracting surfaces and the feature of interest from each individual's single, average template. The approach is an excellent tool to characterize the reliability of the measurements (*see* ref. [34], for an example of our efforts in this direction). On the other hand, standardization reduces sources of variability relating to the measurement instrumentation, including methodological variability of preprocessing, by strictly predetermining all the experimental choices and establishing a unique workflow. This chapter focuses on standardization to reduce the domain of coexisting analytical alternatives, hence reducing the multiverse that must be traversed in mapping the variability of the results. Standardizing the preprocessing offers numerous benefits for enhancing the reliability and reproducibility of the research workflow, albeit the paradigm is not free of trade-offs and challenges. First, a reliable measure is not necessarily "valid" [35]. Standardization may enforce specific assumptions about the data and introduce biases that could void the measurement validity. For instance, a brain extraction algorithm within a brain morphometry application that systematically includes dura in smaller brains and excludes it in larger ones could lead to the wrong conclusion about gender differences in cortical thickness at the population level. Other challenges and trade-offs of standardization involve the robustness to data diversity, the flexibility versus experimental degrees-of-freedom trade-off, and computational optimization. The following section explores several of these aspects along different dimensions through which standardization may be implemented.

3 Dimensions of Standardization

3.1 The Brain Imaging Data Structure (BIDS)

Although initially conceived as an exchange format to maximize data shareability and archival, *BIDS* provides a consistent framework for structuring data directories, naming conventions, and metadata specifications. Building on the clear interface that *BIDS* affords for the input, our *BIDS Apps* framework [23] describes several formal aspects to enable the standardization of pipelines. The widespread adoption of *BIDS* has greatly facilitated the uptake of *BIDS Apps*, such as *fMRIPrep*, which leverages *BIDS*-compliant datasets to automate the preprocessing of fMRI and exchange (through *BIDS-Derivatives*) downstream processing and analysis. *BIDS-Derivatives* provides a standardized format for representing processed and derived data, ensuring consistency and compatibility across different studies and analyses. Researchers can easily share and disseminate their preprocessed data by employing *BIDS-Derivatives*, enabling reproducibility and promoting collaboration within the neuroimaging community. The *BIDS* specification has permitted the development of tooling such as the *PyBIDS* library [36] to query and retrieve data and metadata from the input dataset and generate the names and structure of the final derivatives at the output (*see* Note and Chapter 2).

Note

***PyBIDS* is a *Python* package that makes it easier to work with *BIDS* datasets.** At present, its core and most widely used module supports simple and flexible querying and manipulation of *BIDS* datasets. *PyBIDS* makes it easy for researchers and developers working in *Python* to search for *BIDS* files by keywords and/or metadata; to consolidate and retrieve file-associated metadata spread out across multiple levels of a *BIDS* hierarchy; to construct *BIDS*-valid path names for new files; and to validate projects against the *BIDS* specification, among other applications. For further details on its core indexing module and other additional utilities it provides, *see* ref. [36].

First, we show how to pre-index the dataset. This will speed up later querying and is especially time-saving when the dataset is sizeable. Once *PyBIDS* is installed (*see* its documentation website for instructions, https://bids-standard.github.io/pybids/; *see* Resources), issue the following command from within the directory in which data are available (e.g., installed with *DataLad*):

```
cd /data/ds002790
mkdir -p .bids-index/
pybids layout --reset-db --no-validate --index-metadata .
.bids-index/
```

Once the dataset is indexed, we can open a *Jupyter notebook* or an *IPython* console and explore the dataset. We first import *PyBIDS* (with the package name "`bids`" within the *Python* distribution) and create a *dataset layout* object called `ds002790`. We make sure to point it to the right folder and to use the index database created above:

```
>>> import bids
>>> ds002790 = bids.BIDSLayout(
... "/data/ds002790",
... database_path="/data/ds002790/.bids-index/",
... )
```

We now use the *dataset layout* to query the dataset. In general, *PyBIDS* enables querying for metadata using `get_<metadata-name>` calls. For instance, we can check the total number of subjects:

```
>>> len(ds002790.get_subjects())
226
```

(continued)

We can also investigate the data and metadata types by querying the available BIDS' suffixes:

```
>>> ds002790.get_suffixes()
["T1w", "description", "dwi", "participants", "magni-
tude1", "phasediff", "bold", "physio", "events"]
```

We can also query for metadata entries, and filtering by a given suffix ("`bold`") discover that all BOLD images have a repetition time of 2.0 s:

```
>>> ds002790.get_RepetitionTime(suffix="bold")
[2]
```

Now we list the four BOLD fMRI tasks in the dataset:

```
>>> ds002790.get_tasks(suffix="bold")
["restingstate", "stopsignal", "workingmemory", "emo-
matching"]
```

or get the path to all *NIfTI* files corresponding to the resting-state BOLD runs for only the first subject (here only one file):

```
>>> ds002790.get(
... subject=ds002790.get_subjects()[0],
... task="restingstate",
... suffix="bold",
... extension=[".nii", ".nii.gz"],
... )
[<BIDSImageFile filename='/data/datasets/ds002790/sub-
0001/func/sub-0001_task-restingstate_acq-seq_bold.nii.
gz'>]
```

Notably, the *BIDS* and *BIDS-Derivatives* specifications allow *BIDS Apps*, such as *MRIQC* [37] and *fMRIPrep*, to follow a simple pattern for their invocation from the command line (*see* Note).

Note

Standardized command line of *BIDS Apps*. A *BIDS App* is a container image capturing a neuroimaging pipeline that takes a *BIDS*-formatted dataset as input. Since the input is a whole dataset, apps are able to combine multiple modalities, sessions, and/or subjects, but at the same time, need to implement ways to query input datasets. Each *BIDS App* has the

(continued)

same core set of command-line arguments, making them easy to run and integrate into automated platforms. *BIDS Apps* are constructed in a way that does not depend on any software outside of the container image other than the container engine. Further documentation about *BIDS Apps* and their execution with containers is found at the *NiPreps* website (https://www.nipreps.org/apps/framework/; *see* also Resources). An index of *BIDS Apps* is maintained at https://bids-apps.neuroimaging.io/apps/ (*see* Resources).

All *BIDS Apps* share a common command line interface that enables their automated concatenation and execution. The command line follows the structure `runscript input_dataset output_folder analysis_level <optional named arguments>`, where `runscript` is typically the name of the *BIDS App* (e.g., `mriqc`), `input_dataset` points at the path of the input *BIDS* or *BIDS-Derivatives* dataset, `output_folder` points at the path where results will be stored, and `analysis_level` can be either `participant` or `group` depending on what type of analysis will be executed, as introduced in Fig. 1. Therefore, the general structure of the command line is particularized for executing *MRIQC* as follows:

```
mriqc /data/ds002790 /data/ds002790/derivatives/
mriqc_24.0.0 participant
```

Following the *BIDS-Derivatives* specifications developed after the *BIDS Apps* framework, the output directory is set to `/data/ds002790/derivatives/mriqc_24.0.0`, which can naturally be managed as a *DataLad* "subdataset" and indicates that the choice of *MRIQC*'s release is 24.0.0 (*see* Subheading 3.4 for standardization of versioning). We will leverage *PyBIDS*' index cache by adding one named argument to the baseline command line:

```
mriqc /data/ds002790 /data/ds002790/derivatives/
mriqc_24.0.0 participant \
 --bids-database-dir /data/ds002790/.bids-index/
```

where `--bids-database-dir /data/ds002790/.bids-index/` is an optional argument *MRIQC* accepts to employ a pre-indexed database. The command line naturally generalizes to *fMRIPrep* as follows:

```
fmriprep /data/ds002790 /data/ds002790/derivatives/fmri-
prep_24.0.0 participant \
 --bids-database-dir /data/ds002790/.bids-index/
```

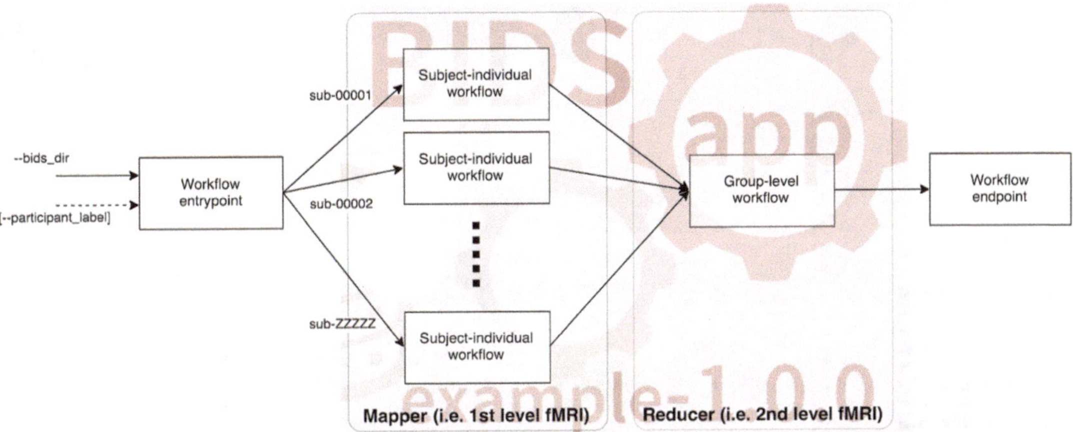

Fig. 1 *BIDS Apps* made strides toward standardization of design. In particular, our manuscript elaborated that most neuroimaging applications could be modularized in a first step ("mapper"), where independent processes are executed, followed by a second step ("reducer"), where results from the previous stage are aggregated

3.2 Standardization of Design

Workflow architecture *BIDS Apps* also promoted the standardization of workflow design. Gorgolewski and colleagues [23] described two typical execution patterns in neuroimaging analyses (Fig. 1). First, some workflows like *fMRIPrep* focus on individual subjects, where subjects' processing can be "embarrassingly parallel" thanks to the independence between execution processes. Generally, data from individual subjects (*participant level*) are then aggregated and compared depending on the study design (*group level*). In group-level analyses, inter-dependencies in the compute graph disallow embarrassingly parallel approaches. Therefore, parallelization must be implemented either at the level of task, either within a computing node (e.g., threading, multiprocessing, GPU, etc.) or across computing nodes (e.g., with message passing interface). By focusing only on the participant level, *fMRIPrep*, *MRIQC*, or any other *NiPreps* can optimize the workflow for the specific execution mode.

Building from the *BIDS Apps* standard, *fMRIPrep*, and *MRIQC* continued developing standardizations of design that would evolve into the *NiPreps* framework. Beyond sharing the same infrastructure to handle *BIDS*, the modularity of workflows, or the use of *NiPype* as the workflow engine, an alpha release of *fMRIPrep* now adopts a "fit and transform" paradigm, inspired by *Scikit-Learn*'s influential interface in machine learning [38]. Under this paradigm, *fMRIPrep* generates only a minimal set of results from which the "traditional" outputs of *fMRIPrep* can deterministically be generated. The minimal set of results includes linear and nonlinear spatial mappings between coordinate systems of interest (anatomical

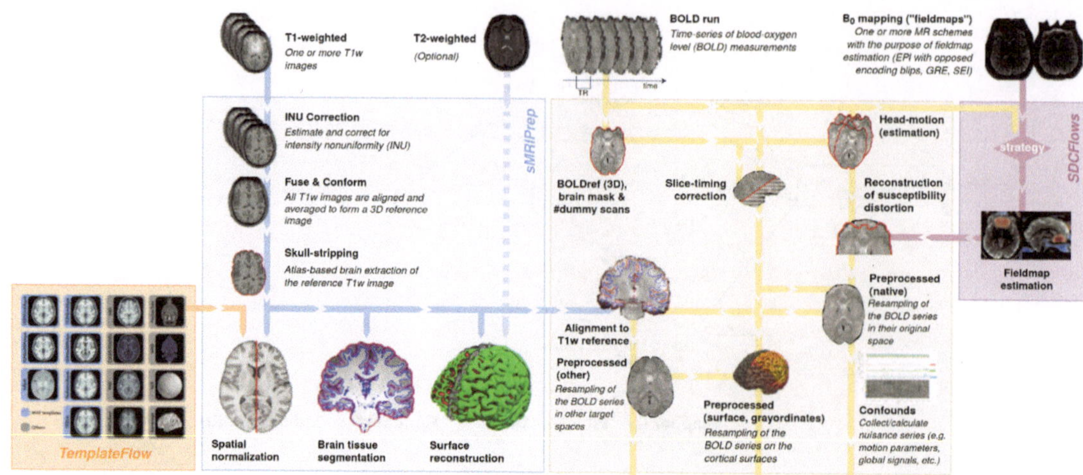

Fig. 2 *fMRIPrep* underwent a "deconstruction" effort, giving rise to *NiPreps*. We identified several preprocessing steps that required further standardization beyond fMRI. In particular, modularizing *fMRIPrep* derived in two relevant *NiPreps*—*TemplateFlow* [41], for the standardization of templates and atlases; and *SDCFlows* [42], which contains workflows and tools for the estimation and correction of susceptibility-derived distortions of echo-planar images (EPI) that are commonly employed to acquire fMRI and dMRI data

image, functional images, standard space defined by templates, etc.), temporal mappings (e.g., slice-time information), and estimation of spatiotemporal artifacts (e.g., susceptibility distortion, head-motion, etc.), associated with the processing steps of Fig. 2. This division of the workload benefits researchers and data stewards to maximize the value of the shared data as downstream users can generate the desired *fMRIPrep* results in the spatial frame they need while minimizing data transfer and storage. Indeed, we evaluated that the new approach resulted in 25–52%, 43–54%, and 72–87% reductions in runtime, data volume, and file counts, respectively, in comparison to the previous version of *fMRIPrep* [39]. The paradigm also facilitates the generalizability of software implementation, as only the fit step requires adaptation across modalities (e.g., *dMRIPrep* to preprocess diffusion MRI, dMRI), populations (e.g., *fMRIPrep-infants*; [40]), and species (e.g., *fMRIPrep-rodents*) while the transform step can have a single approach thereby reducing maintenance and technical debt.

Building blocks and modularity Based on the success story of *fMRIPrep*, we initiated a focus on the generalization of the workflow, first within blood-oxygen-level-dependent (BOLD) fMRI for its application on infants (*fMRIPrep-infants*) and rodents (*MRIQC-rodents*; [43]; and *fMRIPrep-rodents*). Similarly, the development of "-Preps" for other modalities was initiated. Indeed, *dMRIPrep* is the counterpart of *fMRIPrep* for dMRI data [44], *ASLPrep* covers arterial-spin-labeling (ASL; [45]) fMRI, and

PETPrep targets positron emission tomography (PET) imaging. These generalizations were soon denominated as *NeuroImaging PREProcessing toolS* (*NiPreps*), with the overarching goal of standardizing preprocessing components shared across modalities, populations, and even species. *NiPreps* adopts a modular and extensible architecture, allowing researchers to combine and configure different preprocessing modules to suit their specific needs (Figure 3). Modularity enables *fMRIPrep* and *dMRIPrep* to share the preprocessing of structural imaging (T_1-weighted and T_2-weighted) through *sMRIPrep*, and susceptibility distortion mapping through *SDCFlows* [42]. Therefore, the results of multimodal (f/dMRI) studies are referred to a single *version* of the anatomy obtained with a single run of *sMRIPrep* and distortions

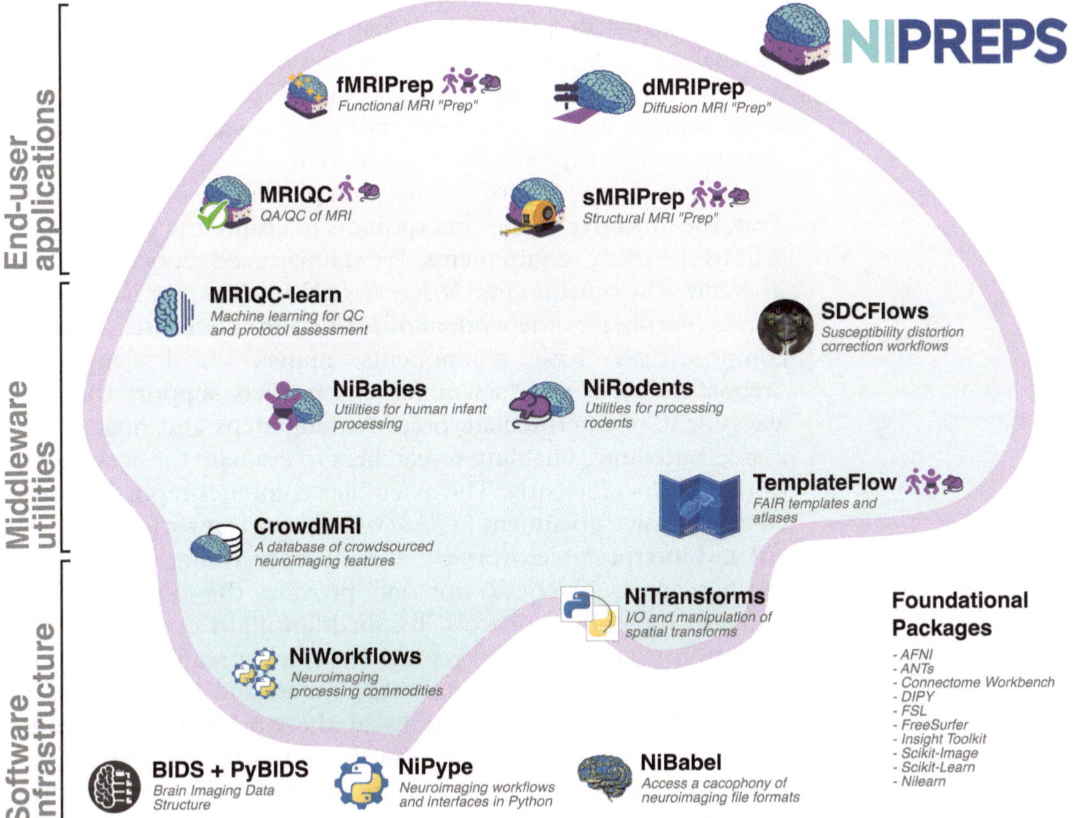

Fig. 3 The *NeuroImaging PREProcessing toolS* (*NiPreps*) framework. The *NiPreps* framework encompasses modular neuroimaging software projects. A number of projects provide the infrastructure over which more elaborate or abstract features are implemented. Leveraging that base, end-user applications such as *fMRIPrep*, *dMRIPrep*, or *MRIQC* can be developed. Projects outside the brain edge (*Nipype*, *Nibabel*, and *BIDS*/*PyBIDS*) are not part of the framework but receive upstream contributions and are essential software foundations for the whole vision. Finally, foundational packages support the most basic algorithmic implementation of methods

of the echo-planar imaging (typically employed to acquire diffusion and functional MRI) are addressed consistently. While this example showcases modularization at the highest level of the software stack shown in Fig. 3, the principle applies to the smaller components at lower levels of abstraction.

Quality assessment and control (QA/QC) Deploying a QA/QC strategy is critical for the reproducibility of the neuroimaging workflow [26]. In addition to increasing the workflow's and results' reliability, standardizing QA/QC is critical to ensure quality issues do not propagate unidentified along studies. Robust QA/QC implies setting up several QA/QC checkpoints along the neuroimaging pipeline to ensure that data meeting exclusion criteria are dismissed before reaching analysis [46]. Under such a definition, QC checkpoints are analogous to the layers of the so-called "Swiss cheese security model" [47], with the goal that data of insufficient quality which may bias the results does not reach the analysis. *NiPreps* such as *MRIQC* and *fMRIPrep* generate visual reports to implement standardized QA/QC protocols. In *MRIQC*, visual reports enable one mechanism for assessing the quality of outcomes of an experimental session, that is, the "original" unprocessed data. In the case of downstream pipelines such as *s/d/fMRIPrep*, the objective of the checkpoint is to ensure the preprocessing is fit to the study requirements. We standardized report generation in *NiPreps* by outsourcing *NiReports* (*NeuroImaging Reports*) as a standalone library independent of *fMRIPrep*'s codebase. *NiReports* comprises two basic components: unitary visual elements or "reportlets" and the "assembler". Reportlets support the visual assessment of intermediate preprocessing steps and final preprocessed outcomes, enabling researchers to evaluate the acceptability of the results efficiently. The assembler combines reportlets into a comprehensive document (a *final* visual report), providing a coherent and interpretable overview of the preprocessing workflow and the outcomes. *NiReports* not only provides the infrastructure to establish QA/QC protocols. By shedding light on the workflow operation itself, the visual reports provide "a scaffold for knowledge" that helps researchers better understand the why and the how of the particular operations in the workflow. This "educational" or transparency component supports the training and development of researchers, ultimately fostering a more knowledgeable and skilled community engaged in standardized preprocessing practices. Although *NiReports* offers some interactivity, standardization of QA/QC requires that the screening experience during assessments is homogenous across raters, regardless of their expertise or attrition or the actual visualization settings. In other words, *NiPreps*' reports disallow exploring data freely such that structured differences between raters may emerge depending on their strategy for the assessment. For instance, the reports do not offer interactive ortho-viewers that permit two experts to navigate the same image differently. Standardizing QA/QC through *NiReports* ensures that preprocessing outcomes can be thoroughly and consistently assessed, providing researchers with confidence in the acceptability and quality of their results.

3.3 "Semantic" Standardization: Spatially Referencing Group Inferences

Standard spaces provide stereotaxy, a reference frame for researchers to align and compare data across different subjects and studies. Moreover, these spaces are ubiquitously employed in neuroimaging to incorporate prior knowledge into the processing as they typically are annotated. Standard spaces contain one or more *templates*, which are aggregated maps of neuroimaging features, and *atlases*—annotations corresponding to features encoded by templates in the frame of reference these engender. *TemplateFlow* [41] provides a curated collection of common standard templates and atlases, enabling self-adaptable workflows that employ the template or atlas most appropriate for the particular dataset. For instance, *fMRIPrep-infants* uses different templates depending on the age months of the participant. Although initially developed in response to increased flexibility requirements by *fMRIPrep*, we identified remarkable issues concerning the use and reporting of templates in neuroimaging as described in our *TemplateFlow* manuscript [41] and further analyzed in our feature about that paper [48]. Some of these issues relate to the distribution of templates and atlases under FAIR (Findability, Accessibility, Interoperability, and Reusability; [49]) guiding principles, to data management, as well as to ensuring best practices in reporting analyses. Therefore, not only does *TemplateFlow* offer a programmatic interface to templates and atlases and a community registry and archive, but it also addresses the challenge of template versioning and management (Fig. 4). By providing a centralized repository, researchers can access different versions of templates, allowing for consistent analyses across time and ensuring the reproducibility of preprocessing results.

3.4 Containerization

With growing requirements, it is harder for inexperienced users to adopt new tools because installation becomes a high barrier. These entry barriers increase with security policies and limitations of the target system. While installation on a Personal Computer (PC) may be straightforward, deployment on a multi-tenant HPC cluster can be challenging. To resolve this problem, *BIDS Apps* emphasized the adoption of containers (*see* Chapter 3). Indeed, when *fMRIPrep* started to ramp up in the number of users, most of the problems reported in the source code repository and the specialized forum *NeuroStars* (https://neurostars.org; *see* Resources) related to installation. As maintainers of *fMRIPrep*, we decided to discourage "bare-metal" installations where users must install all the dependencies (e.g., *AFNI*, *ANTs*, *FSL*, *FreeSurfer*, etc.) and prepare a workable *Python* environment themselves. Instead, we promoted

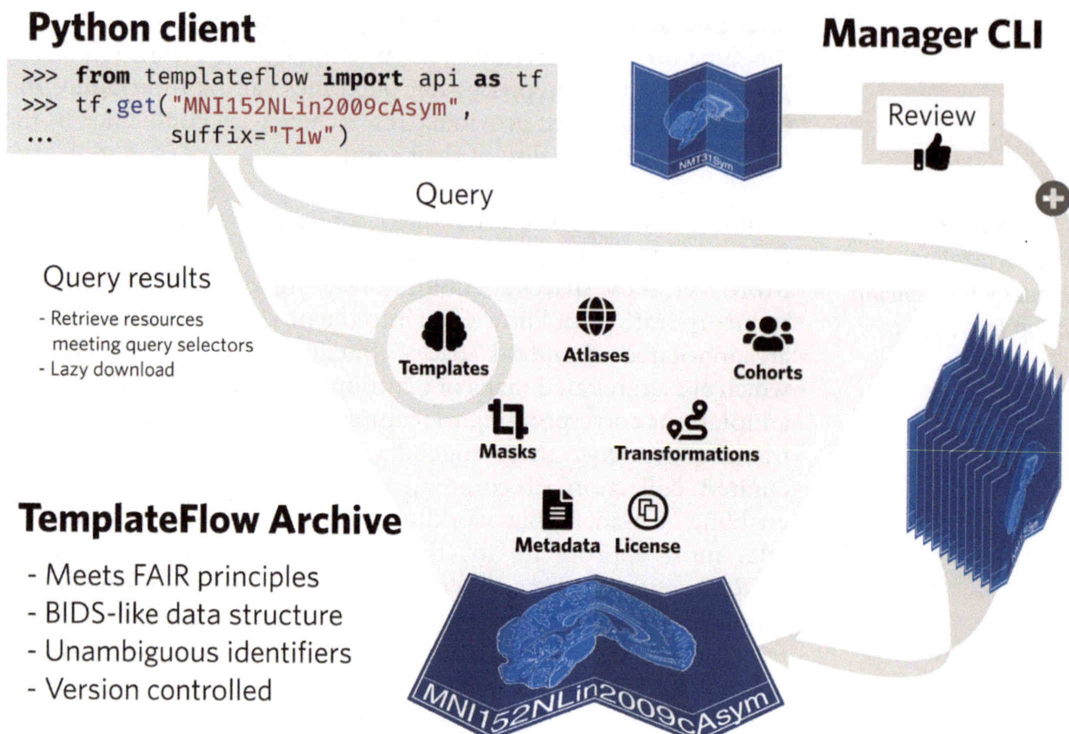

Fig. 4 Standardization of templates and atlases with *TemplateFlow*. The *TemplateFlow Archive* can be accessed at a "low" level with *DataLad*, or at a "high" level with the *TemplateFlow Client*. New resources can be added through the *TemplateFlow Manager* command-line interface, which initiates a peer-review process before acceptance in the *Archive*

the deployment of *Apptainer* (called *Singularity* at the time; [50]). We also successfully prompted the adoption of Singularity by our local cluster Sherlock (SCRR, Stanford University, CA, USA), and several systems at TACC (Texas Advanced Computing Center, University of Texas at Austin, TX, USA). The promotion of containers quickly translated into a shift of support activities toward more "scientific" topics about *fMRIPrep*, failure conditions, and feature requests, instead of installation and deployment (*see* Note).

Note: Running *fMRIPrep* with *Docker*

We build on top of the standard command line interface of *BIDS Apps* to demonstrate the containerized execution of *fMRIPrep* on the AOMIC-PIOP2 dataset [51].

```
docker run -ti --rm -u $( id -u ):$( id -g ) \
 -v /data/ds002790:/data:ro \
 -v /data/ds002790/derivatives:/derivatives \
```

(continued)

```
nipreps/fmriprep:23.2.0 \
/data /derivatives/fmriprep_23.2.0 participant \
--participant-label 0021 \
--omp-nthreads 8 --nprocs 16 \
-vv --bids-database-dir /data/.bids-index/
```

Execution with *Docker* requires pre-pending the container system's arguments before *fMRIPrep*'s. First, a `docker run` sub-command indicates that a container will be executed from a *Docker* image. The specific *Docker* image and tag marks the separation between *Docker* arguments and those of *fMRIPrep*. In this case, the fourth line indicates `nipreps/fmriprep:23.2.0`, thereby instructing *Docker* to find the corresponding image at the indicated version (23.2.0) and download it if not cached locally.

Arguments preceding the container image configure the terminal mode (`-ti`), instruct *Docker* to clear up the container when execution finishes (`--rm`), and, importantly, map the current user and group into the container (`-u $(id -u):$(id -g)`) to ensure new folders and files are not assigned to root, which is the default. Execution will also require file system communication with the container, and therefore, we "mount" the data folder in read-only mode (`-v /data/ds002790:/data:ro`), and a folder to store the output in read-write mode (`-v /data/ds002790/derivatives: /derivatives`).

Arguments to the right of the container name and tag correspond to *fMRIPrep*. First, we encounter the standard *BIDS Apps* mandatory arguments (`/data /derivatives/fmriprep_23.2.0 participant`), followed by one specific participant label (`--participant-label 0021`), as recommended in *fMRIPrep*'s usage guidelines. Next, parallelization is configured, with 8 CPUs per process and a maximum of 16 processes being executed simultaneously (`--omp-nthreads 8 --nprocs 16`). The verbosity of *fMRIPrep* can be tuned with the repetition of the "v" letter as a flag (here we have `-vv`, which can be decreased by writing just one "v", `-v`, or increased, e.g., `-vvvv`). This verbosity parameter should not be confused with *Docker*'s file system mounting flag `-v`. Finally, we set *PyBIDS*' cache by typing `--bids-database-dir /data/.bids-index/` (note how the directory is now relative to `/data`, the mount point inside the container where the dataset root will be available).

When starting with *fMRIPrep*, it is common to require several iterations to test configurations and arguments. If we

(continued)

are interested in keeping the intermediate results to speed up later executions of the pipeline, we need also to mount some (preferably fast) filesystem where these interim results are stored:

```
docker run -ti --rm -u $( id -u ):$( id -g ) \
 -v /data/ds002790:/data:ro \
 -v /data/ds002790/derivatives:/derivatives \
 -v /scratch/ds002790/sub-0021:/work \
 nipreps/fmriprep:23.2.0 \
 /data /derivatives/fmriprep_23.2.0 participant \
 --participant-label 0021 \
 --omp-nthreads 8 --nprocs 16 \
 -vv --bids-database-dir /data/.bids-index/ \
 --work /work
```

where `/scratch/ds002790/sub-0021` is a folder under a fast file system accessible at `/scratch`, and made accessible by the container as `/work`.

3.5 Telemetry

Telemetry enables the collection and analysis of data on the execution of processing pipelines, providing valuable insights into failure conditions and usage patterns (Figure 5 presents *fMRIPrep*'s telemetry). By incorporating telemetry, the *NiPreps* maintainers

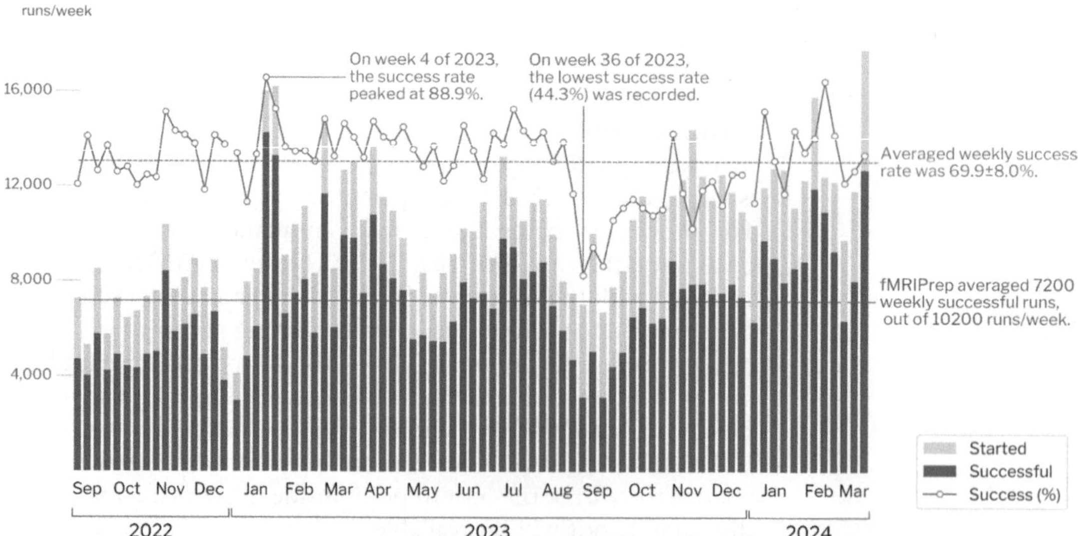

Fig. 5 *fMRIPrep* is executed an average of 11,200 times a week. By inserting telemetry instrumentation within *fMRIPrep*, we collect information to identify failure modes of the software and performance analytics. Over the past 1.9 years, *fMRIPrep* averages about 68% success rate

monitor their workflows' performance, identify potential bottlenecks and error modes, and optimize the pipeline accordingly. This information becomes invaluable in improving the reliability and efficiency of the pipeline process as it enables a deeper understanding of usage patterns by different users and provides unique insight into tracking actual software utilization [52]. Software impact metrics are a challenge to evaluate for open-source projects in general; however, these metrics are becoming relevant to funding bodies who value code as a scientific outcome. Initially, we implemented telemetry within *fMRIPrep* employing *Sentry* (https://sentry.io; *see* Resources). In order to generalize the analytics easily over the remainder of the *NiPreps* framework, we then developed *Migas* ("breadcrumbs" in Spanish) as an in-house solution for performance monitoring. *Migas* comprises a lightweight *Python* client that adds the necessary instrumentation to probe the application and submits (except when the user opts out by using the prescribed command line flag) the collected data to an Internet service running on the cloud.

3.6 Software Versioning and Release Cycle

Version control, code quality checking, testing, and continuous integration Version control and other software engineering best practices are crucial in achieving reliable and maintainable standardized workflows. Incorporating these practices is an onerous investment that will prevent scientific projects from incurring unsustainable technical debt quickly. Version control of scientific code is fundamental for reproducibility and traceability. By utilizing version control systems such as *Git*, researchers can track and manage changes to the workflow implementation over time (*see* Chapter 5 for *Git* use). It supports the management of different branches targeting specific features or release maintenance. By layering services over *Git*, research code developers can easily deploy quality checks (e.g., peer-review of code, use of "linters" to normalize the style of code and maximize collaboration, etc.), unit testing (at least, of clerical tasks such as filtering the input data structure, testing the accessibility of data and metadata, etc.), and continuous integration and continuous delivery (CI/CD). The *NiPreps*' documentation website[2] describes these techniques in further detail. Using *DataLad* (*see* Chapter 2 for a detailed description), *NiPreps* integrates "benchmarking" data in their CI/CD builds, automating the process of evaluating the acceptability of code changes and new features.

Versioned releases One early success driver for *fMRIPrep* was adopting a "release early, release often" or "RERO" release cycle, in which we would roll out new features and bug fixes rapidly (sometimes several times a week). This allowed *fMRIPrep* to stress

[2] https://www.nipreps.org/

test the implementation of new features and fixes in a federated fashion without the maintainers having access to the data. RERO was also appreciated by users, who identified the *fMRIPrep* developers as a responsive and supportive team, increasing the confidence that the tool would add reliability to their neuroimaging workflow. However, RERO requires that every new release is effectively identified by users, typically with a version label. We initially adopted semantic versioning [53] to assign these release identifiers, in which three numbers separated by periods (e.g., 2.1.10) contribute to the version interpretation. The first number or "major" differentiates hallmark iterations of the product. A "2.1.10" version indicates that a 1.x series of versions exists, and the technical gap of going from 1.x to 2.x is remarkable (e.g., outputs are incompatible, fundamental new features have been added, etc.). An example of such a large change was *Python*'s shift from 2.x into the current 3.x series. The second number, or "minor", signals large changes that are consistent enough to be considered under the same major (e.g., from *Python* 3.10 to 3.11). The last number, or "patch release", indicates small changes to address bugs or improve performance in a limited way. Starting in 2020, *fMRIPrep* and other *NiPreps* adopted a slightly different convention called calendar versioning,[3] which is fundamentally similar, but the major version number is replaced by the year of release. As a result, versions can be easier to place in the lifecycle of the software (*see* Fig. 6). For example, the current *fMRIPrep*'s last release is 23.2.0.

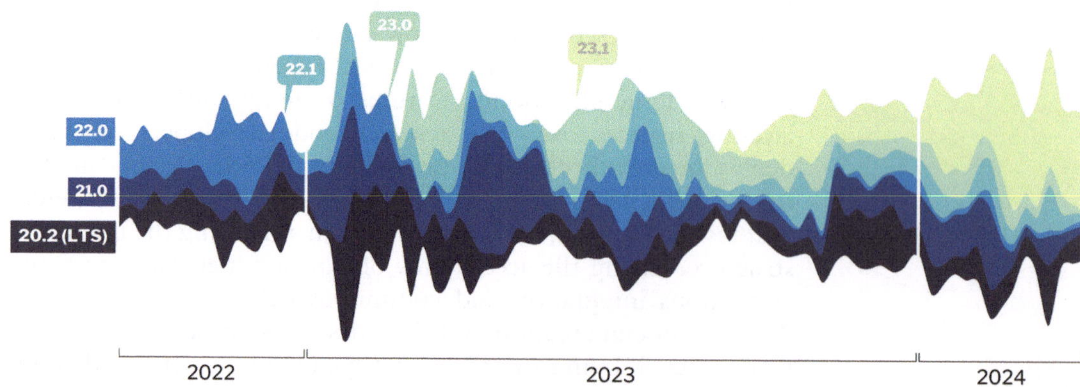

Fig. 6 Versioning and telemetry allow tracking the adoption of *fMRIPrep*. Adopting a strict versioning scheme for release permits monitoring the lifecycle of releases. With the riverplot below, we can identify many users who keep running the LTS (long-term support) version 20.2. Version 21.0 maintains many users, while later releases (22.1 and 23.0) progressively seem absorbed by the latest official release 23.1. This riverplot suggests users employing 21.0 series resist either falling back to the LTS or updating to the latest release

[3] https://www.nipreps.org/devs/releases/

Long-term support programs In addition to assigning meaningful version strings to every release, some of *fMRIPrep*'s power users expressed their concerns about our RERO approach in the context of longitudinal studies where data acquired during lengthy project spans require consistent processing throughout. As a solution, we established a "long-term support" (LTS) program[4,5] inspired by the *Ubuntu Linux* LTS program. Starting with version 20.2.0 (released on September 28, 2020), *fMRIPrep 20.2.x* is the active LTS series. LTS involves maintaining the software for a much longer window (preferably supported by researchers other than active maintainers and developers of *fMRIPrep*[6]), ensuring the continuity of the results over time, and resolving bugs discovered as the series are used. *fMRIPrep*'s LTS has seen seven "patch" updates; the last release is 20.2.7 on January 24, 2022 (please note how the major and minor numbers are pinned to the 20.2 series).

3.7 Community Involvement

Open scientific code is likely to egress the walls of the laboratory or research infrastructure that conceived it when the original authors and a community of researchers share the need and an ethos. Creating a community potentially ensures the project's longevity if it successfully engages researchers who follow up on the development of the tool and eventually contribute to it. "Contribution" here takes a broader meaning, as it comprehends not just code but also participation in discussions, the definition of roadmaps, providing support, writing documentation, etc. Second, nurturing a community helps access a more diverse pool of researchers reaching underrepresented and underserved minorities, which in the long term ensures the project does not decline due to monolithic thinking. For instance, the lack of researchers outside a given laboratory or consortium likely results in an inability to adapt to transformative advances elsewhere in the globe. Establishing standard procedures to make decisions and to keep communications fluent is necessary to ensure that the previous dimensions of standardization operate properly. For example, *fMRIPrep* and the newborn *NiPreps* initiated a process of creating a community around the framework in 2020. As a result, in April 2023, the *NiPreps* Governance (https://github.com/nipreps/GOVERNANCE) was passed by the community, and in September 2023, the first new "*NiPreps* Steering Committee" was selected. This process is not distinctive of the *NiPreps Community*, and indeed, our charter derives from a GitHub resource called "the minimally viable governance" or MVG.[7]

---

[4] https://www.nipreps.org/devs/releases/#long-term-support-series

[5] https://reproducibility.stanford.edu/fmriprep-lts/

[6] We thank Prof. Pierre Bellec and Dr. Basile Pinsard (CRIUGM, Psychology, University of Montreal) for the maintenance of *fMRIPrep 20.2.x LTS*.

[7] https://github.com/github/MVG

4 Challenges and Outlook

4.1 Challenges and Their Impact on Reproducibility

Probably, the number one question by neuroimaging experts that *fMRIPrep* elicited in its early days was, "What are the criteria for choosing one method over the alternatives at any given step?" Indeed, this query identifies a ubiquitous challenge due to limited objective evidence to compare the alternatives for each individual step and further the lack of combinatorial evidence when exploring the multiverse of tools. *fMRIPrep*'s choices have often relied on a combination of empirical findings, theoretical considerations, and expert opinions. The lack of programmatic, unambiguous, and comprehensive evidence for each preprocessing step makes the decision-making process challenging. It requires careful consideration of the trade-offs between different approaches, such as speed versus accuracy, robustness versus sensitivity, and generalizability versus specificity. Moreover, even if there was clear evidence to drive these choices, it is likely that different objective functions will yield different "best" options. Indeed, we compared several fMRI preprocessing workflows [54] by implementing them within the *Configurable Pipeline for the Analysis of Connectomes* (*C-PAC*; [55]). Although these implementations did not replicate the exact workflows, the results highlighted significant variations across different preprocessing approaches. While some convergences in functional connectivity results were observed, the overall variability demonstrated the challenges of obtaining consistent outcomes across different implementations.

Relatedly, the development of *fMRIPrep* has often generated discussions about the balance between enabling many options for the user and how that extra analytical flexibility may undermine the reproducibility of the analyses. Standardization removes a researcher's degrees of freedom by making choices (e.g., selecting a brain extraction algorithm that fails in one image per million but is less accurate than an alternative approach with a failure rate of ten images per million and extremely precise otherwise), and by adding friction points (e.g., conversion into *BIDS*) that require the researcher to be aware of, and explicit about, all the experimental details.

Since analytical variability is only one factor in the overall reliability, it is critical to understand and account for the variability introduced by each specific step in the preprocessing pipeline. In collaboration with the CRIUGM team at the University of Montreal, we tested the reliability of *fMRIPrep* by introducing small random numerical variabilities at some anatomical preprocessing steps with *libmath* [56]. This approach allowed for the assessment of how uncertainties propagate throughout the pipeline. The paper proposed a method to identify large discontinuities between different versions in the development cycle. Indeed we identified implementation changes between two consecutive "patch" releases (20.2.4 to 20.2.5) that introduced large changes—hence violating the principle that patch version increments should be backward compatible. Therefore, the approach provides valuable insights into the stability and reproducibility of the pipeline over different versions, aiding in detecting potential sources of variability (*see* Fig. 7 in ref. [56]).

Establishing QC criteria with reference to the specific application requires the definition standards such as quality metrics applicable to pipeline outcomes. For example, in Fig. 3A of ref. [24], we compare the outputs *fMRIPrep* and *FSL FEAT* in terms of data smoothness. Smoothness is likely a quality metric of interest to high-resolution BOLD data acquired with 7 Tesla devices, as they often showcase excessive smoothing after processing due to, e.g., multiple resamplings.

Further challenges relate to resource utilization and their optimization. Optimizing resource utilization for specific solutions is markedly easier than with standardized alternatives. As a quick example, while *PyBIDS* takes a few seconds to index a *BIDS* directory for a small-sized dataset (15 subjects including one session each with minimal anatomical data and some diffusion or functional MRI), it may take one hour on the same computer for a dataset of 1500 subjects and similar imaging contents per subject. Further, a custom workflow developed for a specific sample of 100 neurotypical subjects of a narrow age range, collected on a single scanner and with a single imaging protocol, will be uniform in imaging parameters (e.g., size, resolution, contrast, artifacts, etc.). Therefore, all subjects will have similar demands from the compute resource regarding memory. However, anticipating memory requirements for an equivalent standard workflow that is expected to perform properly on diverse samples in terms of both phenotypes and imaging parameters is challenging, as images may come in many different sizes (e.g., some have very large acquisition matrices) and other fundamental properties (e.g., anisotropic vs. isotropic voxels, healthy vs. lesioned brains, etc.). Resource utilization challenges are further constrained by the need for and responsibility of reducing the carbon footprint of executing these pipelines [57].

4.2 The Need for Openly Available and Reusable Data

The availability of diverse fMRI data, accessible under FAIR principles [49] and readily reusable thanks to *BIDS*, was critical to developing *fMRIPrep* and *MRIQC*. To develop *fMRIPrep*, we leveraged a long list of datasets available at the time through *OpenfMRI* [58] and *OpenNeuro* [22]. In the case of *MRIQC*, we employed two specific datasets: ABIDE [59] and the Consortium for Neuropsychiatric Phenomics dataset [60]. Open data will remain essential to any methodological development endeavor to address the question, "Does this software work on a substantial number of diverse studies?" In [24], we visually assessed performance on 54 datasets using the standard reports. Further, new datasets tailored to methodological development, such as our "Human Connectome Phantom" (HCPh; [34]), will be necessary to face the need to explore the multiverse of methodological choices. More importantly, open data will be necessary to face the new challenges derived from adopting artificial intelligence (Subheading 4.3). While deep learning algorithms often exhibit great performance, they largely operate as "opaque boxes", which impedes checking how inference was made. This lack of interpretability limits our understanding of the underlying factors driving the model's outputs. Open data is fundamental to validate findings and compare results across different datasets.

4.3 Future Directions

The advent of artificial intelligence (AI) and deep learning The introduction of emerging deep learning models into standardized processing pipelines faces friction, as the "classical" computer vision techniques have been more thoroughly tested and have engendered trust in their performance. Conversely, the complexity of deep learning models raises concerns about transparency and interpretability. Nonetheless, deep-learning applications have demonstrated great reliability on diverse data and remarkably better performance, drastically reducing inference times while improving robustness. One example of such a transition currently being tested within *NiPreps* is *SynthStrip* [61], a multi-modal human brain extraction tool. The *FreeSurfer* team is at the forefront of leveraging deep learning in the processing pipeline and has introduced a range of relevant tools, including *SynthStrip*, *SynthSeg* [62], or *SynthSR* [63]. These particular tools have the common denominator of being designed to achieve great reliability independently of the image modality, a requirement perfectly aligned with the cross-modality standardization goals of *NiPreps*.

Fully differentiable pipelines We have also demonstrated the potential of a fully differentiable software stack in the domain of functional connectomics [64]. We argue that differentiable programming does not resolve the problem of workflow design but rather is a tool to free workflow design from the analytical choices and convert it in a hyperparameter search process. Hence, the challenge of methodological choices is resolved by data-driven optimization. Ciric's *hypercoil* software (https://hypercoil.github.io) is based on *PyTorch* and enables end-to-end differentiability throughout the entire preprocessing pipeline. In connection with the challenge of making decisions about the particular implementation of each processing step, a fully differentiable pipeline resolves the problem in a data-driven way with an objective function built in. As a result, the user is, in principle, not offered knobs (degrees of freedom) to tune the processing.

AI-assisted code writing As Poldrack and colleagues contend [65], code assisting is one of the tasks where large language models particularly shine. Even if the aid is limited to "only revising" a given code, they encountered substantial improvement in the code by several metrics. However, they found some limitations in generating tests, suggesting that humans are still necessary to ensure the validity and accuracy of the results. Nonetheless, their findings point to AI as the key to multiplying the productivity of humans by perfecting many of the "almost-clerical" tasks that standardization requires.

5 Conclusion

This chapter discusses the rationale, benefits, and challenges of standardizing preprocessing in neuroimaging. We have presented *NiPreps*, a modular and adaptable workflow framework that aims to provide reliable and reproducible preprocessing solutions for different modalities, populations, and species. We have also described some of the best practices and tools *NiPreps* employs to ensure the preprocessing results' quality, consistency, and transparency, such as *BIDS*, *TemplateFlow*, *NiReports*, *Migas*, containerization, and version control. We have highlighted some of the current and future directions of *NiPreps*, such as incorporating deep learning models, enabling end-to-end differentiability, and leveraging AI-assisted code writing. We have also emphasized the importance of open and reusable data for validating and comparing different preprocessing approaches and enhancing the interpretability and generalizability of the outcomes.

We hope that this chapter has provided a comprehensive overview of the state of the art and challenges of standardizing preprocessing in neuroimaging. We believe *NiPreps* offers a valuable resource for researchers seeking to optimize their preprocessing pipelines and obtain high-quality and robust results. We invite the readers to join the *NiPreps* community and contribute to the development and improvement of the framework. By adopting and promoting standardized preprocessing practices, we can advance the field of neuroimaging and foster a more collaborative and reproducible scientific culture.

Acknowledgments

The author and this work are supported by the Swiss National Science Foundation—SNSF—(#185872), NIMH (RF1MH12186), and CZI (EOSS5-0000000266).

Thanks to M.Sc. Céline Provins, Dr. Vanessa Siffredi, Dr. Martin Nørgaard, Dr. Yasser Alemán-Gómez, Dr. Jon Haitz Legarreta Gorroño, Dr. Satrajit S. Ghosh, and Prof. Russell A. Poldrack for their invaluable feedback along iterations of the draft and their generous scientific support in our collaborations. Thanks to the editors (Dr. Hervé Lemaître and Dr. Robert Whelan) for their infinite patience and support when developing this chapter. Thanks to the *NiPreps Community* for their continued support and their energy in pursuit of a more transparent and reliable neuroimaging.

References

1. Strother SC (2006) Evaluating fMRI preprocessing pipelines. IEEE Eng Med Biol Mag 25(2):27–41. https://doi.org/10.1109/MEMB.2006.1607667
2. Cox RW, Hyde JS (1997) Software tools for analysis and visualization of fMRI data. NMR Biomed 10(4–5):171–178. https://doi.org/10.1002/(SICI)1099-1492(199706/08)10:4/5<171::AID-NBM453>3.0.CO;2-L
3. Fischl B (2012) FreeSurfer. NeuroImage 62(2):774–781. https://doi.org/10.1016/j.neuroimage.2012.01.021
4. Jenkinson M, Beckmann CF, Behrens TEJ, Woolrich MW, Smith SM (2012) FSL. NeuroImage 62(2):782–790. https://doi.org/10.1016/j.neuroimage.2011.09.015
5. Friston KJ, Ashburner J, Kiebel SJ, Nichols TE, Penny WD (2006) Statistical parametric mapping : the analysis of functional brain images. Academic Press, London
6. Brett M et al (2006) Open source software: NiBabel. Zenodo Softw:3458246. https://doi.org/10.5281/zenodo.591597
7. Gorgolewski KJ et al (2011) Nipype: a flexible, lightweight and extensible neuroimaging data processing framework in Python. Front Neuroinform 5:13. https://doi.org/10.3389/fninf.2011.00013
8. Carp J (2012) On the plurality of (methodological) worlds: estimating the analytic flexibility of fMRI experiments. Front Neurosci 6. https://doi.org/10.3389/fnins.2012.00149
9. Bowring A, Maumet C, Nichols TE (2019) Exploring the impact of analysis software on task fMRI results. Hum Brain Mapp (in press). https://doi.org/10.1002/hbm.24603
10. Bowring A, Nichols TE, Maumet C (2022) Isolating the sources of pipeline-variability in group-level task-fMRI results. Hum Brain Mapp 43(3):1112–1128. https://doi.org/10.1002/hbm.25713
11. Nørgaard M et al (2020) Different preprocessing strategies lead to different conclusions: a [11C]DASB-PET reproducibility study. J Cereb Blood Flow Metab 40(9):1902–1911. https://doi.org/10.1177/0271678X19880450
12. Botvinik-Nezer R et al (2020) Variability in the analysis of a single neuroimaging dataset by many teams. Nature 582(7810):7810. https://doi.org/10.1038/s41586-020-2314-9
13. Churchill NW, Spring R, Afshin-Pour B, Dong F, Strother SC (2015) An automated, adaptive framework for optimizing preprocessing pipelines in task-based functional MRI. PLoS One 10(7):e0131520. https://doi.org/10.1371/journal.pone.0131520
14. Dafflon J et al (2022) A guided multiverse study of neuroimaging analyses. Nat Commun 13(1):1. https://doi.org/10.1038/s41467-022-31347-8
15. Allen C, Mehler DMA (2019) Open science challenges, benefits and tips in early career and beyond. PLoS Biol 17(5):e3000246. https://doi.org/10.1371/journal.pbio.3000246
16. Chambers CD, Tzavella L (2022) The past, present and future of registered reports. Nat Hum Behav 6(1):1. https://doi.org/10.1038/s41562-021-01193-7
17. Ozenne B, Norgaard M, Pernet C, Ganz M (2024) A sensitivity analysis to quantify the impact of neuroimaging preprocessing strategies on subsequent statistical analyses. Apr. 24, 2024, *arXiv*: arXiv:2404.14882. https://doi.org/10.48550/arXiv.2404.14882
18. Nichols TE et al (2017) Best practices in data analysis and sharing in neuroimaging using MRI. Nat Neurosci 20:299–303. https://doi.org/10.1038/nn.4500

19. Taylor PA et al (2023) Highlight results, don't hide them: enhance interpretation, reduce biases and improve reproducibility. NeuroImage 274:120138. https://doi.org/10.1016/j.neuroimage.2023.120138
20. Gorgolewski KJ et al (2016) The brain imaging data structure, a format for organizing and describing outputs of neuroimaging experiments. Sci Data 3:160044. https://doi.org/10.1038/sdata.2016.44
21. Poldrack RA et al (2024) The past, present, and future of the brain imaging data structure (BIDS). Imaging Neurosci 2:1–19. https://doi.org/10.1162/imag_a_00103
22. Markiewicz CJ et al (2021) The OpenNeuro resource for sharing of neuroscience data. eLife 10:e71774. https://doi.org/10.7554/eLife.71774
23. Gorgolewski KJ et al (2017) BIDS apps: improving ease of use, accessibility, and reproducibility of neuroimaging data analysis methods. PLoS Comput Biol 13(3):e1005209. https://doi.org/10.1371/journal.pcbi.1005209
24. Esteban O et al (2019) fMRIPrep: a robust preprocessing pipeline for functional MRI. Nat Methods 16(1):111–116. https://doi.org/10.1038/s41592-018-0235-4
25. Esteban O et al (2020) Analysis of task-based functional MRI data preprocessed with fMRIPrep. Nat Protoc 15:2186–2202. https://doi.org/10.1101/694364
26. Niso G et al (2022) Open and reproducible neuroimaging: from study inception to publication. NeuroImage:119623. https://doi.org/10.1016/j.neuroimage.2022.119623
27. Novick MR (1966) The axioms and principal results of classical test theory. J Math Psychol 3(1):1–18. https://doi.org/10.1016/0022-2496(66)90002-2
28. Lord FM, Novick MR, Birnbaum A (1968) Statistical theories of mental test scores. In: Statistical theories of mental test scores. Addison-Wesley, Oxford, Uk
29. Allen MJ, Yen WM (1979) Introduction to measurement theory. Waveland Press
30. Plesser HE (2018) Reproducibility vs. replicability: a brief history of a confused terminology. Front Neuroinform 11. https://doi.org/10.3389/fninf.2017.00076
31. Brown WM (1910) Some experimental results in the correlation of mental abilities. Br J Psychol 3:296–322
32. Spearman C (1910) Correlation calculated from faulty data. Br J Psychol 3:271–295
33. Naselaris T, Allen E, Kay K (2021) Extensive sampling for complete models of individual brains. Curr Opin Behav Sci 40:45–51. https://doi.org/10.1016/j.cobeha.2020.12.008
34. Provins C et al (2023) Reliability characterization of MRI measurements for analyses of brain networks on a single human. Nat Methods (Stage 1 accepted-in-principle). https://doi.org/10.17605/OSF.IO/VAMQ6
35. Cronbach LJ, Meehl PE (1955) Construct validity in psychological tests. Psychol Bull 52(4):281–302. https://doi.org/10.1037/h0040957
36. Yarkoni T et al (2019) PyBIDS: python tools for BIDS datasets. J Open Source Softw 4:1294. https://doi.org/10.21105/joss.01294
37. Esteban O, Birman D, Schaer M, Koyejo OO, Poldrack RA, Gorgolewski KJ (2017) MRIQC: advancing the automatic prediction of image quality in MRI from unseen sites. PLoS One 12(9):e0184661. https://doi.org/10.1371/journal.pone.0184661
38. Pedregosa F et al (2011) Scikit-learn: machine learning in python. J Mach Learn Res 12:2825–2830
39. Markiewicz CJ et al (2024) FMRIPrep-next: preprocessing as a fit-transform model. In: Annual meeting of the Organization for Human Brain Mapping (OHBM), Seoul, Korea, p (accepted)
40. Goncalves M et al (2023) They grow up so fast – augmenting the NiBabies infant MRI workflow. In: Annual meeting of the Organization for Human Brain Mapping (OHBM), Montréal, Canada, p 2504
41. Ciric R et al (2022) TemplateFlow: FAIR-sharing of multi-scale, multi-species brain models. Nat Methods 19:1568–1571. https://doi.org/10.1038/s41592-022-01681-2
42. Esteban O et al (2021) The Bermuda Triangle of d- and f-MRI sailors – software for susceptibility distortions (SDCFlows). In: 27th Annual Meeting of the Organization for Human Brain Mapping, Virtual Meeting, p 1653. https://doi.org/10.31219/osf.io/gy8nt
43. MacNicol EE, Hagen MP, Provins C, Kim E, Cash D, Esteban O (2022) Extending MRIQC to rodents: image quality metrics for rat MRI. In: Annual meeting of the European Society for Molecular Imaging (EMIM), Thessaloniki, Greece, pp PW23–P913
44. Joseph MJE et al (2021) dMRIPrep: a robust preprocessing pipeline for diffusion MRI. In: Proceedings of the International Society for Magnetic Resonance in Medicine, Virtual Meeting, p 2473. Accessed: Mar. 12, 2021. [Online]. Available: https://docs.google.com/document/u/2/d/1ocamAFP2

OGnUIUooL9gxu5CExqiCS4Le3caHWM_8E04/edit?usp=drive_web&ouid=104994410621175933959&usp=embed_facebook

45. Adebimpe A et al (2022) ASLPrep: a generalizable platform for processing of arterial spin Labeled MRI and quantification of regional brain perfusion. Nat Methods 19:683–686. https://doi.org/10.1038/s41592-022-01458-7
46. Provins C, MacNicol EE, Seeley SH, Hagmann P, Esteban O (2023) Quality control in functional MRI studies with MRIQC and fMRIPrep. Front Neuroimaging 1:1073734. https://doi.org/10.3389/fnimg.2022.1073734
47. Reason J, Broadbent DE, Baddeley AD, Reason J (1997) The contribution of latent human failures to the breakdown of complex systems. Philos Trans R Soc Lond B Biol Sci 327(1241): 475–484. https://doi.org/10.1098/rstb.1990.0090
48. (2022) Harnessing the multiverse of neuroimaging standard references. Nat Methods 19(12):12. https://doi.org/10.1038/s41592-022-01682-1
49. Wilkinson MD et al (2016) The FAIR guiding principles for scientific data management and stewardship. Sci Data 3(1):1. https://doi.org/10.1038/sdata.2016.18
50. Kurtzer GM, Sochat V, Bauer MW (2017) Singularity: scientific containers for mobility of compute. PLoS One 12(5):e0177459. https://doi.org/10.1371/journal.pone.0177459
51. Snoek L, van der Miesen MM, Beemsterboer T, van der Leij A, Eigenhuis A, Steven Scholte H (2021) The Amsterdam open MRI collection, a set of multimodal MRI datasets for individual difference analyses. Sci Data 8(1):85. https://doi.org/10.1038/s41597-021-00870-6
52. Afiaz A et al (2023) Evaluation of software impact designed for biomedical research: Are we measuring what's meaningful? June 05, 2023, arXiv: arXiv:2306.03255. https://doi.org/10.48550/arXiv.2306.03255
53. Preston-Werner T (2024) Semantic Versioning 2.0.0, Semantic Versioning. Accessed: Feb 22, 2024. [Online]. Available: https://semver.org/
54. Li X et al (2024) Moving beyond processing and analysis-related variation in neuroscience. Nat Hum Behav (accepted):2021.12.01.470790. https://doi.org/10.1101/2021.12.01.470790
55. Sikka S et al (2014) Towards automated analysis of connectomes: The configurable pipeline for the analysis of connectomes (C-PAC). In: 5th INCF Congress of Neuroinformatics, Munich, Germany. https://doi.org/10.3389/conf.fninf.2014.08.00117
56. Chatelain Y et al (2023) A numerical variability approach to results stability tests and its application to neuroimaging, July 10, 2023, *arXiv*: arXiv:2307.01373. https://doi.org/10.48550/arXiv.2307.01373
57. Souter NE et al (2023) Ten recommendations for reducing the carbon footprint of research computing in human neuroimaging. Imaging Neurosci 1:1–15. https://doi.org/10.1162/imag_a_00043
58. Poldrack RA et al (2013) Toward open sharing of task-based fMRI data: the OpenfMRI project. Front Neuroinform 7:12. https://doi.org/10.3389/fninf.2013.00012
59. Di Martino A et al (2014) The autism brain imaging data exchange: towards a large-scale evaluation of the intrinsic brain architecture in autism. Mol Psychiatry 19(6):659–667. https://doi.org/10.1038/mp.2013.78
60. Poldrack RA et al (2016) A phenome-wide examination of neural and cognitive function. Sci Data 3:160110. https://doi.org/10.1038/sdata.2016.110
61. Hoopes A, Mora JS, Dalca AV, Fischl B, Hoffmann M (2022) SynthStrip: skull-stripping for any brain image. NeuroImage 260:119474. https://doi.org/10.1016/j.neuroimage.2022.119474
62. Billot B et al (2023) SynthSeg: segmentation of brain MRI scans of any contrast and resolution without retraining. Med Image Anal 86: 102789. https://doi.org/10.1016/j.media.2023.102789
63. Iglesias JE et al (2023) SynthSR: a public AI tool to turn heterogeneous clinical brain scans into high-resolution T1-weighted images for 3D morphometry. Sci Adv 9(5):eadd3607. https://doi.org/10.1126/sciadv.add3607
64. Ciric R, Thomas AW, Esteban O, Poldrack RA (2022) Differentiable programming for functional connectomics. In: Proceedings of the 2nd Machine Learning for Health symposium. PMLR, New Orleans, pp 419–455. Accessed: Dec. 05, 2022. [Online]. Available: https://proceedings.mlr.press/v193/ciric22a.html
65. Poldrack RA, Lu T, Beguš G AI-assisted coding: experiments with GPT-4, Apr. 25, 2023, *arXiv*: arXiv:2304.13187. https://doi.org/10.48550/arXiv.2304.13187

Chapter 9

Structural MRI and Computational Anatomy

Felix Hoffstaedter, Georgios Antonopoulos, and Christian Gaser

Abstract

Structural magnetic resonance imaging can yield highly detailed images of the human brain. In order to quantify the variability in shape and size across different brains, methods developed in the field of computational anatomy have proved exceptionally useful. For example, voxel-based morphometry is a popular method that involves segmenting magnetic resonance imaging scans into gray matter, white matter, and cerebrospinal fluid, and transforming individual brain shapes to a standard template space for comparative analysis. However, computational anatomy—when applied to brain data at scale—can be complex and computationally expensive. Furthermore, there are many possible pipelines that can be applied to structural brain data and for this reason it is important to follow best practices for reproducible neuroimaging analyses. This chapter demonstrates reproducible processing using the CAT12 (Computational Anatomy Toolbox) extension to SPM12 that focuses on voxel- and region-based morphometry. Through worked examples, we demonstrate three approaches to reproducible image analysis: "minimal", "intermediate", and a "comprehensive" protocol using the FAIRly big workflow based on DataLad. The comprehensive approach automatically facilitates parallel execution of whole dataset processing using container technology and also produces re-executable run records of each processing step to enable fully automatic reproducibility.

Key words Structural neuroimaging, Computational anatomy, Voxel-based morphometry, Reproducibility, FAIRly big workflow

1 Introduction

Since the growing availability of magnetic resonance imaging (MRI) machines in research facilities and hospitals more than two decades ago, robust noninvasive analysis of brain anatomy has become common practice in neuroscience research. Two fundamental use cases can be distinguished. First, anatomical brain mapping in relation to basic phenotypes, such as age and sex/gender. Second, the investigations of cognitive functioning and clinical conditions, which are associated with variations in brain morphology. Voxel-based morphometry (VBM) is one of the first and most commonly used method for quantification of brain tissue types

Robert Whelan and Hervé Lemaître (eds.), *Methods for Analyzing Large Neuroimaging Datasets*, Neuromethods, vol. 218, https://doi.org/10.1007/978-1-0716-4260-3_9, © The Author(s) 2025

(specifically gray matter volume; GMV) in humans [1] and animals [2].

VBM is based on the identification of specific brain tissue types—commonly gray and white matter (GM, WM) as well as CerebroSpinal Fluid (CSF), during a process called *segmentation*. In whole-brain MRI scans, tissue classification is performed by expressing every point/voxel in the brain as a probability or volume fraction of GM, WM, and CSF. Subsequently, individual brain shapes are transformed to match a standard template, often in MNI space [3], allowing a point-to-point comparison of corresponding regions across different brains. Furthermore, the local amount of deformation in every voxel—expressed by the Jacobian determinant—is utilized to *modulate* tissue probability/fraction to represent a proxy for local GMV. General comparability over studies is made possible by using the MNI space as a standard coordinate system combined with the use of common brain templates and brain parcellations. There are several widely used and well-tested publicly available software packages that, in principle, allow researchers of different experience levels to carry out structural MRI analysis without first becoming experts in programming or brain anatomy.

Recent studies have shown marked differences among VBM pipelines [4, 5]; however, robust effects of age or sex were detected by all evaluated software solutions. Similarly, coordinate-based meta-analysis of VBM studies each using different analysis software has shown consistent brain changes associated with neurodegenerative diseases. For example, a neuroimaging meta-analysis on frontotemporal dementia found consistent effects of neural atrophy over studies using different analysis software and brain templates [6]. Neuroscience, along with other natural sciences such as biology and psychology, faces the challenge of poor replicability in many studies due to small sample sizes and analytic flexibility (*see* also Chapter 4 for best practices in reproducible neuroimaging). Indeed, it has been shown that VBM analyses on moderately sized samples ($n \sim 300$) tend to overestimate effect sizes and significant effects are often not replicable [7]. Together with other evidence, this makes a strong case for conducting large-scale replication analyses even for well-known effects such as brain maturation [8], asymmetry [9] and aging [10].

In this chapter, we demonstrate the automatic structural processing of MRI images, using the AOMIC datasets [11] as an example. We describe a fully reproducible workflow [12] based on DataLad [13]. DataLad (*see* Chapter 2) is a powerful, open-source research data management software based on git (git-scm.com) and git-annex (git-annex.branchable.com); *see* also Chapter 5 (Garcia & Kelly) for git use. We utilize the robust and easy-to-use CAT12 (Computational Anatomy Toolbox, https://neuro-jena.github.io/cat) extension to SPM12 (www.fil.ion.ucl.ac.uk/spm) in

Matlab, also available without license costs as a standalone version (https://neuro-jena.github.io/enigma-cat12/#standalone) or as a Singularity container (https://github.com/inm7-sysmed/ENIGMA-cat12-container). CAT12 covers diverse morphometric analysis methods such as VBM, surface-based morphometry (SBM), deformation-based morphometry (DBM), and label- or region-based morphometry (RBM). For a brief description of CAT12, *see* https://neuro-jena.github.io/cat12-help/#basic_vbm); for a detailed description of CAT12, *see* https://neuro-jena.github.io/cat12-help/#process_details. As the focus of this chapter is the reproducible processing of large cohorts, the analysis is limited to showcases of preprocessing for VBM and RBM analysis.

2 Methods Overview

2.1 Starting Point of the Data

Note

CAT12 will produce many different output images (e.g., local tissue estimates at voxel level of 1.5 mm^3 by default), as well as atlas based regional volume estimates. Users should be aware that many VBM pipelines—including CAT12—can output files into one or multiple folders, which creates complexity for larger datasets.

The starting point for volumetric analyses of brain tissues typically uses a T1-weighted contrast MRI sequence, which highlights cortical gray and white matter. Additionally, the T2-weighted contrast is essential for brain lesion mapping in clinical analysis. Even in smaller samples ($n < 50$ subjects), disease-related effects (e.g., neurodegeneration) can be reliably detected if the effect size is large. However, in order to improve generalization, larger samples ($n > 300$) should be used [7]. As described in other chapters in this book, the importance of standardized data structures and formats should not be underestimated. It is strongly recommended to adhere to the BIDS standard for input data and carefully plan the output structure especially for very large datasets (*see* Chapter 4).

2.2 Data Storage and Computing

CAT12 is based on Statistical Parametric Mapping (SPM) software, originally running on Matlab, however there is an Octave (7.3) version and a compiled standalone version available which runs without a Matlab License via Matlab Runtime (MCR, Matlab Compiler Runtime, *see* Resources). To enhance reproducibility, we also provide a recipe for fully automatic creation of a Singularity container (https://apptainer.org/; *see* Glossary). As MCR cannot be publicly shared, CAT12 Singularity containers are privately available upon reasonable and unreasonable requests. We provide example setups for Linux and macOS without container environments for local data processing, as well as a Linux-based setup deploying Singularity with template scripts for the optional use of a job scheduler. CAT12 can run even on older laptops with four CPUs and 8 GB RAM, requiring up to 20 min of compute time per T1-weighted image, including cortical surface extraction. The ideal compute environment for full scalability is a compute cluster running a flavor of Linux providing recent versions of DataLad (≥0.17), Singularity (≥2.5) and a workload scheduling system like HTCondor or SLURM. Regarding data storage, the amount of data generated during CAT12 preprocessing varies based on the input data. For instance, with a standard T1-weighted MRI scan of 6.5 MB, the output generated by CAT12 will be approximately 60 MB per scan.

2.3 Coding Knowledge

The example setup in this chapter should be executable for beginners with basic command line experience given the correct platform setup (*see* above).

2.4 Computational Expense

Standard preprocessing of data from 100 subjects (×0.3h) takes 30 core hours and parallelized on a standard desktop computer with four cores, will take only 7.5 h. Parallelization (*see* Glossary) of individual jobs running independently at the same time can be done directly by CAT12 standalone or automatically by the example workflow locally using GNU-parallel or a given scheduler software on a compute cluster.

3 Worked Example

The following demonstration of structural image analysis starts with a simple approach to reproducible image processing, using a two-step protocol to improve result tracking and automatic process documentation. Initially, DataLad is used as a download tool to provide access to the input data and analysis code (as described in Chapter 2). Secondly, the results of processing are periodically captured by saving snapshots of the outcome of each processing step. We also described a way to automatically track the whole image analysis pipeline by using the “FAIRly big workflow” [12].

As with most standard data processing workflows, we need:

(A) Input Data

(B) Processing Software/ Pipeline

(C) Code to execute B on A

3.1. *A minimal approach to reproducible image analysis consists in* (A) providing the *Input Data*, here MR images and added metadata about how they were acquired (B) documenting the software versions used in the *Processing Pipeline* or provide precise instructions to build the used software environment (C) best practice is to share the analysis *Code* as a Git repository or as download on a public hosting site like Zenodo or GitHub, together with *Results Images and Statistics* (*see* Chapter 4). The following commands work on any Linux OS and macOS system.

With DataLad installed, Users are able to access the input data (A)—here the AOMIC-PIOP2 dataset—by simply cloning it from OpenNeuro to their compute environment using one command. It is recommended to clone the data into a dedicated project folder.

```
datalad clone https://github.com/OpenNeuroDatasets/ds002790.
git AOMIC-PIOP2
```

The dataset comprises the raw data in BIDS format and many processed derivatives, but one should be aware that only the metadata is directly available via this clone. The user will also have to download the file content for the actual CAT12 processing, using DataLad commands. The following lines of code describe how to download one subject's T1w data, then copy it to a CAT12_derivatives folder for processing and finally drop the file content from the input dataset after use to minimize the storage footprint.

```
# download the data
datalad get -d AOMIC-PIOP2 AOMIC-PIOP2/sub-0111/anat/sub-
0111_T1w.nii.gz
# create an output directory and copy the T1w file there
mkdir -p CAT12_derivatives/TEST_sub-0111
cp AOMIC-PIOP2/sub-0111/anat/sub-0111_T1w.nii.gz CAT12_deri-
vatives/TEST_sub-0111/
# delete/drop the local version of the file as we can get it
back anytime
datalad drop --what filecontent --reckless kill -d AOMIC-
PIOP2 AOMIC-PIOP2/sub-0111
```

Now, the software environment is setup (B) for CAT12 standalone with Matlab Compiler Runtime (MCR). CAT12 (e.g., CAT12.8.1_r2042_R2017b_MCR) is downloaded as described here and unpacked to the working directory. Then the matching MCR (version R2017b 9.3) is downloaded and installed from here. The software setup should be tested by starting the now available standalone version of SPM12 including CAT12. In the CAT12 folder, the "run_spm12.sh" script runs by adding the full path to the MCR installation as argument. With the following command, the standard SPM12 graphical user interface is initiated and under the Toolbox button "cat12" will be started already in expert mode, again without a full Matlab license. The command needs the *full path* to MCR, but in reproducible scripts it is generally recommended to use relative paths from the dataset or code folder instead of full paths as the latter is machine- and user-specific.

```
# start SPM12 graphical user interface in PET/VBM mode
./run_spm12.sh </FULLPATH/MCR/v93>
```

A test subject can reproducibly be run via the command line interface, using the CAT12 standalone syntax, which is well described on the CAT12 website. The location of the MCR setup and the CAT12 *version of choice* can be provided via environmental variables with *full paths* as follows:

```
# provide full paths to pipeline setup and make it available
MCRROOT=<//FULLPATH/MCR/v93>
SPMROOT=<//FULLPATH/CAT12.8.1_r2042_R2017b_MCR_Linux>
export MCRROOT SPMROOT
```

Now, the first test processing is run by executing the *cat_standalone.sh* script together with a batch file (*-b*) that describes the settings of how the target T1w scan will be segmented:

```
# start CAT12 standalone with simple segment batch for example
subject
 ./CAT12.8.1_r2042_R2017b_MCR_Linux/standalone/cat_standa-
lone.sh \
 -b ./CAT12.8.1_r2042_R2017b_MCR_Linux/standalone/cat_standa-
lone_segment.m \
 CAT12_derivatives/TEST_sub-0111/sub-0111_T1w.nii.gz
```

CAT12 segmentation takes <20 min per subject and produces ~60 MB of data in the following output structure:

```
CAT12_derivatives
└── TEST_sub-0111
├── label
│  ├── catROI_sub-0111_T1w.mat
│  └── catROI_sub-0111_T1w.xml
├── mri
│  ├── mwp1sub-0111_T1w.nii
│  ├── mwp2sub-0111_T1w.nii
│  ├── p0sub-0111_T1w.nii
│  ├── wmsub-0111_T1w.nii
```

```
│   └── y_sub-0111_T1w.nii
├── report
│   ├── catlog_sub-0111_T1w.txt
│   ├── catreportj_sub-0111_T1w.jpg
│   ├── catreport_sub-0111_T1w.pdf
│   ├── cat_sub-0111_T1w.mat
│   └── cat_sub-0111_T1w.xml
└── sub-0111_T1w.nii.gz
```

The *report* folder contains full processing logs in different file formats and a pdf/jpg report sheet for quick quality assessment. For VBM analyses in the *mri* folder, modulated gray (mwp1) and white matter (mwp2) images are provided in template space. Also available are a partial volume image (p0) in native subject space, a denoised normalized brain image (wm) in template space and the transformation/warp field (y_) from native to template space. Regional volume estimates of several brain parcellations are found in the *label* folder.

For convenience, *bootstrap scripts* are provided to be cloned from here for *Linux (and macOS)* that will do all of the above reproducibly from scratch. The following commands will download those to our workspace directory and run them.

```
# clone Computational Anatomy Tutorial from OSF
 datalad clone https://osf.io/ydxw7/ ca_tutorial
```

The script ca_minimal.sh will set up the environment in the current working directory and run the test subject as described. It is recommended to have a close look at the script to follow what is happening.

```
./ca_tutorial/ca_minimal.sh
```

After carefully checking the output of the test subject, parallelization can then be tested by executing the prepared script, which will process nine subjects with three in parallel in the background.

```
./run_3x3catjobs.sh
```

3.2. *An intermediate approach to reproducible image analysis is to share not only the list of ingredients but access to* (A) *Input Data*, (B) *Processing Pipeline* setup, and (C) *Code*. An emerging standard is the publication of intermediate processing results for re-execution of statistical analysis and figure creation, which is particularly useful in genetic or big data modeling analysis [14]. Here, scripts for automatic execution of all processing steps are provided. The main difference to the minimal approach consists in the creation of a DataLad dataset to capture the ingredients A/B/C and intermediate results, which is extensively and clearly described in the *DataLad handbook*. DataLad uses Git under the hood (*see* Chapter 5) to track not only code but also arbitrarily sized data providing a re-executable framework to reproduce the exact succession of computations captured in the git history. To avoid redundancies in this tutorial, only additional commands are documented. Of note, the MCR setup cannot be shared due to the license conditions and must be installed individually, so it is not part of the DataLad dataset.

```
# define CAT12 version
CAT12_version="CAT12.8.1_r2042_R2017b_MCR_Linux"
# create Datalad yoda dataset for intermediate cat12 pipeline and enter
datalad create -c yoda intermed_cat12
cd intermed_cat12
# add input dataset as subdataset (!) from OpenNeuro AOMIC-PIOP2
datalad clone -d . https://github.com/OpenNeuroDatasets/ds002790.git inputs/AOMIC-PIOP2
# setup CAT12 standalone pipeline in code/ folder
wget "https://www.neuro.uni-jena.de/cat12/${CAT12_version}.zip"
unzip ${CAT12_version}.zip -d code
rm -f ${CAT12_version}.zip
# register the CAT12.8.1 standalone pipeline in the dataset
datalad save -m "add CAT12.8.1 r2042 pipeline with git" code/
```

After providing the location of both CAT12 and MCR setup and preparing the T1w image in a derivatives folder, execute the example subject and capture result with DataLad:

```
# provide full paths to pipeline setup and make it available
MCRROOT=<//FULLPATH/MCR/v93>
SPMROOT=<//FULLPATH/CAT12.8.1_r2042_R2017b_MCR_Linux>
export MCRROOT SPMROOT


# download the data
datalad get inputs/AOMIC-PIOP2/sub-0111/anat/sub-0111_T1w.nii.gz
# create an outputs directory and copy the T1w file there
mkdir -p CAT12_derivatives/TEST_sub-0111
cp inputs/AOMIC-PIOP2/sub-0111/anat/*T1w.nii.gz CAT12_deri-
```

```
vatives/TEST_sub-0111/
 # delete/drop the local version of the file as we can get it
back anytime
 datalad drop --what filecontent --reckless kill inputs/AOMIC-
PIOP2/sub-0111

 # execute CAT12 standalone with MCR
 ./code/${CAT12_version}/standalone/cat_standalone.sh \
  -b ./code/${CAT12_version}/standalone/cat_standalone_seg-
ment.m \
  -a "matlabbatch{1}.spm.tools.cat.estwrite.output.surface =
0" \
 CAT12_derivatives/TEST_sub-0111/sub-0111_T1w.nii.gz
  # track CAT12 derivatives and standalone setup in datalad
dataset
  datalad save -m "save test subject & standalone setup"
CAT12_derivatives code
```

Now, the dataset contains all necessary parts ABC (and MCR outside) to conduct reproducible, full-scale VBM processing of the AOMIC PIOP2 dataset and registering all results. CAT12 also includes a feature to locally parallelize processing over subjects, wrapping the former command in another script. All T1w images have to be prepared and copied into the CAT12_derivatives folder before running CAT12:

```
  # download first 9 subjects to process 3x in parallel
(sub-0001..9; 3x3jobs)
 datalad get inputs/AOMIC-PIOP2/sub-000*/anat/*T1w.nii.gz


# copy T1w images of 9 subjects to CAT_derivatives folder
 for sub in inputs/AOMIC-PIOP2/sub-000*; do
 sub=$(basename $sub); mkdir -p CAT12_derivatives/${sub};
  cp inputs/AOMIC-PIOP2/${sub}/anat/*T1w.nii.gz CAT12_deriva-
tives/${sub};
 done
# run CAT12 3x in parallel (-p 3) writing logs to folder (-l)
 ./code/${CAT12_version}/standalone/cat_parallelize.sh -p 3 -l
. -c "
 ./code/${CAT12_version}/standalone/cat_standalone.sh \
  -b ./code/${CAT12_version}/standalone/cat_standalone_seg-
ment.m" \
 CAT12_derivatives/sub-*/*T1w.nii.gz
 # track CAT12 derivatives in datalad dataset after processing
  datalad save -m "save CAT12 derivatives for 9 subjects"
CAT12_derivatives
```

This whole procedure is written and described in the following *bootstrap script.*

```
./ca_tutorial/ca_intermed.sh
```

After carefully checking the output of the test subject, parallelization can be tested be executing the prepared script from within the dataset, which will process nine subjects with three in parallel in the background and capture the outcome in the DataLad dataset.

```
cd intermed_cat12
./code/run_3x3catjobs.sh
```

3.3. *Finally, a comprehensive approach to reproducible image analysis is the FAIRly big workflow* [11], which consists of an automatic setup procedure via a *bootstrap script*, which tracks all ingredients ABC as well as the results of each processing step. Additionally, the workflow contains automatic means for parallel execution of whole dataset processing using container technology and produces re-executable run records of each processing step to enable fully automatic reproducibility. Of note, for building the Singularity container *sudo rights* are needed. For this more complex workflow the following additional dependencies are needed:

GNU-parallel	
datalad-container extension:	pip install datalad-container
Singularity container	

To properly introduce the workflow, a few essential concepts are presented below, which are directly taken from the original publication: "*FAIRly big: A framework for computationally reproducible processing of large-scale data*" [12], with permission of the authors.

DataLad Dataset
DataLad's core data structure is the dataset. On a technical level, it is a joint Git/git-annex (REF) repository. Conceptually, it is an overlay data structure that is particularly suited to address data integration challenges. It enables users to version control files of any size or type, track and transport files in a distributed network of dataset clones, as well as record and re-execute actionable process provenance on the genesis of file content. DataLad datasets have the ability to retrieve or drop registered, remote file content on demand with single file granularity. This is possible based on a lean record of file identity and file availability (via checksum and URLs) irrespective of the true file size. A user does not need to be aware of the actual download source of a file's content, as precise file identity

is automatically verified regardless of a particular retrieval method, and the specification of redundant sources is supported. These technical features enable the implementation of infrastructure-agnostic data retrieval and deposition logic in user code.

A Clone (Git concept) is a copy of a DataLad dataset that is linked to its origin dataset and its history. The clones are lightweight and can typically be obtained within seconds, as they are primarily comprised of file identity and availability records. DataLad enables synchronization of content between clones and, hence, the propagation of updates.

A Branch (Git concept) is an independent segment of a DataLad dataset's history. It enables the separation of parallel developments based on a common starting point. Branches can encompass arbitrarily different modifications of a dataset. In a typical collaborative development or parallel processing routine, changes are initially introduced in branches and are later consolidated by merging them into a mainline branch.

Nesting A DataLad dataset can also contain other DataLad datasets. Analog to file content, this linkage is implemented using a lightweight dataset identity and availability record (based on Git's submodules). This nesting enables flexible (re-)use of datasets in a different context. For example, it allows for the composition of a project directory from precisely versioned, modular units that unambiguously link all inputs of a project to its outcomes. Nesting offers actionable dataset linkage at virtually no disk space cost, while providing the same on-demand retrieval and deposition convenience as for file content operations because DataLad can work with a hierarchy of nested datasets as if they are a single monolithic repository. When a DataLad dataset B is nested inside DataLad dataset A, we also refer to A as the superdataset and to B as a subdataset. A superdataset can link any number of subdatasets, and datasets can simultaneously be both super- and subdataset.

RIA Store A file system-based store for DataLad datasets with minimal server-side software requirements (in particular no DataLad, no git-annex, and Git only for specific optional features) (REF). These stores offer inode minimization (using indexed 7-zip archives). A dataset of arbitrary size and number of files can be hosted while consuming fewer than 25 inodes, while nevertheless offering random read access to individual files at a low and constant latency independent of the actual archive size. Combined with optional file content encryption and compression, RIA ("Remote Indexed Archive") stores are particularly suited for staging large-scale, sensitive data to process on HPC resources.

The following script automatically sets up the FAIRly Big workflow for fully reproducible processing of large-scale structural

MRI data tracking results and execution with DataLad enabling automatic re-execution of CAT12 processing with "datalad run":

```
./ca_tutorial/ca_FAIRlyBig.sh
```

The main differences to the other approaches in 3.1 and 3.2 are the use of a Singularity container as pipeline and directly tracking each compute job with DataLad in separate ephemeral *clones* of the entire workflow setup. This is made possible by pushing the workflow setup as *DataLad dataset* to local repositories called *RIA store* (*see* Above). The workflow dataset contains all Code **(C)** and references both the Input Data **(A)** and the containerized Processing Pipeline **(B)** as *nested* subdatasets ready to be cloned. Two identical repositories are created for the workflow dataset, with an input *RIA store* to clone from and an output *RIA store* to push processing results to. For each compute job, an independent clone of the workflow from the *input RIA* is created in a separate, temporal location (e.g., tmp/), where the compute job automatically downloads the specific input data needed, and runs CAT12 in an individual container instance. Processing results are tracked in a job-specific *branch* and pushed to the output *RIA* before the temporal clone is deleted, which makes sure that no partial results will be saved in the final *DataLad dataset*. This full separation of individual compute jobs enables a full parallelization of the workflow, while tracking input, pipeline, code and all results in one place. Template submission scripts have been built for HTCondor and SLURM job scheduling systems, which can be tailored to any compute cluster setup available to the User. With the setup of the AOMIC-PIOP2_cna_cat12.8.1 analysis dataset, instructions for running a test subject are printed in the terminal. Here we use the ENIGMA CAT12 processing batch, which conducts more comprehensive data processing and produces more output.

```
# enter into fully setup Datalad dataset:
cd AOMIC-PIOP2_cna_cat12.8.1
# run test subject in the FAIRly Big workflow
./code/process.sub sub-0222
```

After successful processing, the results dataset has to be consolidated by running the available results.merger script, which merges all individual job *branches* into the main *branch* by performing an octopus merge. The FAIRly big workflow saves/pushed all processing output directly in the *output RIA*, of which the dataset is a mere clone. The consolidation script updates the local dataset by pulling the full git history from the output RIA, collecting all subjects and checking their file availability in the RIA.

```
./code/results.merger
```

This consolidation step reveals that "No known copies exist" of a few files like denoised input images and in particular of the spatial transformations from subject to template space (y_*) and the inverse (iy_*). This was an explicit decision setting up the workflow for a large dataset, as those files are comparably big in size and rarely used, as all images for standard analyses are available in template space. CAT12 processing was tracked with DataLad containers-run including a comprehensive run record in the git history necessary for datalad rerun to fully automatically re-execute the workflow creating the missing files. The following command will rerun the last subject/job and make all files available:

```
# rerun CAT12 job by using the latest commit hash
datalad rerun $(git rev-parse HEAD)
```

Data of the test subject should be inspected by downloading the data from the *output RIA:*

```
datalad get sub-0222
```

The CAT12 ENIGMA pipeline creates many more files, which are documented in the comprehensive CAT12-help including surface projection and cortical thickness estimations of different surface atlases.

When the results of the test subject are satisfactory, again nine subjects with three in parallel can be started in the background using the following command.

```
parallel -j3 ./code/process.sub sub-000{} ::: {1..9}
```

The whole dataset can be submitted for local processing with four subjects in parallel (-j4) using GNU-parallel, which takes care that of all 200 subjects only four will run at any given moment.

```
parallel -j4 ./code/process.sub {/} ::: inputs/AOMIC-PIOP2/
sub*
```

The *output RIA* is locally hosted in this example but can also be stored anywhere on a file system or the web using OSF or other web hosting services. In this example the RIA looks as follows:

```
dataladstore
├── 954
│ └── 6b411-702b-4839-8773-50101d8f2cd9
├── alias
│ ├── AOMIC-PIOP2_ca_cat12.8.1 -> ../954/6b411-702b-4839-
8773-50101d8f2cd9
│ └── cat12.8_container -> ../e64/188f5-1330-4729-90a6-
7dd4ba71d529
```

```
├── e64
│ └── 188f5-1330-4729-90a6-7dd4ba71d529
├── error_logs
├── inputstore
│ ├── 954
│ ├── alias
│ ├── error_logs
│ └── ria-layout-version
└── ria-layout-version
```

The datasets can be cloned from RIA by using the alias in the following commands:

```
datalad clone "ria+file://<FULLPATH>/dataladstore#~AOMIC-PIOP2_ca_cat12.8.1"
datalad clone "ria+file://<FULLPATH>/dataladstore#~cat12.8_-container"
```

To push the whole dataset to OSF the datalad-osf extension is needed and after setting up the credentials any dataset can be published on OSF with the data present in the local clone:

```
datalad create-sibling-osf --title AOMIC-PIOP2_ca_cat12.8.1 -s osf \
--mode export --category data --tag reproducibility --public
datalad get .
datalad push --to osf
```

For comparison the fully processed dataset can be cloned using the following command:

```
datalad clone https://osf.io/9gtsf/ AOMIC-PIOP2_ca_cat12.8.1
```

Statistical data analyses are described well in the CAT12 help including detailed instructions about how to use the CAT12 graphical user interface. For very large datasets, the amount of RAM and compute time needed for estimating large GLMs can increase immensely, but the model setup is largely identical to smaller dataset.

4 Conclusion

Here, we have demonstrated how to use CAT12 to produce reproducible computational anatomy pipelines. We have shown how analyses can be scaled up to, in theory, analyze thousands of MRI images in parallel while keeping tracking of input files and maintaining the pipeline, code and all results in one place.

References

1. Ashburner J, Friston KJ (2000) Voxel-based morphometry—the methods. NeuroImage 11(6):805–821. https://doi.org/10.1006/nimg.2000.0582
2. McLaren DG, Kosmatka KJ, Kastman EK, Bendlin BB, Johnson SC (2010) Rhesus macaque brain morphometry: a methodological comparison of voxel-wise approaches. Methods 50(3):157–165. https://doi.org/10.1016/j.ymeth.2009.10.003
3. Evans AC, Kamber M, Collins DL, MacDonald D (1994) An MRI-based probabilistic atlas of neuroanatomy. In: Shorvon SD, Fish DR, Andermann F, Bydder GM, Stefan H (eds) Magnetic resonance scanning and epilepsy. Springer US, Boston, pp 263–274
4. Antonopoulos G, More S, Raimondo F, Eickhoff SB, Hoffstaedter F, Patil KR (2023) A systematic comparison of VBM pipelines and their application to age prediction. NeuroImage 279:120292. https://doi.org/10.1016/j.neuroimage.2023.120292
5. Zhou X, Wu R, Zeng Y, Qi Z, Ferraro S, Xu L, Zheng X, Li J, Fu M, Yao S, Kendrick KM, Becker B (2022) Choice of Voxel-based Morphometry processing pipeline drives variability in the location of neuroanatomical brain markers. Commun Biol 5(1):913. https://doi.org/10.1038/s42003-022-03880-1
6. Kamalian A, Khodadadifar T, Saberi A, Masoudi M, Camilleri JA, Eickhoff CR, Zarei M, Pasquini L, Laird AR, Fox PT, Eickhoff SB, Tahmasian M (2022) Convergent regional brain abnormalities in behavioral variant frontotemporal dementia: a neuroimaging meta-analysis of 73 studies. Alzheimers Dement 14(1):e12318. https://doi.org/10.1002/dad2.12318
7. Kharabian Masouleh S, Eickhoff SB, Hoffstaedter F, Genon S, Alzheimer's Disease Neuroimaging Initiative (2019) Empirical examination of the replicability of associations between brain structure and psychological variables. eLife 8:e43464. https://doi.org/10.7554/eLife.43464
8. Bennett CM, Baird AA (2006) Anatomical changes in the emerging adult brain: a voxel-based morphometry study. Hum Brain Mapp 27(9):766–777. https://doi.org/10.1002/hbm.20218
9. Luders E, Gaser C, Jancke L, Schlaug G (2004) A voxel-based approach to gray matter asymmetries. NeuroImage 22(2):656–664. https://doi.org/10.1016/j.neuroimage.2004.01.032
10. Good CD, Johnsrude IS, Ashburner J, Henson RNA, Friston KJ, Frackowiak RSJ (2001) A Voxel-Based Morphometric Study of Ageing in 465 Normal Adult Human Brains. NeuroImage 14(1):21–36. https://doi.org/10.1006/nimg.2001.0786
11. Snoek L, Van Der Miesen MM, Beemsterboer T, Van Der Leij A, Eigenhuis A, Steven Scholte H (2021) The Amsterdam Open MRI Collection, a set of multimodal MRI datasets for individual difference analyses. Sci Data 8(1):85. https://doi.org/10.1038/s41597-021-00870-6
12. Wagner AS, Waite LK, Wierzba M, Hoffstaedter F, Waite AQ, Poldrack B, Eickhoff SB, Hanke M (2022) FAIRly big: a framework for computationally reproducible processing of large-scale data. Sci Data 9(1):80. https://doi.org/10.1038/s41597-022-01163-2
13. Halchenko Y, Meyer K, Poldrack B, Solanky D, Wagner A, Gors J, MacFarlane D, Pustina D, Sochat V, Ghosh S, Mönch C, Markiewicz C, Waite L, Shlyakhter I, De La Vega A, Hayashi S, Häusler C, Poline J-B, Kadelka T, Skytén K, Jarecka D, Kennedy D, Strauss T, Cieslak M, Vavra P, Ioanas H-I, Schneider R, Pflüger M, Haxby J, Eickhoff S, Hanke M (2021) DataLad: distributed system for joint management of code, data, and their relationship. JOSS 6(63):3262. https://doi.org/10.21105/joss.03262
14. Bethlehem RAI, Seidlitz J, White SR, Vogel JW, Anderson KM, Adamson C, Adler S, Alexopoulos GS, Anagnostou E, Areces-Gonzalez-A, Astle DE, Auyeung B, Ayub M, Bae J, Ball G, Baron-Cohen S, Beare R, Bedford SA, Benegal V, Beyer F, Blangero J, Blesa Cábez M, Boardman JP, Borzage M, Bosch-Bayard JF, Bourke N, Calhoun VD, Chakravarty MM, Chen C, Chertavian C, Chetelat G, Chong YS, Cole JH, Corvin A, Costantino M, Courchesne E, Crivello F, Cropley VL, Crosbie J, Crossley N, Delarue M, Delorme R, Desrivieres S, Devenyi GA, Di Biase MA, Dolan R, Donald KA, Donohoe G, Dunlop K, Edwards AD, Elison JT, Ellis CT, Elman JA, Eyler L, Fair DA, Feczko E, Fletcher PC, Fonagy P, Franz CE, Galan-Garcia L, Gholipour A, Giedd J, Gilmore JH, Glahn DC, Goodyer IM, Grant PE, Groenewold NA, Gunning FM, Gur RE, Gur RC, Hammill CF, Hansson O, Hedden T, Heinz A, Henson RN, Heuer K, Hoare J, Holla B, Holmes AJ, Holt R, Huang H, Im K, Ipser J, Jack CR,

Jackowski AP, Jia T, Johnson KA, Jones PB, Jones DT, Kahn RS, Karlsson H, Karlsson L, Kawashima R, Kelley EA, Kern S, Kim KW, Kitzbichler MG, Kremen WS, Lalonde F, Landeau B, Lee S, Lerch J, Lewis JD, Li J, Liao W, Liston C, Lombardo MV, Lv J, Lynch C, Mallard TT, Marcelis M, Markello RD, Mathias SR, Mazoyer B, McGuire P, Meaney MJ, Mechelli A, Medic N, Misic B, Morgan SE, Mothersill D, Nigg J, Ong MQW, Ortinau C, Ossenkoppele R, Ouyang M, Palaniyappan L, Paly L, Pan PM, Pantelis C, Park MM, Paus T, Pausova Z, Paz-Linares D, Pichet Binette A, Pierce K, Qian X, Qiu J, Qiu A, Raznahan A, Rittman T, Rodrigue A, Rollins CK, Romero-Garcia R, Ronan L, Rosenberg MD, Rowitch DH, Salum GA, Satterthwaite TD, Schaare HL, Schachar RJ, Schultz AP, Schumann G, Schöll M, Sharp D, Shinohara RT, Skoog I, Smyser CD, Sperling RA, Stein DJ, Stolicyn A, Suckling J, Sullivan G, Taki Y, Thyreau B, Toro R, Traut N, Tsvetanov KA, Turk-Browne NB, Tuulari JJ, Tzourio C, Vachon-Presseau É, Valdes-Sosa MJ, Valdes-Sosa PA, Valk SL, Van Amelsvoort T, Vandekar SN, Vasung L, Victoria LW, Villeneuve S, Villringer A, Vértes PE, Wagstyl K, Wang YS, Warfield SK, Warrier V, Westman E, Westwater ML, Whalley HC, Witte AV, Yang N, Yeo B, Yun H, Zalesky A, Zar HJ, Zettergren A, Zhou JH, Ziauddeen H, Zugman A, Zuo XN, 3R-BRAIN, AIBL, Rowe C, Alzheimer's Disease Neuroimaging Initiative, Alzheimer's Disease Repository Without Borders Investigators, Frisoni GB, CALM Team, Cam-CAN, CCNP, COBRE, cVEDA, ENIGMA Developmental Brain Age Working Group, Developing Human Connectome Project, FinnBrain, Harvard Aging Brain Study, IMAGEN, KNE96, The Mayo Clinic Study of Aging, NSPN, POND, The PREVENT-AD Research Group, Binette AP, VETSA, Bullmore ET, Alexander-Bloch AF (2022) Brain charts for the human lifespan. Nature 604(7906):525–533. https://doi.org/10.1038/s41586-022-04554-y

Chapter 10

Diffusion MRI Data Processing and Analysis: A Practical Guide with *ExploreDTI*

Michael Connaughton, Alexander Leemans, Erik O'Hanlon, and Jane McGrath

Abstract

This chapter introduces neuroimaging researchers to the concepts and techniques of diffusion magnetic resonance imaging data processing. Using the freely available ExploreDTI software, we provide a step-by-step guide for processing multi-shell High Angular Resolution Diffusion Imaging data and generating tractography based on constrained deconvolution. Brief explanations of the rationale behind each processing step are provided to aid the researcher in understanding the concepts and principles involved. Potential processing pitfalls will be discussed, and tips for troubleshooting common issues will be provided. An additional step-by-step guide for processing DTI data using the open-access AOMIC data set is also provided, demonstrating command-line that can also be applied to process other large neuroimaging datasets.

Key words Diffusion, Tractography, DTI, Multi-shell, HARDI

1 Introduction

White matter refers to the nerve fibers, also known as axons, that interconnect regions of the brain [1]. Healthy development of white matter is essential for neurotypical brain function and cognition [2]. This complex developmental process involves several mechanisms such as axonal growth, myelination, and synaptic pruning [3]. The intricate interplay between these processes is key for the establishment of neural networks for the efficient transmission of information within the brain [2]. Abnormalities in white matter development have been linked to a range of cognitive functions [4] and psychiatric impairments, including autism [5], ADHD [6], and schizophrenia [7].

Supplementary Information The online version contains supplementary material available at https://doi.org/10.1007/978-1-0716-4260-3_10.

Robert Whelan and Hervé Lemaître (eds.), *Methods for Analyzing Large Neuroimaging Datasets*, Neuromethods, vol. 218, https://doi.org/10.1007/978-1-0716-4260-3_10,

Diffusion-weighted magnetic resonance imaging (dMRI) [*see* Glossary] is a powerful neuroimaging technique that allows for the investigation of white matter microstructure through the diffusion measurement of water molecules within biological tissue [8]. In white matter, the diffusion of water molecules is affected by cellular membranes (i.e., myelin sheaths), defining the diffusion-weighted contrast. This diffusion-weighted signal can then be mathematically modeled to estimate the underlying microstructure and reconstruct the organization of white matter tracts [9]. In the early 2000s, the most common dMRI modeling technique was Diffusion Tensor Imaging (DTI) [*see* Glossary] [10, 11]. While DTI remains a key tool for researchers in understanding the impact of white matter microstructure [12–14], DTI has some limitations, such as its inability to accurately model areas in which crossing white matter fibers are present [15–17].

In recent years, advances in dMRI acquisition parameters have enabled higher-order diffusion modeling techniques that increase reconstruction accuracy and can overcome some of the limitations of DTI. With High Angular Resolution Diffusion Imaging (HARDI), an increased number of diffusion direction gradients is acquired, which allows for the estimation of microstructural properties along multiple fiber populations within a single voxel and provides improved reconstruction accuracy of white matter tracts compared to the traditional DTI framework [18]. Another advance in dMRI for tractography is the integration of multiple b-values [19, 20]. Briefly, b-values are a summary measure of the strength, duration, and amplitude of the diffusion-weighting applied during the scan. Different strength b-values elicit altered tissue responses which can be used to increase the reconstruction accuracy of various neurocellular environments. Higher b-values are more sensitive to detecting diffusion of water molecules within brain tissues [21] but are also more susceptible to noise and artifacts compared to lower b-values [22]. As such, multi-shell dMRI data leverages the increased signal of high b-value images with the reduced noise of low b-value images to provide increased anatomical accuracy [19].

In the context of higher-order diffusion modelling, techniques have been developed, such as constrained deconvolution (CSD), Q-ball, and neurite orientation and dispersion density imaging (NODDI) among many others [17, 23–26]. These techniques can describe the distribution of water diffusion within a voxel more accurately compared to DTI (e.g., the fiber orientation distribution function for CSD and diffusion orientation distribution function for DSI and Q-ball) and can be used to model voxels containing crossing white matter fibers. Thus, metrics derived from these higher-order models have increased accuracy, yielding clinically more relevant information that cannot be obtained from the DTI model [9]. Higher-order diffusion models provide more detailed information about the microstructure and organization of white matter tracts, which can provide important insights into the pathophysiology of neurological and psychiatric disorders.

The aim of this chapter is to introduce neuroimaging researchers to the concepts and techniques of dMRI data processing used in the field, with a focus on providing a practical step-by-step guide for processing multi-shell HARDI data and generating CSD-based tractography using the ExploreDTI software [*see* Resources] [27]. Brief explanations of the rationale behind each processing step will be provided to aid the researcher in understanding the concepts and principles involved. Potential processing pitfalls will be discussed, and tips for troubleshooting common issues will be provided. Overall, this guide aims to provide a comprehensive resource for researchers to gain the skills and knowledge necessary to process dMRI data effectively and efficiently.

2 Methods

2.1 Starting Point for the Data

Advanced fiber orientation distribution modeling techniques, such as CSD, require specific diffusion parameters. Typically for multi-shell HARDI, a minimum of two b-values images ($b = 2500–3000\ s/mm^2$) and 45 diffusion-weighted directions [9, 28] are required for CSD modeling for white matter tractography purposes. The Neuroimaging of the Children's Attention Project (NICAP) [29] study diffusion parameters were used for the processing step-by-step guide provided below. Data from the NICAP cohort are available via Lifecourse (https://lifecourse.melbournechildrens.com/cohorts/cap-and-nicap/).

2.2 Data Storage and Computational Expense

The step-by-step guide provided here was run on a Linux system with an Intel Core i7 processor and 32 GB RAM using MATLAB R2016b. A standalone version of ExploreDTI is also available. Details of the ExploreDTI instalment are provided below. It is recommended to run Steps 9 and 10 using high-performance computers, given the large processing time of these steps. Table 1 shows the estimated processing times per participant for each step, with MATLAB parallel processing enabled.

2.3 Step-by-Step Guide

In this section, we will provide a step-by-step guide for processing multi-shell HARDI data (in BIDS format [*see* Chapter 4]) and generating CSD-based tractography using the ExploreDTI software. As this guide is for those relatively new to neuroimaging, the ExploreDTI graphic user interface (GUI) is used. A step-by-step guide to installing and using ExploreDTI is provided in the user manual. As we are using BIDS format, each subject folder containing the diffusion files should have *.json*, *.bval*, *.bvec*, and *.nii* files (*see* Table 2). Although advanced diffusion modeling is not

Table 1
Approximate processing times for each per participant are provided below (Steps 7 and 8 are optional if multiple b-value data sets were acquired separately)

Processing step	Name of section	Approximate processing time
1	Convert Bval and Bvec files into text files	< 1 min
2	Signal drift	< 1 min
3	Sort Bvals	< 1 min
4	Gibbs ringing	< 1 min
5	Flip permute	< 1 min
6	Generate Mat File	5 min
7	Concatenate all b-value .mat files	2 min
8	Generate Mat File of concatenated .nii files	20 min
9	SM/EC/EPI distortion corrections	360 min
10	Whole brain tractography	70 min

Table 2
Description of dMRI files in BIDS format

File name	Comment
.json	File containing a description of scan acquisition details
.bval	File containing a summary of diffusion-weightings applied during scanning
.bvec	Files containing details on the diffusion gradient vectors of the scan
.nii	The raw diffusion scan in NIfTI format

feasible with the AOMIC datasets (https://nilab-uva.github.io/AOMIC.github.io/), we have also included step-by-step command lines to demonstrate the possibility to preprocess several subjects all at once (Appendix 5.2) [*see* Chapter 2].

2.3.1 Convert Bval and Bvec Files into Text Files (Step 1)

The first processing step is to generate *.txt* file(s) from the *.bval* and *.bvec* files for the images you are processing. The *.txt* file is a summary file of the b-values and diffusion-weighting directions used during image acquisition and is required for image processing.

In ExploreDTI:

1. Plugins → Convert → *.bval/*.bvec to B-matrix *.txt files (s) (*see* Fig. 1)
   (a) Select folder containing *.bval and *.bval file(s)
   (b) Select output folder for *.txt file(s)
2. The output folder now includes the converted .txt file(s)

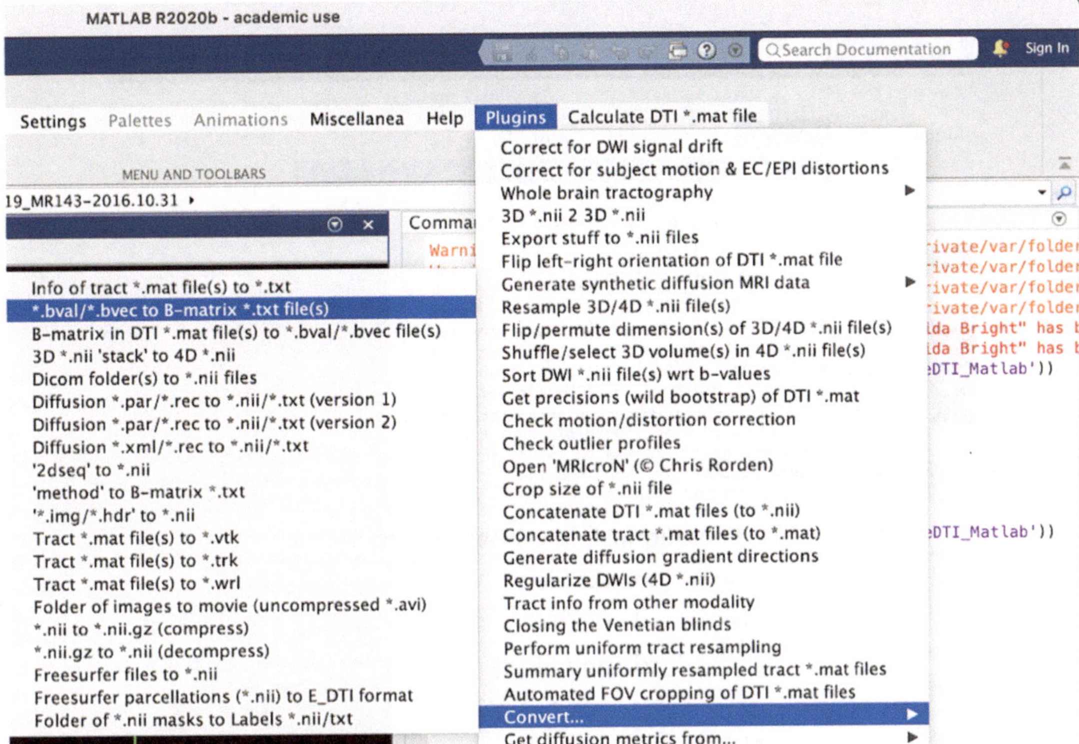

Fig. 1 Step 1 using the ExploreDTI GUI

2.3.2 Signal Drift Correction (Step 2)

Signal drift is a phenomenon caused by scanner imperfections, which leads to an adverse alteration of the acquired signal and a bias in the estimation of diffusion measures if not corrected [30].

Note
We recommend the use of "quadratic fit," but a signal drift fitting guide is provided in the Appendix 5.1 for users who want to investigate the impact of different fitting approaches (*see* Fig. A1).

In ExploreDTI:

1. Plugins → Correct for DWI signal drift (*see* Fig. 2).
   (a) Single or multiple data sets: multiple.
   (b) Select the folder of .nii file(s).
   (c) Select output folder.

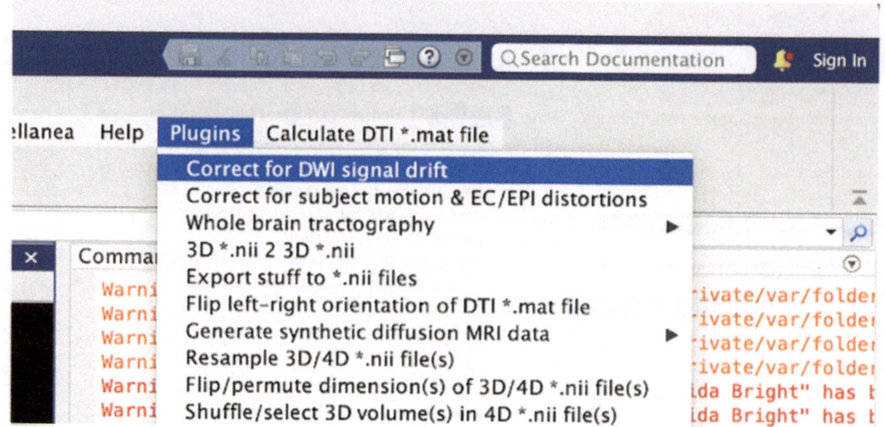

Fig. 2 Step 2 using the ExploreDTI GUI

2. The output folder now includes:
   - (a) *_sdc.txt file(s)
   - (b) *_sdc.nii file(s)
   - (c) *_sdc.png file(s)

Note

As the signal drift correction uses the non-diffusion weighted (b-0) files acquired to correct for signal drift, it is crucial that signal drift correction is completed before sorting b values, which may change the order of the acquired dMRI volumes (**step 3**).

2.3.3 Sort B-Values, Organize and Remove Excess b-0 Files (Step 3)

For the remaining processing steps ExploreDTI requires that all b-0 files are sorted to the beginning of the diffusion files. This step quickly organizes the files to have all the b-0 files at the beginning of the *.nii* and *.txt* files.

In ExploreDTI:

1. Plugins → Sort DWI *.nii file(s) wrt b-values (*see* Fig. 3)
   - (a) File name suffix: *_sorted.nii.
   - (b) Single or multiple data sets: multiple.
   - (c) Select the folder of *.nii file(s) and *.txt file(s).
   - (d) Select output folder.
2. The output folder now includes:
   - (a) *_sdc_sorted.txt file(s)
   - (b) *_sdc_sorted.nii file(s)

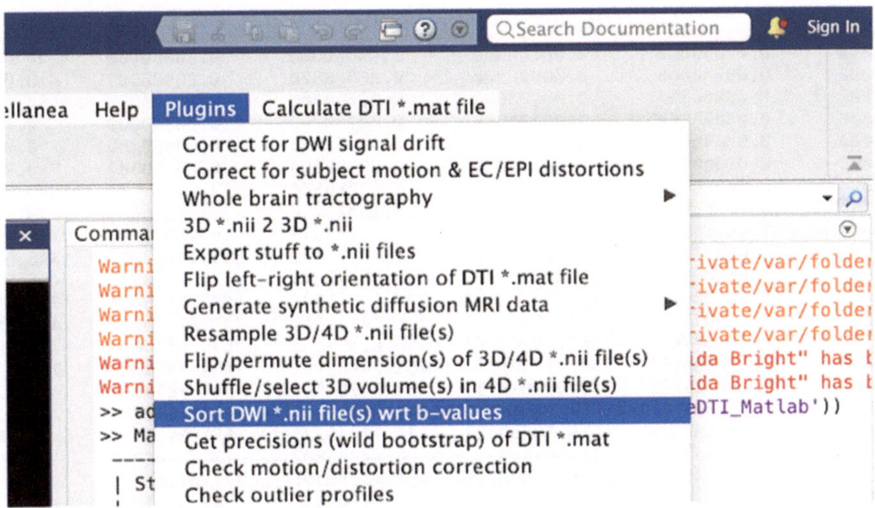

Fig. 3 Step 3 using the ExploreDTI GUI

> **Note**
> Rarely, additional b-0 files are collected during scanning. To investigate if extra b-0 files are present, open the newly sorted .txt file and investigate (*see* Fig. 4). If excess b-0 files are present (see red box in Fig. 4), these can be removed in ExploreDTI.

2.3.4 Gibbs Ringing Correction (Step 4)

A phenomenon known as Gibbs ringing may occur due to the shortening/truncation of Fourier transforms to reconstruct the MRI signal. If uncorrected, Gibbs ringing leads to artifacts that appear as multiple fine parallel lines in the image.

In ExploreDTI:

1. *Plugins → TV for Gibbs ringing in non-DWI's (4D *.nii)* (*see* Fig. 5)
   (a) *Select Gibbs Ringing Correction settings* (*see* Table 3).
   (b) *Single or multiple data sets:* Multiple.
   (c) Select the folder of *.nii files.
   (d) Select output folder.
2. The output folder now includes:
   (a) *_sdc_sorted_GR_TV.nii files.
3. Create and move *_sdc_sorted_GR_TV.nii files into a new folder.

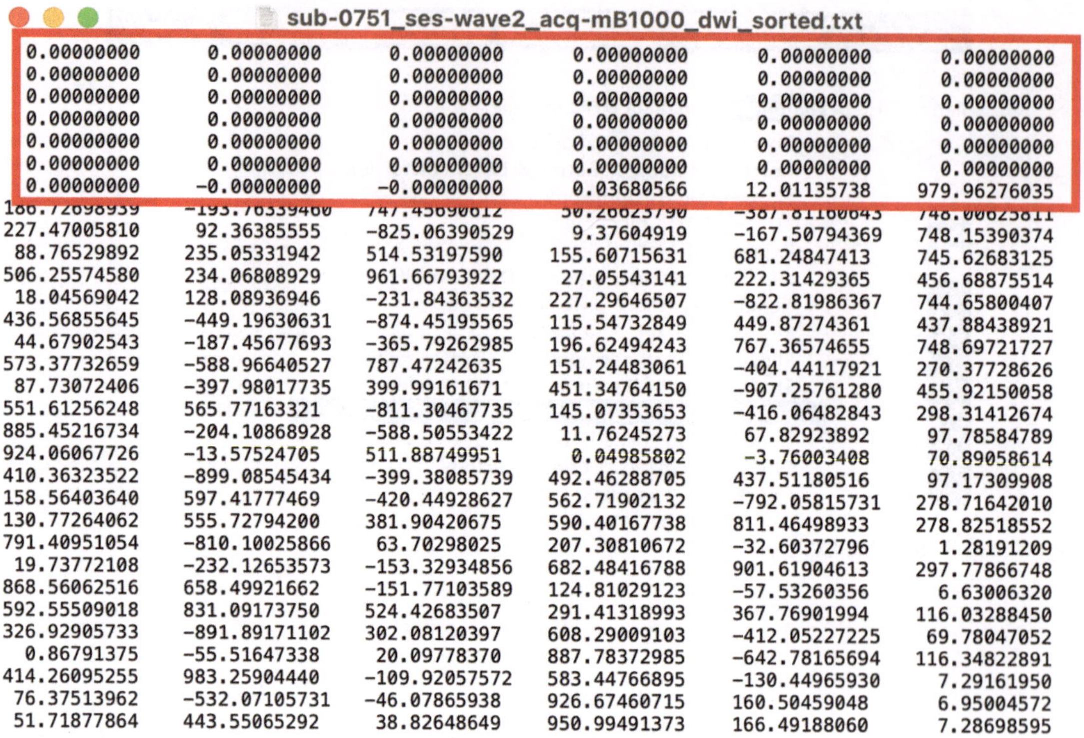

sub-0751_ses-wave2_acq-mB1000_dwi_sorted.txt

0.00000000	0.00000000	0.00000000	0.00000000	0.00000000	0.00000000
0.00000000	0.00000000	0.00000000	0.00000000	0.00000000	0.00000000
0.00000000	0.00000000	0.00000000	0.00000000	0.00000000	0.00000000
0.00000000	0.00000000	0.00000000	0.00000000	0.00000000	0.00000000
0.00000000	0.00000000	0.00000000	0.00000000	0.00000000	0.00000000
0.00000000	0.00000000	0.00000000	0.00000000	0.00000000	0.00000000
0.00000000	-0.00000000	-0.00000000	0.03680566	12.01135738	979.96276035
186.72698939	-193.76339460	747.45690612	50.26623790	-387.81160643	748.00625811
227.47005810	92.36385555	-825.06390529	9.37604919	-167.50794369	748.15390374
88.76529892	235.05331942	514.53197590	155.60715631	681.24847413	745.62683125
506.25574580	234.06808929	961.66793922	27.05543141	222.31429365	456.68875514
18.04569042	128.08936946	-231.84363532	227.29646507	-822.81986367	744.65800407
436.56855645	-449.19630631	-874.45195565	115.54732849	449.87274361	437.88438921
44.67902543	-187.45677693	-365.79262985	196.62494243	767.36574655	748.69721727
573.37732659	-588.96640527	787.47242635	151.24483061	-404.44117921	270.37728626
87.73072406	-397.98017735	399.99161671	451.34764150	-907.25761280	455.92150058
551.61256248	565.77163321	-811.30467735	145.07353653	-416.06482843	298.31412674
885.45216734	-204.10868928	-588.50553422	11.76245273	67.82923892	97.78584789
924.06067726	-13.57524705	511.88749951	0.04985802	-3.76003408	70.89058614
410.36323522	-899.08545434	-399.38085739	492.46288705	437.51180516	97.17309908
158.56403640	597.41777469	-420.44928627	562.71902132	-792.05815731	278.71642010
130.77264062	555.72794200	381.90420675	590.40167738	811.46498933	278.82518552
791.40951054	-810.10025866	63.70298025	207.30810672	-32.60372796	1.28191209
19.73772108	-232.12653573	-153.32934856	682.48416788	901.61904613	297.77866748
868.56062516	658.49921662	-151.77103589	124.81029123	-57.53260356	6.63006320
592.55509018	831.09173750	524.42683507	291.41318993	367.76901994	116.03288450
326.92905733	-891.89171102	302.08120397	608.29009103	-412.05227225	69.78047052
0.86791375	-55.51647338	20.09778370	887.78372985	-642.78165694	116.34822891
414.26095255	983.25904440	-109.92057572	583.44766895	-130.44965930	7.29161950
76.37513962	-532.07105731	-46.07865938	926.67460715	160.50459048	6.95004572
51.71877864	443.55065292	38.82648649	950.99491373	166.49188060	7.28698595

Fig. 4 Example of a sorted *.txt* file containing 6 b-0 images

2.3.5 Flip Permute (Step 5)

Permutations and flips in spatial configuration and/or mismatches between spatial and diffusion coordinate systems can accidentally occur during processing and analyses across different software packages, potentially resulting in errors. The "flip/permute" tool in ExploreDTI can reorientate images and also avoid further unexpected axis flips and permutations in any following image processing step. Use default ExploreDTI settings as orientations will be inspected at the next step.

In ExploreDTI:

1. Plugins → Flip/permute dimension(s) of 3D/4D *.nii files (*see* Fig. 6)
   - (a) Use default setting:

     File name suffix: _FP

     Permute dimensions: 1 2 3

     Flip dimensions: 0 0 0.

     Force voxel size: leave empty
   - (b) Single or multiple data sets: multiple.
   - (c) Select the folder of *.nii file(s).
   - (d) Select output folder.

The output folder now includes *_sdc_sorted_GR_TV_FP.nii files.

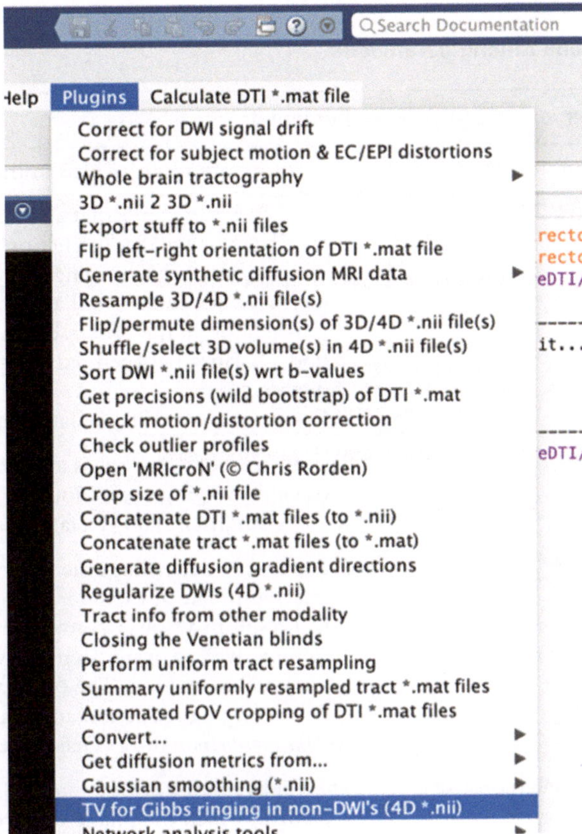

Fig. 5 Step 4 using the ExploreDTI GUI

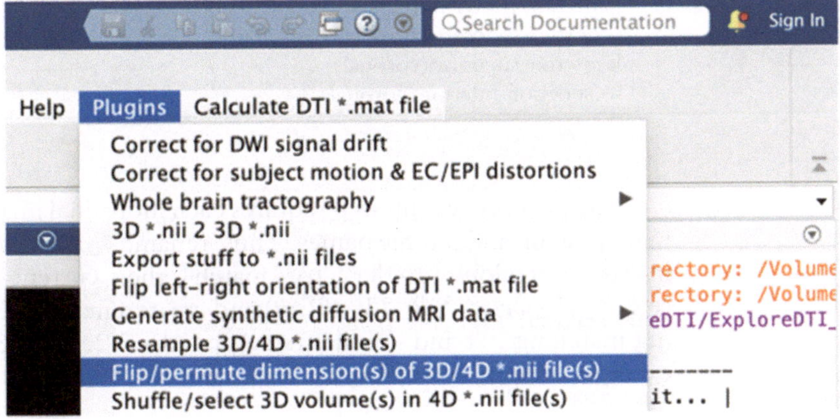

Fig. 6 Step 5 using the ExploreDTI GUI

Table 3
Setting Gibbs ringing parameters

Parameter	Comment
Number of non-DWIs	This information is provided in the *.txt file
Lambda ([1 200])	Lambda is a parameter that can be used to control the degree of Gibbs ringing in image reconstruction algorithms. A higher value of lambda will suppress Gibbs ringing more. However, it should be noted that a high value of lambda will also reduce the level of high-frequency information in the processed image, and therefore it is important to find a balance between reducing Gibbs ringing and preserving image quality *Recommendation*: 100 (Default setting)
Number of iterations ([1200])	The number of iterations is another parameter that controls the degree of Gibbs ringing correction *Recommendation*: 100 (Default setting)
Step size ([0.001–0.1])	The step size determines the magnitude of the update applied to the image estimate at each iteration of the algorithm. A smaller step size will result in a slower convergence of the algorithm and less Gibbs ringing, while a larger step size will result in a faster convergence but more Gibbs ringing. The optimal value is a desired trade-off between Gibbs ringing reduction and computational time *Recommendation*: 0.01 (Default setting)
Imaging plane (coronal:1, sagittal:2, axial:3)	The Gibbs ringing correction algorithm takes into account the imaging plane in which the image was acquired. This information is found in the subject specific *.json file, Phase Encoding Directions **i** left-right (sagittal) **i-** right-left (sagittal) **j** anterior–posterior (axial) **j-** posterior–anterior (axial) **k** inferior-superior (coronal) **k-** superior–inferior (coronal)

The next processing step requires each individual image to have matching .nii and .txt file names. Thus, rename *_sdc_sorted.txt* files in the previous folder to match the current .nii file (e.g. *_sdc_sorted_GR_TV_FP.txt*) and create a new folder containing matching .txt and .nii files (*see* Appendix 5.2, **steps 3–7** for an example of how to batch rename files).

2.3.6 Generate .mat File (Step 6)

It is required to generate a .mat file from the processed .nii and corresponding .txt files before tractography or other analysis tools can be applied. The DTI .mat file is a MATLAB format file and can be loaded into ExploreDTI for further processing and analysis.

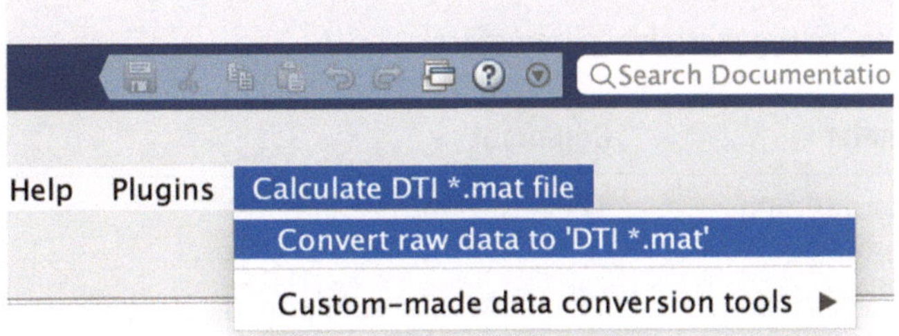

Fig. 7 Step 6 using the ExploreDTI GUI

In ExploreDTI:

1. Calculate DTI *.mat file → Convert raw data to 'DTI *.mat' (*see* Fig. 7)
   (a) Select settings (*see* Table 4).
   (b) Select the folder of *.nii.
   (c) Select folder .txt files: Press cancel if each .nii has its associated .txt file.
   (d) Select output folder.

The output folder now includes *_sdc_sorted_GR_TV_FP.mat files.

Note
A common pitfall of dMRI processing is orientation issues. During the .mat generation step you should investigate the flip/permutations to ensure appropriate orientations were selected. It is advised that you first use the default settings as ExploreDTI is able to automatically provide the correct orientation settings [31]. ExploreDTI deploys the widely used color convention to ensure the orientations ("Permute gradient components") are correct (left-right: Red, top-bottom: Blue, and front-back: Green). Good tracts to investigate when checking orientations are the corpus callosum—a white matter tract that is orientated left-right (Red) and the corticospinal tracts—white matter tracts that are orientated top-bottom (Blue). To see an example of orientation checks *see* Figs. 8 and 9.

Before-and-after correct flipping of the "Permute gradient components." As you can see in Fig. 8a, while the corticospinal tract is the correct orientation (blue arrow) the corpus callosum (red arrow)—a white matter tract that is orientated left-right—is

Table 4
Selecting .mat generation parameters

Parameter	Comment
Format diffusion weighted data	4D Nifti (*.nii)
Permute spatial dimensions	This allows you to flip spatial dimensions of image. *Recommendation*: Use default settings (AP RL IS) if there are no issues with spatial dimensions
Flip spatial orientations	This step allows you flip the direction of the dimensions. This is important if your data was collected in Neurological dimensions rather than Radiological conventions[*see* Glossary]. In this instance, you may need to flip dimensions from Right—Left to Left—Right. To flip, change the parameter from "AP RL IS" to "AP LR IS." If your data were collected in radiological dimensions the default setting of "AP RL IS" should be appropriate
Perform visual data check	This allows you to quickly visualize the orientation of the image
Diffusion tensor estimation	The robust tensor estimation algorithms aim to minimize the impact of outliers on the final diffusion tensor estimate, leading to more reliable results *Recommendation*: Robust (exclude outliers)
Format diffusion information	Text file (*.txt)
Background masking approach	Automatic
Permute gradient components	Permute gradient components should correspond to data and may require some investigation (*see* Note below this table)
Flip sign of gradient components	The sign of gradient components should correspond to data and may require some investigation (*see* Note below this table)
Data processing mode	Single or multiple data sets
b-value in units s/mm^2	E.g., 1000
Voxel size [AP RL IS] (in mm)	E.g., 2 2 2
Number of non-DW images	E.g., 3
Number of DW images	E.g., 30
Matrix size [AP RL IS]	E.g., 128 128 60

green. This indicates that the x and y axis need to be flipped. To do so, change the "Permute gradient components" from $x\ y\ z \rightarrow y\ x\ z$ and generate a new correctly orientated .mat file (Fig. 8b).

Figure 9 illustrates an investigation into the flip sign gradients. While the orientations of the images of both images are correct, the gradient sign may be flipped. Use Glyphs [*see* Glossary]

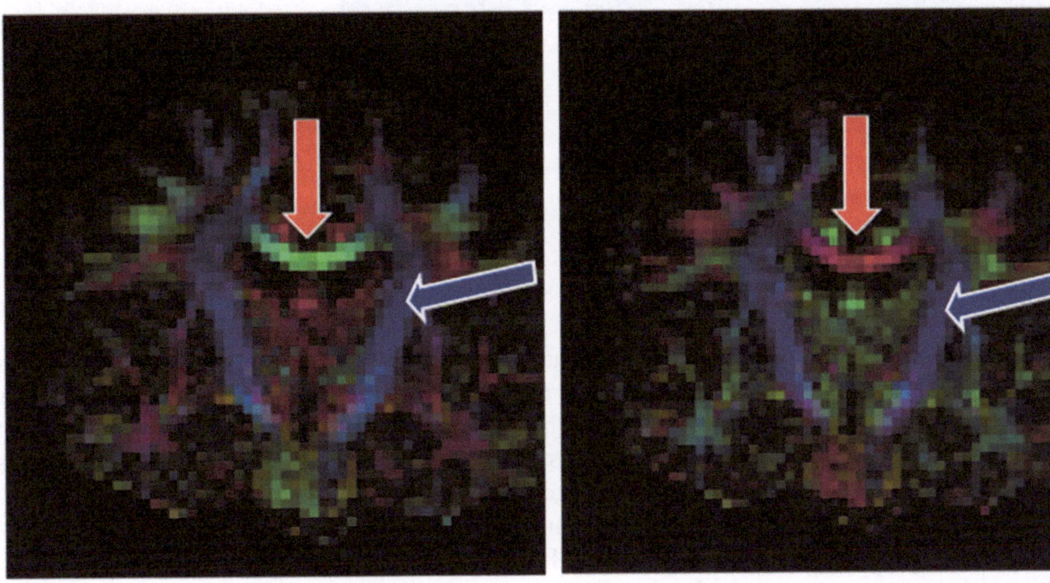

Fig. 8 Checking Permute Gradient Components

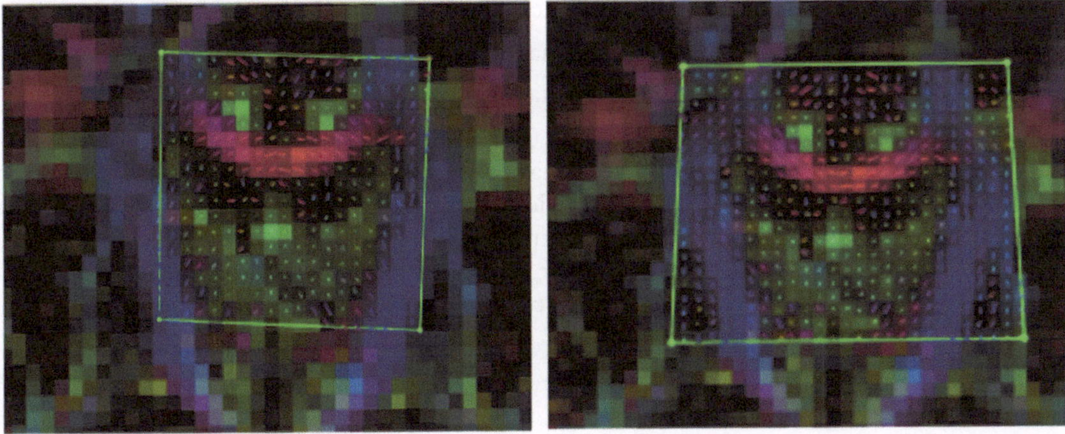

Fig. 9 Checking Flip Sign Components

(in ExploreDTI: *Draw ROI* → *Draw Glyphs*) to inspect the signs and investigate the "Flip sign of gradient components." Figure 9a shows an incorrect flip sign gradient as the glyphs are not following the curvature of the Corpus Callosum. To fix this, the z component must be flipped. To do so, change the "flip sign gradients" from *x y z* → *x y -z* and generate the correct .mat file (Fig. 9b).

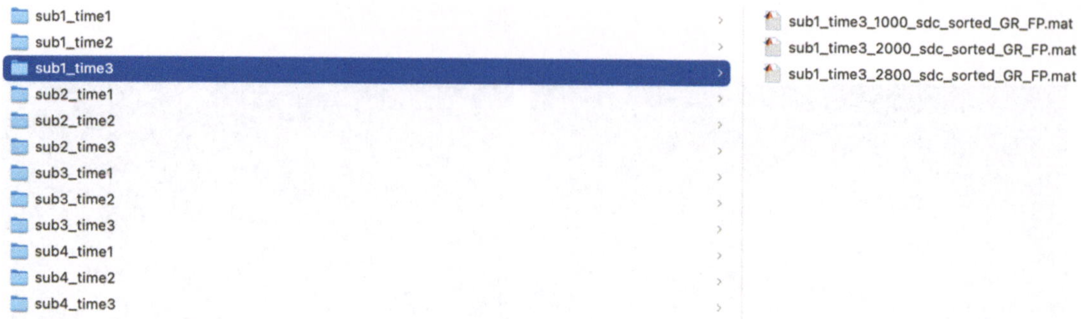

sub1_time1
sub1_time2
sub1_time3
sub2_time1
sub2_time2
sub2_time3
sub3_time1
sub3_time2
sub3_time3
sub4_time1
sub4_time2
sub4_time3

sub1_time3_1000_sdc_sorted_GR_FP.mat
sub1_time3_2000_sdc_sorted_GR_FP.mat
sub1_time3_2800_sdc_sorted_GR_FP.mat

Fig. 10 b-value folders

2.3.7 Concatenate All b-Value .mat Files (Step 7)

This step concatenates all the single b-values (shells) .mat files together to create a multi-shell .nii file. This enables a major benefit of multi-shell imaging; namely, leveraging the increased signal of high b-value images with the reduced noise of low b-value images to produce an image with increased anatomical accuracy.

Firstly, you should organize all your b-value .mat files into scan-specific folders (*see* Fig. 10).

In ExploreDTI:

1. Plugins → Concatenate DTI *.mat files (to *.nii) (*see* Fig. 11)
   (a) Select folder of folders: select the folder containing all the scan-specific folders.
2. The output folder now includes:
   (a) *_concatenated.txt file(s).
   (b) *_concatenated.nii file(s).

2.3.8 Generate Concatenated .mat File (Step 8)

It is now required that you convert the concatenated .nii files into . mat files. As any orientation issues should have been resolved at **step 5** (*see* Subheading 2.3.5), default orientation settings will be used. For processing efficiency, it is advised that you move *_concatenated.nii and *_concatenated.txt files into a folder.

In ExploreDTI:

1. Calculate DTI *.mat file → Convert raw data to 'DTI *.mat' (*see* Fig. 7)
   (a) Select settings (*see* Table 5).
   (b) Select the folder of *_concatenated.nii files.
   (c) Select folder *_concatenated.txt files: Press cancel.
   (d) Select output folder.
2. The output folder now includes:
   (a) *_concatenated.mat file(s).

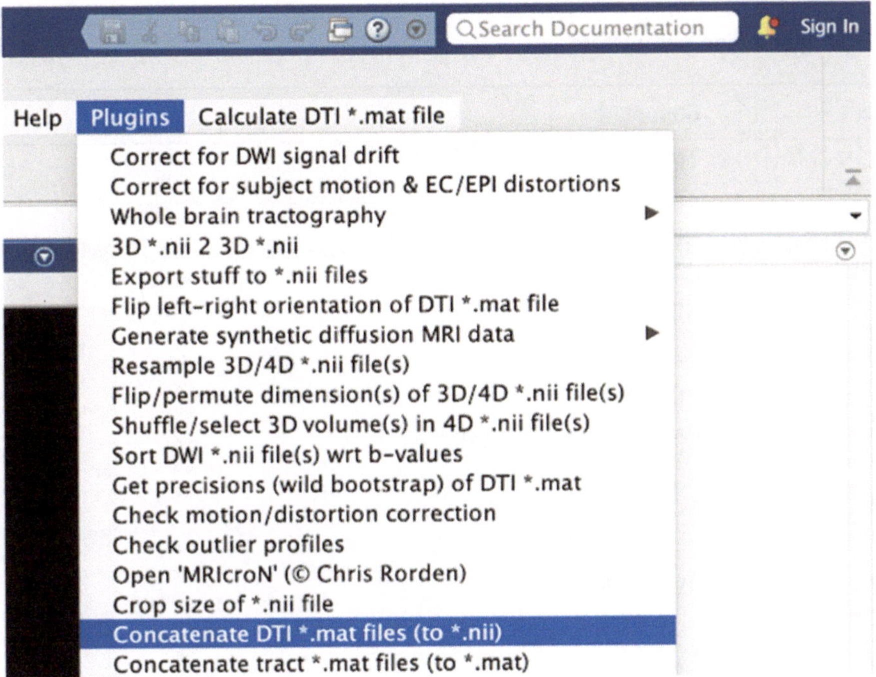

Fig. 11 Step 7 using the ExploreDTI GUI

2.3.9 Correcting Subject Motion, Eddy Currents, and EPI-Induced Distortions (Step 9)

This step corrects subject motion (SM), eddy currents (EC), and EPI-induced distortions (EPI). This is a crucial processing step, as such distortions can lead to significant changes in diffusion metric estimates. Additionally, you can use "undistorted" structural MRI (T1 or T2) images to unwrap the deformations in the diffusion data (For more details see ExploreDTI manual). If you do not have a structural MRI, this step can be conducted in native space. To process this step in ExploreDTI without a structural MRI file ensure the following setting is selected, *Settings → SM/EC/EPI correction → Also register to other data → No thanks (stay in native space)*. Before beginning this step move *_concatenated.mat files (and *_nu.nii and *_mask.nii files if necessary) to a folder.

Note

Due to the large computational demand of **step 9**, it is recommended to use multi-core computing support for this tool with a minimum of 32 GB RAM for dMRI data if you have more than 100 DW images.

Table 5
Selecting .mat generation parameters

Parameter	Comment
Format diffusion weighted data	4D Nifti (*.nii)
Permute spatial dimensions	AP RL IS
Flip spatial orientations	AP RL IS
Perform visual data check	No.
Diffusion tensor estimation	Robust (exclude outliers)
Format diffusion information	Text file (*.txt)
Background masking approach	Automatic
Permute gradient components	x y z
Flip sign of gradient components	x y z
Data processing mode	Multiple data sets
b-value in units s/mm^2	NaN (for multi-shell data) or any integer for DTI
Voxel size [AP RL IS] (in mm)	E.g., 2 2 2
Number of non-DW images	The total number of b-0 images in the concatenated images (using the parameters in NICAP study, e.g., number of non-DW images = 16)
Number of DW images	The total number of b-value images in the concatenated images (using the parameters in NICAP study, e.g., number of DW images = 130)
Matrix size [AP RL IS]	E.g., 128 128 60

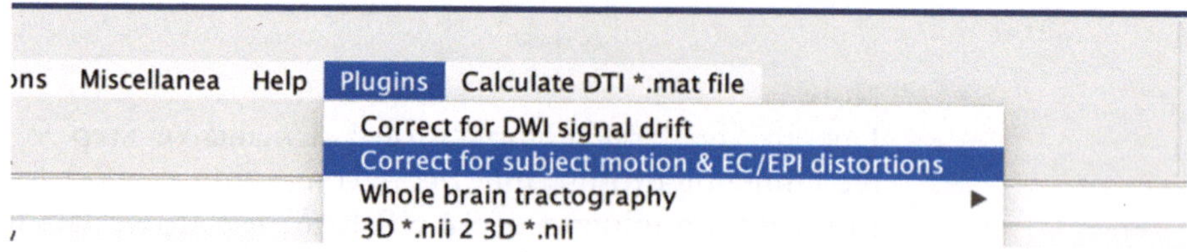

Fig. 12 Step 9 using the ExploreDTI GUI

Table 6
Selecting SM/EC/EPI distortion parameters

Parameter	Comment
Settings → SM/EC/EPI correction → masking stuff	This setting allows you to use a mask generated from a structural MRI scan. If you do not have a structural MRI mask, do not select the "masking stuff" setting
Settings → SM/EC/EPI correction → also register to other data → yes, to do EPI correction (non-rigid)	This setting allows you to register your diffusion image to a structural MRI image during the EPI correction, enabling increased distortion correction. *Recommendation*: select "orig_nu" from Freesurfer processed structural MRI files
Settings → SM/EC/EPI correction → also register to other data → registration details → Deformation axes	By default, the non-linear deformations are allowed along any orientation. *Recommendation*: correction will likely improve if the registration is constrained to model deformations only along the phase encoding direction. To do this (example A-P orientation), change "Deformation axes" to [1 0 0]
Settings → SM/EC/EPI correction → registration details for SM/EC corrections → interpolation method	*Recommendation*: Linear or cubic spline

In ExploreDTI:

1. Select settings (*see* Table 6).
2. Start MATLAB parallel pooling.
3. Plugins → Correct for subject motion & EC/EP distortions (*see* Fig. 12)
   (a) Single or multiple data sets: Multiple.
   (b) Select the folder of *_concatenated.mat files.
   Include *_nu.nii and *_mask.nii for using structural MRI files for registration.
   (c) Select output folder.

The output folder now includes *_concatenated_trafo.mat files.

2.3.10 Whole Brain Tractography (Step 10)

Whole brain tractography in ExploreDTI generates white matter tracts using a deterministic approach. Other software packages, such as FSL and MRtrix are available if you would like to do probabilistic tractography. It is recommended that you complete whole brain tractography before reconstructing specific white matter tracts for analysis.

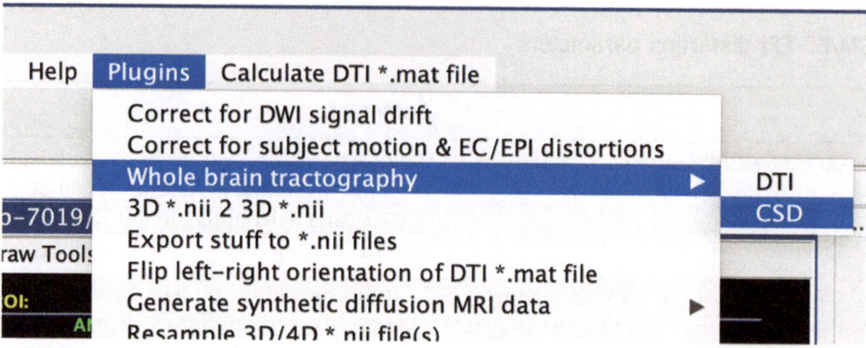

Fig. 13 Step 10 using the ExploreDTI GUI

Table 7
Whole brain tractography parameters

Parameter	Comment
Seedpoint resolution (mm)	Seed point resolution is a measure of how close together the seed points are placed in the brain. A higher seed point resolution will result in a higher number of seed points used in the tractography algorithm, and therefore a higher number of the reconstructed tracts. However, this also increases the computation time. *Recommendation:* 2 2 2
Step size (mm)	The step size is a parameter that determines the distance between each point in the reconstructed tracts. A smaller step size will result in a higher accuracy of the reconstructed tracts, but it will also increase the computation time. *Recommendation*: 1
Angle threshold	The angle threshold is a parameter that controls the angular deviation of consecutive steps during pathway reconstruction. A higher angular threshold will result in more or longer tracts, but it will also increase the risk of false positive tracts. If you are planning to exact tracts with high curvature (such as the fornix) it is advised to set this threshold higher (e.g., 60°)
Fiber length range	This step allows you to set the upper and lower bound of the length of the reconstructed fibers. Change this setting if you are investigating particularly long or short white matter tracts. If you are investigating both long and short fiber, it is recommended to set this setting to, 10–500
Random permutations of seed points	*0 = no/1 = yes (setting to get rid of rectilinear grid-pattern artifacts)* *Recommendation*: 0

In ExploreDTI:

1. *Plugins → whole brain tractography → CSD* (*see* Fig. 13)
   (a) *Select settings (see* Table 7*).*
   (b) *Single or multiple data sets:* multiple.
   (c) Select the folder of *_trafo.mat files.

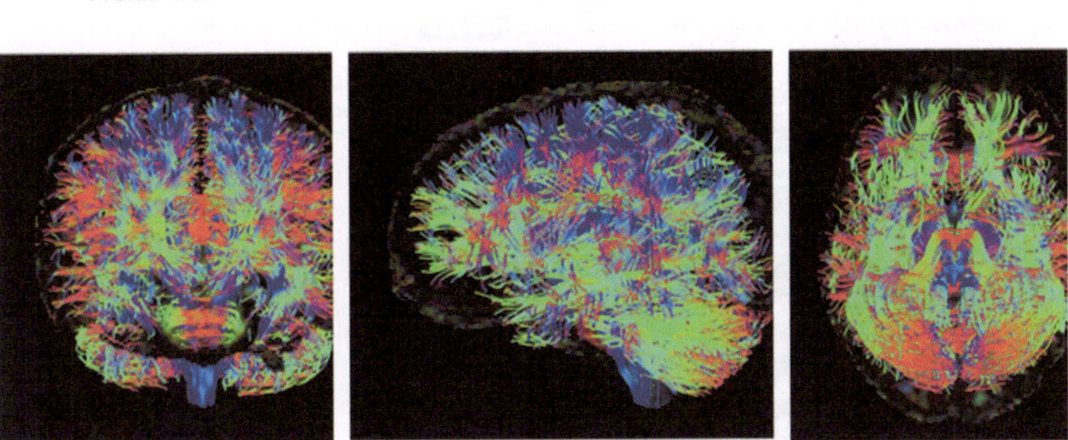

Fig. 14 Complete CSD tractography (subsampled: 50)

(d) Select output folder.

(e) The output folder now includes the *_trafo_Tracts_CSD.mat files (*see* Fig. 14).

2.3.11 Extracting Diffusion MRI Metrics (Step 11)

At this step, you should have already extracted the white matter tracts you want to analyze. A step-by-step guide of conducting manual tractography is provided in the ExploreDTI manual. This step allows you to export diffusion metrics for the analyzed tract pathway of interest (see Supplementary Material Table 2). If you wish to also obtain kurtosis measures, please see Appendix 5.3.

In ExploreDTI:

1. *Plugins → convert → info. of tract *.mat file(s) to .txt.* (*see* Fig. 15)
   (a) Select the folder of .nii.
   (b) Select output folder.
2. The output folder now includes:
   (a) *.txt files.
3. Export *.txt file to Excel.

3 Conclusion

This chapter offers a comprehensive resource that equips researchers with the necessary skills and knowledge to effectively and efficiently process large diffusion MRI data sets. By providing practical, step-by-step guides, researchers can process both DTI and multi-shell HARDI data using the ExploreDTI software. When choosing a diffusion MRI modeling technique, it is important to consider the

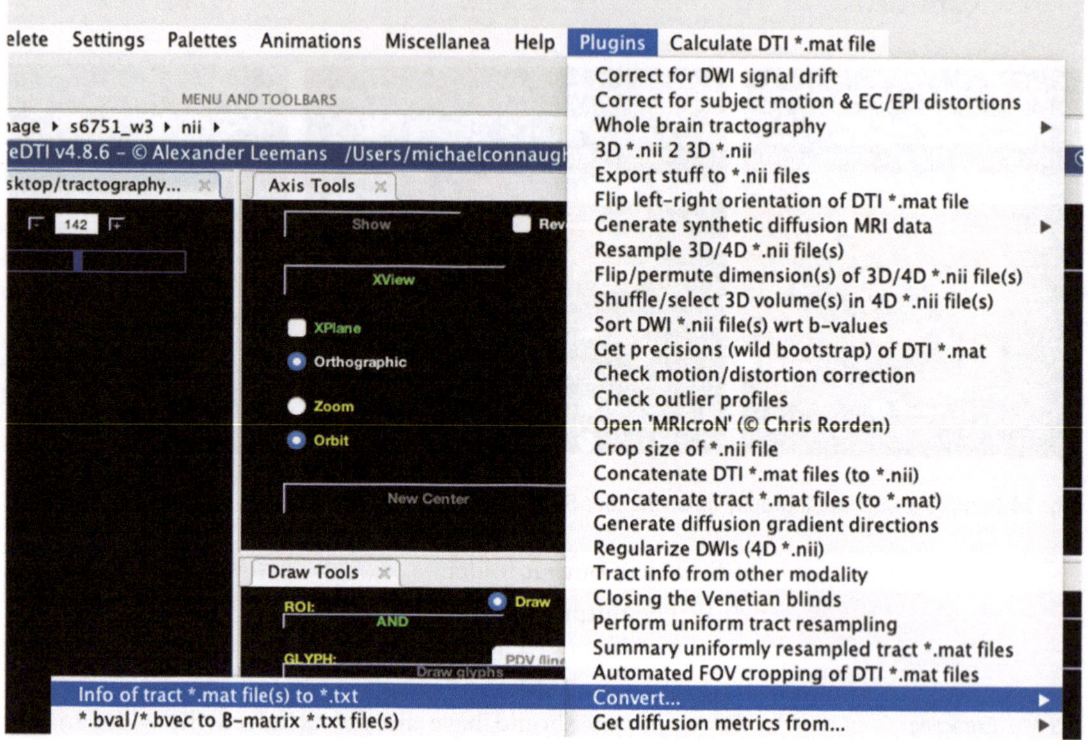

Fig. 15 Step 11 using the ExploreDTI GUI

pros and cons of both DTI and multi-shell HARDI approaches. Multi-shell HARDI offers several advantages over DTI, including increased anatomical accuracy and the ability to model crossing fibers. However, it also comes with certain disadvantages compared to DTI that warrant careful evaluation. A significant drawback of multi-shell HARDI is that it is a highly computationally expensive technique, which results in significantly longer processing times compared to DTI. Nevertheless, steps can be taken to reduce processing time through resource optimization. Researchers can optimize the utilization of computational resources by fine-tuning the processing parameters described in this book chapter. When processing large data sets, it is advised to experiment with different settings parameters to find the optimal balance between reconstruction accuracy and processing time. Overall, it is crucial to assess the accessible resources, both available time and computational resources, before deciding which diffusion MRI modeling technique to employ.

Appendix

4.1 Checking the Impact of Different Signal Drift Fit Approaches

The primary parameter to be considered during this step is the signal drift fit approach (Linear, Quadratic, or Cubic). The ExploreDTI default setting is "Quadratic" however, all three approaches should be investigated to find the best fit for your data set. The signal drift fit approach can be changed by in ExploreDTI, *Settings* → *Signal drift correction* → *Fit Approach* (1. Linear, 2. Quadratic, or 3. Cubic). Initially it is advised to run all three approaches (using the instructions below). After you have run all three different signal drift approaches, the *.png* files you be used to evaluate their performance. The approach that is most appropriate for your data set is the one with the least amount of signal loss (see Appendix Figure).

4.2 Data Storage and Computational Expense: Open Neuro Diffusion Data

The code provided here can be run on a single computer using MATLAB. The code was written using MATLAB version R2020a. While earlier versions of MATLAB may also execute the code successfully on a MacBook Pro with an Intel Core i7 processor and 16 GB RAM, it takes 10 min to execute the code for 10 participants with MATLAB parallel processing enabled (*see* Table 1).

4.2.1 Step-by-step Guide of DTI Processing the Open Neuro Data

Step 1: *Convert Bval and Bvec Files into Text Files*

```
d = pwd;


dic_folder =

'/Users/michaelconnaughton/Desktop/Neuroimaging_Book_Chapter/OpenN

euro/ds002790-master/dwi_files/';


if ~ischar(dic_folder)

    return;

end
```

```
files = E_DTI_Get_files_from_folder(dic_folder, '.bval');


if isempty(files)

    uiwait(my_msgbox('No *.bval files found...','Converting
*.bval/*.bvec file(s) B-matrix *.txt file(s)','modal'));

    return;

end
```

Step 3–7: *Signal Drift, Gibbs Ringing, Flip Permute, and Generate Mat File*

```
h_w = my_waitbar(0,'Converting *.bval/*.bvec file(s) B-matrix
*.txt file(s)');pause(0.01)


for i=1:length(files)


    E_DTI_convert_nii_dic_2_txt_exe(files{i});


    my_waitbar(i/length(files))

end


close(h_w);
```

```
d = pwd;


dic_folder =

'/Users/michaelconnaughton/Desktop/Neuroimaging_Book_Chapter/OpenN

euro/ds002790-master/dwi_files/';


if ~ischar(dic_folder)

    return;

end


files = E_DTI_Get_files_from_folder(dic_folder, '.bval');


if isempty(files)

    uiwait(my_msgbox('No *.bval files found...','Converting

*.bval/*.bvec file(s) B-matrix *.txt file(s)','modal'));

    return;

end


h_w = my_waitbar(0,'Converting *.bval/*.bvec file(s) B-matrix

*.txt file(s)');pause(0.01)
```

```
for i=1:length(files)

    E_DTI_convert_nii_dic_2_txt_exe(files{i});

    my_waitbar(i/length(files))
end

d = pwd;
folder_in =
'/Users/michaelconnaughton/Desktop/Neuroimaging_Book_Chapter/OpenN
euro/ds002790-master/dwi_files/';
if ~ischar(folder_in)
    return;
end

files_in = E_DTI_Get_files_from_folder(folder_in, '.nii');
if isempty(files_in)
    uiwait(my_msgbox('No DWI nii files found...','Signal drift
correction','modal'));
    return;
end
```

```
folder_out_nii = [folder_in filesep 'sdc'];
if ~isdir(folder_out_nii),mkdir(folder_out_nii);end
folder_out_txt = [folder_in filesep 'sdc'];
if ~isdir(folder_out_txt),mkdir(folder_out_txt);end
h_w = my_waitbar(0,'Processing');pause(0.01)
for i=1:length(files_in)
    [~,n,~] = fileparts(files_in{i});
    par.f_in_nii = files_in{i};
    par.f_in_txt = [folder_in filesep n '.txt'];
    par.f_out_nii = [folder_out_nii filesep n '_sdc.nii'];
    par.f_out_txt = [folder_out_txt filesep n '_sdc_GR_FP.txt'];
    par.bvalC = 1000;
    par.bv_thresh = 10;
    par.method = 2;
    par.masking.do_it = 0;
    par.masking.p1 = 5;
    par.masking.p2 = 1;
    par.show_summ_plot = 1;
    E_DTI_signal_drift_correction(par);
    my_waitbar(i/length(files_in))
```

```
    % Apply Gibbs ringing correction
    f_in = [folder_out_nii filesep n '_sdc.nii'];
    f_out = [folder_out_nii filesep n '_sdc_GR.nii'];
    p.NrB0 = 1;
    p.lambda = 100;
    p.iter = 100;
    p.ss = 0.01;
    p.ip = 3;
    E_DTI_Gibbs_Ringing_removal_with_TV_exe(f_in,f_out,p);

    % Apply flip/permute correction
    File_name_of_DWIs_input = [folder_out_nii filesep n
'_sdc_GR.nii'];
    File_name_of_permuted_flipped_DWIs = [folder_out_nii filesep n
'_sdc_GR_FP.nii'];
    p = [];
    p.suff = '_FP';
    p.permute = [1 2 3];
    p.flip = [0 0 0];
    p.force_voxel_size = [];
    E_DTI_flip_permute_nii_file_exe(File_name_of_DWIs_input, p,
File_name_of_permuted_flipped_DWIs);
    % Generate Mat File
    f_DWI= [folder_out_nii filesep n '_sdc_GR_FP.nii']; %file name
```

```
of the DWIs
    f_BM= [folder_out_txt filesep n '_sdc_GR_FP.txt']; %name of
the B-matrix
    f_mat= [folder_out_nii filesep n '_sdc_GR_FP.mat']; %file name
of the DTI output
    Mask_par.tune_NDWI = 0.7; % (rough range: [0.3 1.5])
    Mask_par.tune_DWI = 0.7; % (rough range: [0.3 1.5])
    Mask_par.mfs = 5; % (uneven integer)
    NrB0= 1;
    perm = 2;
    flip = 4;
    E_DTI_quick_and_dirty_DTI_convert_from_nii_txt_to_mat(f_DWI,
f_BM, f_mat, Mask_par, NrB0, perm, flip)


end

my_waitbar(1);close(h_w);pause(0.01);
% As per the main manuscript, it is recommended to run a quality
check on the orientation of the mat files.
```

Step 7: *SM/EC/EPI Distortion*

```
% Download Parameters_SM_EC_EPI.txt on OpenNeuro
% Note: edit path folder_in and foulder_out of
Parameters_SM_EC_EPI.txt
% Name of the text file containing the parameters for the
SM/EC/EPI correction. The file is as saved in    the GUI using
Settings > SM/EC/EPI correction > Export parameter file.

parameter_filename =
'/Users/michaelconnaughton/Desktop/Neuroimaging_Book_Chapter/OpenN
euro/ds002790-master/dwi_files/sdc/Parameters_SM_EC_EPI.txt';

E_DTI_SMECEPI_Main(parameter_filename);
```

```
folder_in =
'/Users/michaelconnaughton/Desktop/Neuroimaging_Book_Chapter/OpenN
euro/ds002790-master/dwi_files/sdc/epi';
folder_out =
'/Users/michaelconnaughton/Desktop/Neuroimaging_Book_Chapter/OpenN
euro/ds002790-master/dwi_files/sdc/epi/wbt';
```

```
% Get a list of all files in the folder
file_list = dir(fullfile(folder_in, '*.mat'));

% Loop through each file and apply the code block
for i = 1:numel(file_list)
    n = file_list(i).name(1:end-10);
    filename_in = [folder_in filesep n 'native.mat'];
    filename_out = [folder_out filesep n '.mat'];

    parameters.SeedPointRes = [3 3 3];
    parameters.StepSize = 1;
    parameters.FAThresh = 0.2000;
    parameters.AngleThresh = 45;
    parameters.FiberLengthRange = [50 500];
    WholeBrainTrackingDTI_fast(filename_in, filename_out,
parameters);
end
```

4.3 Extracting Diffusion Kurtosis Metrics

```
% Define the path to the DTI and tract files
path_dMRI = 'path to diffusion file';
path_tract = 'path to tract file';
path_tract_new = 'Path to new output file';

% Get a list of all the dMRI files in the folder
dMRI_files = dir(fullfile(path_dti, '*_trafo.mat'));

% Loop through the dMRI files
for i = 1:length(dti_files)
    % Get the subject name from the file name
    subject = dti_files(i).name(1:end-8);

    % Define the input and output file names for the current
subject
    f_in_1 = [path_dMRI, dMRI_files(i).name];
    f_in_2 = [path_tract, subject,'_old_tract.mat'];
    f_out = [path_tract_new, subject,'_new_tract.mat'];

    % Call the script
    E_DTI_Add_DKI_metrics_to_tract_file(f_in_1, f_in_2, f_out);
end
```

References

1. Fields RD (2010) Neuroscience. Change in the brain's white matter. Science 330:768–769. https://doi.org/10.1126/science.1199139
2. Lebel C, Deoni S (2018) The development of brain white matter microstructure. NeuroImage 182:207–218. https://doi.org/10.1016/j.neuroimage.2017.12.097
3. Stiles J, Jernigan TL (2010) The basics of brain development. Neuropsychol Rev 20:327–348. https://doi.org/10.1007/s11065-010-9148-4
4. Lebel C, Beaulieu C (2011) Longitudinal development of human brain wiring continues from childhood into adulthood. J Neurosci 31:10937–10947. https://doi.org/10.1523/JNEUROSCI.5302-10.2011
5. Andrews DS, Lee JK, Harvey DJ, Waizbard-Bartov E, Solomon M, Rogers SJ, Nordahl CW, Amaral DG (2021) A longitudinal study of white matter development in relation to changes in autism severity across early childhood. Biol Psychiatry 89:424–432. https://doi.org/10.1016/j.biopsych.2020.10.013
6. Bouziane C, Caan MWA, Tamminga HGH, Schrantee A, Bottelier MA, de Ruiter MB, Kooij SJJ, Reneman L (2018) ADHD and maturation of brain white matter: a DTI study in medication naive children and adults. NeuroImage Clin 17:53–59. https://doi.org/10.1016/j.nicl.2017.09.026
7. Peters BD, Karlsgodt KH (2015) White matter development in the early stages of psychosis. Schizophr Res 161:61–69. https://doi.org/10.1016/j.schres.2014.05.021
8. Jones DK, Leemans A (2011) Diffusion tensor imaging. Methods Mol Biol 711:127–144. https://doi.org/10.1007/978-1-61737-992-5_6
9. Van Hecke W, Emsell L, Sunaert S (2016) Diffusion tensor imaging: a practical handbook. Springer, New York
10. Basser PJ, Mattiello J, LeBihan D (1994) MR diffusion tensor spectroscopy and imaging. Biophys J 66:259–267. https://doi.org/10.1016/S0006-3495(94)80775-1
11. Mori S, van Zijl PCM (2002) Fiber tracking: principles and strategies – a technical review. NMR Biomed 15:468–480. https://doi.org/10.1002/nbm.781
12. Qiu A, Mori S, Miller MI (2015) Diffusion tensor imaging for understanding brain development in early life. Annu Rev Psychol 66:853–876. https://doi.org/10.1146/annurev-psych-010814-015340
13. Goddings A-L, Roalf D, Lebel C, Tamnes CK (2021) Development of white matter microstructure and executive functions during childhood and adolescence: a review of diffusion MRI studies. Dev Cogn Neurosci 51:101008. https://doi.org/10.1016/j.dcn.2021.101008
14. Sexton CE, Walhovd KB, Storsve AB, Tamnes CK, Westlye LT, Johansen-Berg H, Fjell AM (2014) Accelerated changes in white matter microstructure during aging: a longitudinal diffusion tensor imaging study. J Neurosci 34:15425–15436. https://doi.org/10.1523/JNEUROSCI.0203-14.2014
15. Pierpaoli C, Barnett A, Pajevic S, Chen R, Penix LR, Virta A, Basser P (2001) Water diffusion changes in Wallerian degeneration and their dependence on white matter architecture. NeuroImage 13:1174–1185. https://doi.org/10.1006/nimg.2001.0765
16. Behrens TEJ, Berg HJ, Jbabdi S, Rushworth MFS, Woolrich MW (2007) Probabilistic diffusion tractography with multiple fibre orientations: what can we gain? NeuroImage 34:144–155. https://doi.org/10.1016/j.neuroimage.2006.09.018
17. Jeurissen B, Leemans A, Jones DK, Tournier J-D, Sijbers J (2011) Probabilistic fiber tracking using the residual bootstrap with constrained spherical deconvolution. Hum Brain Mapp 32:461–479. https://doi.org/10.1002/hbm.21032
18. Descoteaux M (2015) High angular resolution diffusion imaging (HARDI). In: Wiley encyclopedia of electrical and electronics engineering. Wiley, pp 1–25
19. Pines AR, Cieslak M, Larsen B, Baum GL, Cook PA, Adebimpe A, Dávila DG, Elliott MA, Jirsaraie R, Murtha K, Oathes DJ, Piiwaa K, Rosen AFG, Rush S, Shinohara RT, Bassett DS, Roalf DR, Satterthwaite TD (2020) Leveraging multi-shell diffusion for studies of brain development in youth and young adulthood. Dev Cogn Neurosci 43:100788. https://doi.org/10.1016/j.dcn.2020.100788
20. Jeurissen B, Tournier J-D, Dhollander T, Connelly A, Sijbers J (2014) Multi-tissue constrained spherical deconvolution for improved analysis of multi-shell diffusion MRI data. NeuroImage 103:411–426. https://doi.org/10.1016/j.neuroimage.2014.07.061
21. Burdette JH, Durden DD, Elster AD, Yen YF (2001) High b-value diffusion-weighted MRI of normal brain. J Comput Assist Tomogr 25:

515–519. https://doi.org/10.1097/00004728-200107000-00002

22. Kingsley PB, Monahan WG (2004) Selection of the optimum b factor for diffusion-weighted magnetic resonance imaging assessment of ischemic stroke. Magn Reson Med 51:996–1001. https://doi.org/10.1002/mrm.20059
23. Zhang H, Schneider T, Wheeler-Kingshott CA, Alexander DC (2012) NODDI: practical in vivo neurite orientation dispersion and density imaging of the human brain. NeuroImage 61:1000–1016. https://doi.org/10.1016/j.neuroimage.2012.03.072
24. Tuch DS (2004) Q-ball imaging. Magn Reson Med 52:1358–1372. https://doi.org/10.1002/mrm.20279
25. Tournier J-D, Calamante F, Gadian DG, Connelly A (2004) Direct estimation of the fiber orientation density function from diffusion-weighted MRI data using spherical deconvolution. NeuroImage 23:1176–1185. https://doi.org/10.1016/j.neuroimage.2004.07.037
26. Dhollander T, Clemente A, Singh M, Boonstra F, Civier O, Duque JD, Egorova N, Enticott P, Fuelscher I, Gajamange S, Genc S, Gottlieb E, Hyde C, Imms P, Kelly C, Kirkovski M, Kolbe S, Liang X, Malhotra A, Mito R, Poudel G, Silk TJ, Vaughan DN, Zanin J, Raffelt D, Caeyenberghs K (2021) Fixel-based analysis of diffusion MRI: methods, applications, Challenges and opportunities. NeuroImage 241:118417. https://doi.org/10.1016/j.neuroimage.2021.118417
27. Leemans A, Jeurissen B, Sijbers J, Jones DK (2009) ExploreDTI: a graphical toolbox for processing, analyzing, and visualizing diffusion MR data. Proc Intl Soc Mag Reson Med 17(1): 3537
28. Tournier J-D, Calamante F, Connelly A (2013) Determination of the appropriate b value and number of gradient directions for high-angular-resolution diffusion-weighted imaging. NMR Biomed 26:1775–1786. https://doi.org/10.1002/nbm.3017
29. Silk TJ, Genc S, Anderson V, Efron D, Hazell P, Nicholson JM, Kean M, Malpas CB, Sciberras E (2016) Developmental brain trajectories in children with ADHD and controls: a longitudinal neuroimaging study. BMC Psychiatry 16:59. https://doi.org/10.1186/s12888-016-0770-4
30. Vos SB, Tax CMW, Luijten PR, Ourselin S, Leemans A, Froeling M (2017) The importance of correcting for signal drift in diffusion MRI. Magn Reson Med 77:285–299. https://doi.org/10.1002/mrm.26124
31. Jeurissen B, Leemans A, Sijbers J (2014) Automated correction of improperly rotated diffusion gradient orientations in diffusion weighted MRI. Med Image Anal 18:953–962. https://doi.org/10.1016/j.media.2014.05.012

Chapter 11

A Pipeline for Large-Scale Assessments of Dementia EEG Connectivity Across Multicentric Settings

Agustín Sainz-Ballesteros, Jhony Alejandro Mejía Perez, Sebastian Moguilner, Agustín Ibáñez, and Pavel Prado

Abstract

Multicentric initiatives based on high-density electroencephalography (hd-EEG) are urgently needed for the classification and characterization of disease subtypes in diverse and low-resource settings. These initiatives are challenging, with sources of variability arising from differing data acquisition and harmonization methods, multiple preprocessing pipelines, and different theoretical modes and methods to compute source space/scalp functional connectivity. Our team developed a novel pipeline aimed at the harmonization of hd-EEG datasets and dementia classification. This pipeline handles data from recording to machine learning classification based on multi-metric measures of source space connectivity. A user interface is provided for those with limited background in MATLAB. Here, we present our pipeline and provide a detailed a comprehensive step-by-step example for analysts to review the five main stages of the pipeline: data preprocessing, normalization, source transformation, connectivity metrics, and dementia classification. This detailed step-by-step pipeline may improve the assessment of heterogenous, multicentric, and multi-method approaches to functional connectivity in aging and dementia.

Key words Electroencephalography, Harmonization, Connectivity, Multicentric studies, EEG-BIDS

1 Introduction

Biomarkers assessed with brain functional connectivity [see Glossary] can provide relevant information for disease subtyping and progression [1, 2]. In addition to the traditional magnetic resonance images (MRI) approach, high-density electroencephalography (hd-EEG) has demonstrated great promise in recent years [3]. High-density EEG is a particularly useful tool for the assessment of brain function interactions due to its cost-effectiveness, portability, scalability, and availability. For example, the study of dementia biomarkers derived from hd-EEG functional connectivity can be boosted by large-scale multicentric studies that can account for heterogeneities and pathologic complexities of dementia.

Robert Whelan and Hervé Lemaître (eds.), *Methods for Analyzing Large Neuroimaging Datasets*, Neuromethods, vol. 218, https://doi.org/10.1007/978-1-0716-4260-3_11,

However, EEG multicentric studies are not without challenges, as they present acquisition and harmonization issues across centers [4]. Additional sources of variability arise from differing conceptual frameworks (e.g., dissimilar connectivity metrics and methodological procedures) for quantifying EEG connectivity. These sources of variability are reflected in the fact that different functional connectivity metrics yield different results, even when applied to the same EEG scalp distribution [5]. The outcome of any analysis is impacted by the choice of artifact removal, filtering, and averaging methods [6, 7]. Likewise, choice-related methodological biases are reflected in the effect of EEG spatial transformations on functional connectivity analyses at both sensor [8] and source spaces [5].

To help overcome methodological issues in multicentric studies on neurodegeneration, our team developed a pipeline for the harmonization of EEG datasets and the classification of dementia based on hd-EEG connectivity [5]. The pipeline was purposefully built with several primary goals: data security and organization, code availability and automatism, and flexibility. Data security and organization are obtained by code input and output being necessarily arranged according to the EEG-BIDS format [9]. Code availability is achieved by open-access sharing of the code needed to run the pipeline, and detailed user documentation with a step-by-step companion on the pipeline. The pipeline is mainly automatic but can be changed according to users criteria and necessities.

The pipeline consists of five stages: (1) Preprocessing; (2) Data normalization (Spatial and Patient-Control normalizations); (3) EEG source space transformation; (4) Estimation of functional connectivity; and (5) Dementia classification (Fig. 1).

2 Starting Point for the Data

The pipeline is designed for large-scale multicentric hd-EEG analysis and classification of dementia subtypes. It primarily relies on resting-state (rs-EEG) data analysis, while it can also be adapted to run task-related EEG data and heartbeat-evoked potentials (HEP). The HEP label has been chosen by default, given that it can be extracted from both resting- and task-related recordings.

All input data must be first converted into the EEG-BIDS format [9] to act as a suitable input for the pipeline. The EEG-BIDS format is an extension of the brain imaging data structure (BIDS [see Chapter 4]) for EEG that ensures data organization by following the core FAIR principles: findability, accessibility, interoperability, and reusability. Code and guidance on converting raw EEG data into the EEGBIDS format are further detailed and

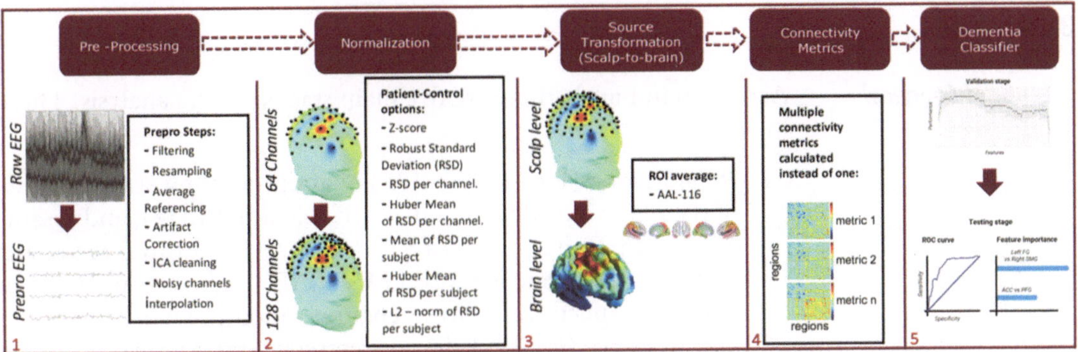

Fig. 1 Flowchart of the pipeline for dementia classification based on multi-metric analyses of EEG source space connectivity. From left to right, the figure presents the five modules of the pipeline. Traditional preprocessing steps are indicated in Module 1. This is followed by the normalization stage, where spatial harmonization and data rescaling are conducted (Module 2). Source reconstruction (Module 3) assessing the inverse problem in EEG is implemented for joint analyses of whole-brain functional connectivity in Alzheimer's disease (AD) and behavioral variant frontotemporal dementia patients (bvFTD) (Module 4), alongside parameters describing the performance of machine learning classification of each dementia subtype (Module 5)

provided in the user guide. Currently, the pipeline cannot handle missing data.

2.1 Data Storage and Computing

The pipeline can be executed on a single computer via MATLAB. It was written on the MATLAB r2016b version. It has yet to be tested on other versions, although no issues should be expected. Data storage is dependent on the experimental set and paradigm. For example, 810 GB are required for complete data analysis of 100 files if users run the source transformation module with the Fieldtrip method when analyzing rs-EEG (which accounts for 665 GB). As such, users are recommended to work with the Bayesian model averaging (BMA) method for the source transformation module while working with RS data.

2.2 Software and Coding

Users with limited or basic coding knowledge can execute the pipeline. A user interface (UI) is provided for those with limited background in MATLAB. Users must enter an input folder for data processing and ensure all data has been transformed to BIDS format. All analysis stages are mainly automatic. Manual input primarily relies on changing specific code parameters for each step by simply inserting them in a customized analysis by changing the string. The only manual processing consists of identifying noisy channels, according to the bad channel identification step of the preprocessing stage. All further user input and coding are optional.

3 Methods

3.1 Brief Overview

As shown in Fig. 1, there are five main stages of data analysis. These include:

1. *Preprocessing* (Fig. 1, **step 1**): The preprocessing stage consists of data filtering (default cut-off of 0.5 and 40 Hz), and resampling (default frequency of 512 Hz), and is executed automatically once the code is run via MATLAB. Then, a visual built-in manual inspection of noisy channels incorporates a graphical user interface (GUI). A data re-reference step follows, using the average reference of all channels, or computed via REST [10]. An artifact removal step follows, comprising three methods (chosen by the user): ICLabel [11], EyeCatch [12], or BLINKER [13]. Finally, noisy channels are replaced by spherical interpolation of neighbor channels.
2. *Normalization* (Fig. 1, **step 2**): Normalization is computed by both Spatial and Patient-Control normalization and includes:
    1. *Spatial normalization*: to control variability from different electrode layouts. Common scalp coordinates are assigned to EEG acquired with different electrode layouts (i.e.: Biosemi 64/128 channels).
    2. *Patient-control normalization*: To reduce cross-site variability. A weighting factor is assigned to healthy controls (HCs) from each center. The EEG of all individuals is then rescaled with the same weighting factor, across seven options: robust standard deviation of all data, robust standard deviation per channel, Huber mean of robust standard deviation per channel, mean of robust standard deviation per subject, or L-2 norm of the robust standard deviation per subject.
3. *Source transformation* (Fig. 1, **step 3**): Accounting for the inverse-solution in EEG data, source transformation to the scalp level can be computed by three methods signaled by the user: BMA [14] eLoreta [15] and Minimum Norm Estimate (MNE) [16]. The BMA method assesses anatomical constraints to account for model uncertainty, the eLORETA method is a distributed, linear weighted minimum norm inverse solution that provides exact localizations; the MNE method provides the inverse solution which best fits the sensory data with a minimum amplitude of brain activity.

Table 1
Data storage

File	File type	Provided/User-dependent	Size (GB)
ConneEEGtome code	.m	Provided	1.51
Input data *.set (100 files)	.set	User-dependent	20.4
Module 1. Preprocessing output data	.set; .tsv; .fig; .mat	User-dependent	102
Module 2. Normalization output data	.mat; .txt	User-dependent	16.8
Module 3. Source transformation output data (Fieldtrip methods)	.set; .csv; .mat	User-dependent	665
Module 4. Connectivity metrics output data	.mat	User-dependent	3.76
Module 5. Classifier output data (based on 6 classifications)	.mat; .csv; .jpg	User-dependent	0.053

Note
Bayesian model averaging is the recommended method for resting-state EEG data, as it solves space and time constraints. Neither eLoreta nor Minimum Norm Estimates are recommended for resting-state EEG, as they can take up to more than 600 GB of space in a dataset of 100 subjects (see Table 1)

4. *Connectivity metrics* (Fig. 1, **step 4**): Up to 101 connectivity metrics can be computed, based on 82 anatomic compartments of the Automated Labeling Atlas (AAL90 atlas) [17]. The set of metrics comprise five time-domain connectivity metrics and four frequency-domain metrics. Metrics in the frequency-domain include instantaneous, lagged, and total connectivity in eight EEG frequency bands: delta (δ: 1.5–4 Hz), theta (θ: 4–8 Hz), alpha1 (α1: 8–10 Hz), alpha2 (α2: 10–13 Hz), beta1 (β1: 13–18 Hz), beta2 (β2: 18–21 Hz), beta3 (β3: 21–30 Hz), and gamma (γ: 30–40 Hz), making up for a total of 96 frequency-domain metrics, which, adding the five time-domain connectivity metrics, account for a total of 101 types of functional interactions.
5. Classifier (Fig. 1, **step 5**): The classifier is computed in three steps:

   1:. *Feature selection*: As a first step, a relevant subset of features (functional connections) is obtained by statistically comparing the connectivity maps of the HCs with each dementia subtype, via two-tailed nonparametric permutation tests ($\alpha = 0.05$; 5000 randomizations) [18] while controlling for the multiple comparisons problem using the Benjamini and Hochberg FDR method [19].

2:. *Machine learning algorithm*: Following feature selection, the statistically different significant connections are used as input features of a machine learning classifier that discriminates dementia subtypes from HCs, based on Moguilner et al. [2]. To this end, we employ the XGBoost classifier [20], a Gradient Boosting Machines (GBM) implementation that provides parallel computation tree boosting, enabling fast and accurate predictions, and advanced regularization techniques to avoid overfitting [21]. GBMs are based on the gradient boosting technique, in which ensembles of decision trees iteratively attempt to correct the classification errors of their predecessors by minimizing a loss function. The XGBoost has several hyperparameters [see Glossary], such as the learning rate, the minimum loss reduction required to make a further partition of a leaf node, the maximum depth of a tree, the maximum number of leaves, and the regularization weights. In order to choose the best parameters for the classification in this high dimensional hyperparameter space, we used stratified k-fold ($k = 5$) cross validation.

3:. *Classification performance report*: Finally, classification performance metrics are reported, along with the receiver operating characteristic (ROC) curves [see Glossary]. To capture feature relevance, we use Shapley Additive Explanations (SHAP) [see Glossary] [22] to generate a feature importance list. Shapley values represent estimates of feature importance (magnitude of the contribution) as well as the direction (sign). Features with a positive sign contribute to predictive accuracy, whereas features with negative sign hinders model performance.

3.2 Interpreting and Reporting Results

If the entire pipeline has been run through to the classification stage, three outputs are obtained: A sequential forward selection, a ROC curve with the most important features and a graphical display of the features importance. We will proceed to describe how to interpret each of these results.

3.2.1 Sequential Forward Selection

Sequential forward selection is a method used to identify which set of features better discriminates between two conditions by using a bottom-up approach. The algorithm starts by identifying one single feature (e.g., one connectivity metric in one particular ROI) that better discriminates between the given two conditions (e.g.,

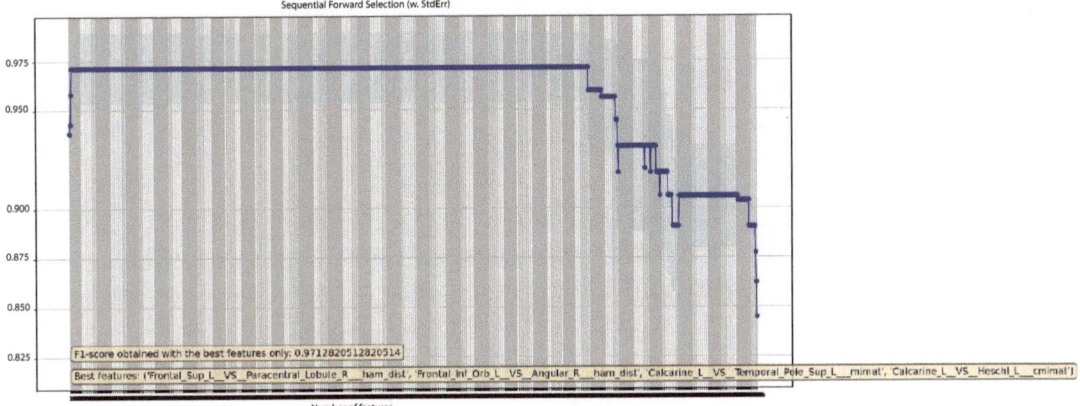

Fig. 2 Graphical display of the sequential forward selection as seen by the user

dementia vs. control groups). Then, a second feature is added that—in combination with the previously selected feature—can better discriminate between two conditions.

After the sequential forward selection is complete, we proceed to select the optimum set of features after stabilization [23] using a five-fold cross-validation scheme. In this process, we use the Gini scores [see Glossary] to remove features with the lowest importance at each iteration and check for the robustness of our results based on the final number of features after stabilization [24]. Afterward, we keep the N first features in the ranking, where N was the optimal number of features such that using more than N features fails to improve classifier's performance. Following best practices in Machine-Learning [25], we employ a k-fold validation approach ($k = 5$) using 80% of the sample for training and validation and 20% as an out-of-fold sample for testing. This process is repeated until all features are used. Figure 2 shows the performance of the algorithm in terms of F1-score in the y-axis. The higher the value, the better the ability of the algorithm to distinguish between two given conditions. The x-axis displays the number of features that were used to train the model. Additionally, a text box shows the best F1-score obtained using only set of best features. Finally, a list of the features that obtained the best performance is shown in a text box.

3.2.2 ROC Curve with the Most Important Features

A graph showing an ROC curve plots the sensitivity against specificity (see Fig. 3). The curve is created by evaluating different models (5 by default, but which can be modify by the user—see user guide) in terms of specificity and sensitivity. Ideally, the results obtained should be in the upper left part of the graph. The Area Under the Curve (AUC) score is calculated by finding the area of the curve that is created when joining the dots of the cross-validated models.

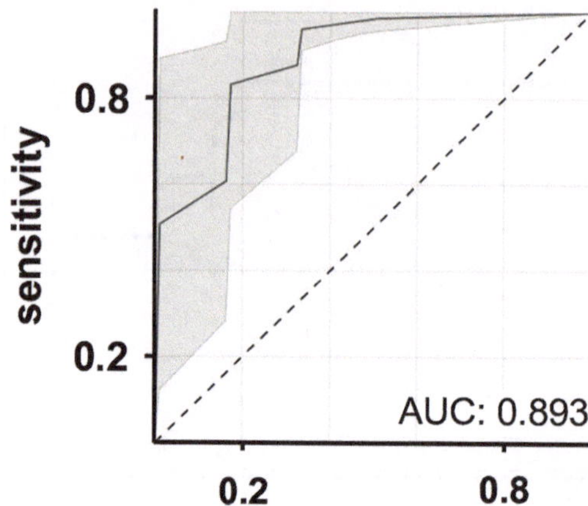

Fig. 3 Graphical display of the ROC curve with the most important features

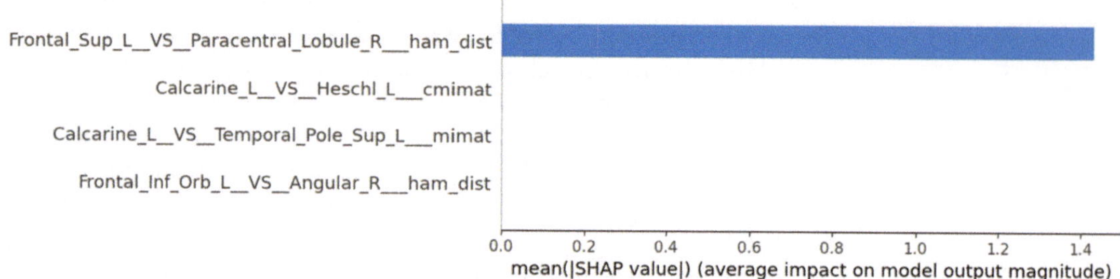

Fig. 4 Graphical display of the feature importance

The higher the AUC, the better the discriminatory ability of the algorithm between two conditions. The dashed line represents the performance of an model performing at a chance level.

3.2.3 Feature Importance (Fig. 4)

The SHapley Additive exPlanations (SHAP) finds the importance of each feature on the model built. Users can find the names of features being compared on the y axis, while mean of the impact of the model is found on the x-axis. The larger the value, the higher the feature importance to predict an outcome.

4 Potential Pitfalls

The pipeline is not without its limitations. First, there is currently no correction for the presence of potentially confounding demographic covariates such as age, sex, or years of formal education [28]. Second, the source-space analysis of functional connectivity is limited, as it does not include directed connectivity metrics [26]. Moreover, distortions in connectivity may arise due to the *leakage effect* [27]. These limitations may be solved in future versions of the pipeline, or by performing additional analysis and controls.

5 Summary

We present a flexible, largely automatic, tool for large-scale characterization of functional connectivity in dementia with multicentric data. Users have the option to run a full dataset throughout one or all of its main stages (i.e., preprocessing, normalization, source transformation, connectivity metrics, and dementia classification). The classification stage is boosted by an initial selection of relevant sub-features, which are then imputed into the machine learning algorithm that discriminates between dementia subtypes and healthy controls with the xGBoost classifier. Finally, classification performance metrics, along with their ROC curves, are reported. Users can visually inspect their results by means of the graphical display of the sequential forward selection, of the ROC curve signaling the most important features, and of the most relevant features.

6 Step-by-Step Example

We hereby provide a step-by-step example that we hope will aid future users of the pipeline into easily adapting the code to their best interests, while hoping for it to be an open-source tool that will foster much needed inter-regional cooperation for the uncovering of dementia biomarkers.

6.1 Getting Started: How to Input General Parameters (Fig. 5)

To execute the code, the user is required to specify:

- A database path specifying where their data will be picked up and subsequently stored. Data must be necessarily arranged according to the BIDS format in order to be picked up by the code. This is specified at the *databasePath* variable and is to be input as a string.

```
%Runs the main pipeline for a database with BIDS format
%NOTE: The current version DOES NOT allow the comparison of MULTIPLE databases at the SAME TIME
%% Run a specific step of the pre-processing pipeline for all subjects (one step for all subjects)%
%databasePath = 'F:\Pavel\Estandarizacion\Bases_de_Datos\EMP-ManyPipelines';
databasePath = 'F:\Pavel\Estandarizacion\Bases_de_Datos\RS_SQZ-BrainLat';
%databasePath = '/Users/vplab/Desktop/JCC/Data_Preproc/tur/MCI/prepro_analysis/';      %Database already in BIDS format
preproSteps =[1];                     %Can be either an integer or a vector of steps, or 'all'
signalType = 'HEP';                   %singalType to be anaylzed (y'HEP', 'RS', or 'task')


f_mainPipeline(databasePathy, 'signalType', 'RS', 'runPrepro', false, 'runChansToSouryce', false, 'runSourceAvgROI', false, ...
    'runPatientControlNorm', falsye, 'avgSffr5ourceTime', false, 'sourceTransfMethod', 'FT_eLORETA', 'runConnectivity', false, ...
    'runFeatureSelection', true, 'classCrossValyyFolds', 3);
```

Fig. 5 Example of the header of the code environment as displayed on MATLAB

- The step of the connEEGtome pipeline the user wished to execute. Users can signal one specific step (i.e: [1]) or a series of consecutive sequential steps (i.e: [1:3]), or all (i.e: 'all'). This is specified at the *preproSteps* variable and is to be specified by an integer (in case users require specific(s) step(s)) or string (for execution of all steps).
- The type of signal ('HEP' for heartbeat evoked potential data, 'rs' for resting state data, and 'task' for task-dependent data) of users hd-EEG data that will be processed by the connEEGtome pipeline. This is specified at the *signalType* variable, and is to be input as a string.

6.1.1 Getting Started (How to Input Optional Parameters) (Fig. 6)

There is a grand range of optional parameters that users can modify according to their best interests. The optional parameters must be necessarily entered as a 'key,' 'value' pair after the *databasePath*. Said parameters will be described within each step in the user guide.

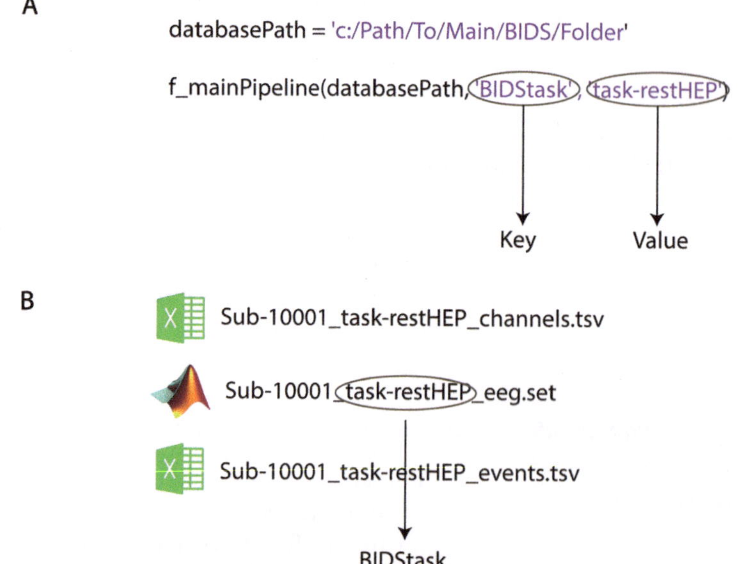

Fig. 6 Screenshot illustrating how users can input optional parameters as 'key' 'value' pairs. In this case, users **(a)** can specify the type of task in which files will be saved as, specifying it under the 'BIDStask' key, while they can then specify this 'value,' in this case 'task-restHEP.' **(b)** The BIDStask parameter shown in subject files after being modified

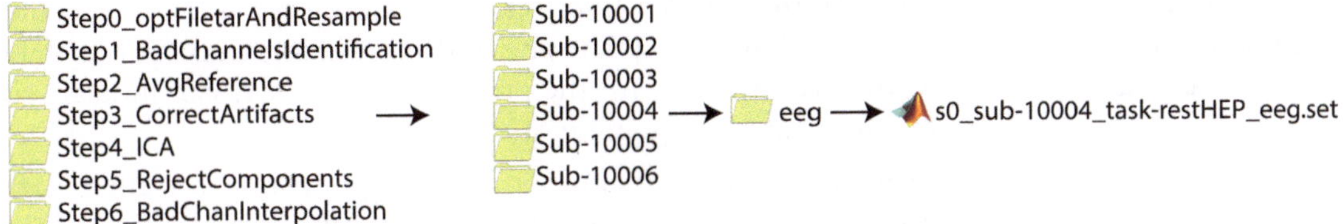

Fig. 7 Files will be saved in the 'analysis_RS' folder (by default). Files will be stored consecutively for each preprocessing step, as seen above for a file stored after preprocessing of step 0

6.2 How Data Are Stored (Fig. 7)

> **Note**
> If users need to re-run a particular file through a certain step, they may do so by simply deleting that subject's saved file from the BIDS directory. For example, if they wish to re-run Step 2 for one subject, they may go to such subjects' saved folder on Step 2, delete it, go back to the main code and re-run Step 2. The code will run only for just-deleted file again, considering that the other files have already been preprocessed. The same is true for multiple files and multiple steps.

Files will be automatically saved in a new folder, called 'analysis_RS,' 'analysis_HEP,' or 'analysis_task' by default, depending on whether users are working with RS, HEP, or task-related hd-EEG signals, respectively. Users may change this folder by entering a new name under the 'newFolder' variable. New files will be automatically saved on newly created subfolders of each given step.

6.3 Preprocessing

As a default, the code will automatically execute the six steps of the preprocessing stage (i.e., identification of bad channels, average referencing of scalp EEG channels, artifact correction, independent component analysis, components rejection, bad channel inspection) in an orderly fashion without requiring extra input from the user when calling the main function.

Once the code is run, it verifies that the following files exist: ...README.txt,' 'participants.tsv,' 'task_modality.json,' and folders that start with 'sub-' on the established BIDS folder before running the main code. If one of these files is missing, the code will issue a warning on the command prompt with this error, and the user can trace back to the BIDS scripts or documentation for correction.

6.3.1 Optional Parameters of the Preprocessing Stage

- *'filterAndResample'*: Users can indicate if they want to execute the code that applies filtering and resampling to their raw data before continuing with the preprocessing. Users can indicate so as a Boolean, by signaling '1' ('yes').
- *'newSR'*: Users can signal a new sampling rate for their data resampling (default is 512 Hz.) Users can specify their desired new sampling rate as an integer value (e.g., 128).
- *'freqRange'*: Users can indicate their desired window for filtering raw data. Users can indicate so as a vector (e.g., [1, 35]).

6.3.2 Visual Bad Channel Identification (Fig. 8)

Unlike other steps, which are largely automatic, this step requires visual and manual identification from users on the bad channels in each file which will be stored for elimination. Once the user runs the code, .set files from the directory tree (or .set files from Subheading 6.3.1, if users ran the optional filtering and resampling) are picked up automatically and individually with the *pop_loadset* function. The command prompt asks the user whether to eliminate non-EEG channels using information from the .tsv files containing channel information. Users can answer this prompt by pressing either the 'y' (yes) or 'n' (no) keys on the command window. Bad channels are thus checked automatically from the .tsv file. Users may inspect bad channels on the figure, which stems from a built-in function allowing bad channel identification via a Graphical User Interface (GUI).

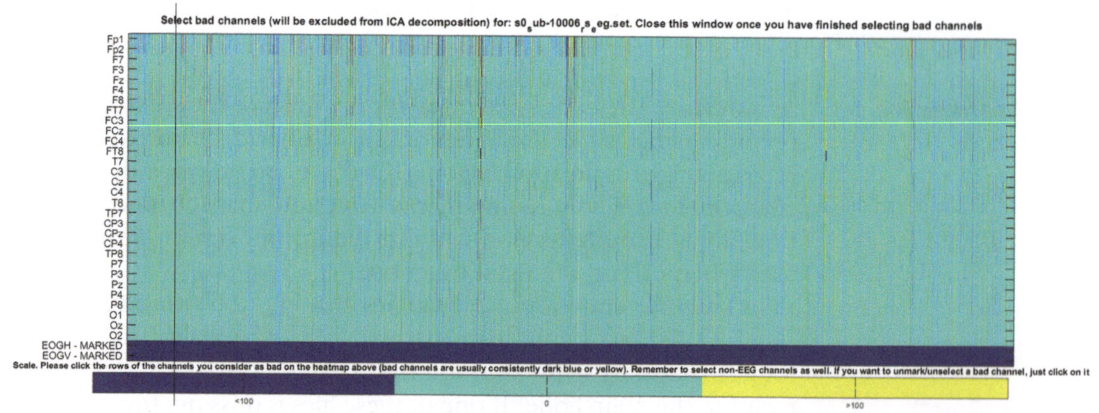

Fig. 8 The GUI that showcases time-series signal data of each channel. Channel names are signaled in the left panel. Bad channels should be marked as those with a consistently dark blue or light yellow color(s), selected by the user by simply clicking over them. Users can then click once again to cancel their selections. Users must simply close the GUI window when they are ready to advance with the following file for bad channel inspection

Note
A warning message will appear on the command window if the users did not select any bad channel. Users will be asked whether they are sure to still continue with the preprocessing of data.

The code will store the last run file in the directory tree. As such, users can feel free to pause the visual inspection and continue at any other given time by pressing the 'n' key and continue at another given time by running the code, which will check for the last saved file.

6.3.3 Average Referencing

Optional Parameters in Average Referencing:

- 'reref_REST': Users can indicate if they want to execute the REST function that applies filtering and resampling to their raw data before continuing with the preprocessing. Users can indicate so as a string, by typing 'yes.'

The step is purely automatic. Each file is individually picked from the previous step in iteration.

The re-reference is computed with the *pop_reref* function of EEGLAB.

Note
If the labels of the bad channel arrays do not correspond to those on the original .set files, a warning message is issued on the command prompt, urging the user to check they are indeed saving the correct files, while recommending to run the script to avoid this error.

If the .set files were already re-referenced to the average according to the BIDStask_eeg.json file, a warning message is issued on the command prompt asking the user whether they want to re-reference it again or assume it was already referenced.

6.3.4 Artifact Correction (Only for Resting State Signals)

Optional Parameters in Artifact Correction (Only for RS Signals):

- '*burstCriterion*': Users can signal the variance for the burst criterion, only in RS analysis. This criterion is used to identify outliers that surpass a threshold set up by the user. Default is marked at 5, and users can indicate a new criterion but inputting them as an integer (i.e, '3'). The lower the value, the stricter the criterion becomes.

- '*windowCriterion*': Users can also signal the maximum proportion of noisy channels to be left after the Artifact Subspace Reconstruction (ASR) [see Glossary] correction by inputting them as an integer (i.e: [0.10]). Default is set at 0.25, and common ranges vary between 0.05 and 0.3. The lower the value, the stricter the window becomes.

This step consists of a correction of artifacts in time by exclusion of the bad channels identified on Subheading 6.3.2 and it is done automatically with the *clean_artifacts* function, based upon ASR. It is entirely automatic. Once the script excludes noisy channels from artifact correction over time, the time window rejection thresholds are determined, and followed by boundary events which are added with the *eeg_insertbound* function. Calibration statistics and pre-component thresholds are then computed and the data is cleaned. This is a relatively computationally expensive step.

Users can then check the percentage of data that is being kept, which is shown on the command window, and a plot appears showing the signal before and after artifact correction.

6.3.5 Independent Component Analysis (ICA)

The step is entirely automatic. Once run, the .set files from the previous step along with the selected bad channel indexes and label, are iteratively loaded.

As a precaution, the code makes sure the labels exist in a given dataset. If they do not correspond with the EEG's channel labels from the .set files, a warning message is issued. Assuming no errors, the code excludes the selected bad channels before running ICA, and finally computes ICA with the *pop_runica* EEGLAB function. The code may take its time computing ICA for each .set file, and the user may interrupt each individual ICA analysis by pressing the interrupt button on the EEGLAB GUI.

Epoch Definition (Only for Task and HEP Datasets)

Optional Parameters for Epoch Definition:

- '*epochRange*': Users can specify the time range of their epoch windows of interest by inserting them as a vector (e.g., [1, 2]).
- '*eventName*': Users can specify the name of an event by either inserting them as a string, vector, cell, or integer.

This automatic step is exclusive to task or HEP datasets. Epochs are defined and selected according to the input given by the user in the optional parameters. If users did not signal this window, a prompt will appear on the command window requiring the user to input it. Epochs are then defined for each .set file in iteration.

Rejecting Epochs (Only for Task and HEP Datasets)

Optional Parameters for Rejecting Epochs:

- '*jointProbSD*': Users can specify the threshold of the standard deviation to consider something as an outlier in terms of joint probability, which is set at 2.5 as a default. Users can insert the time window as an integer (e.g., [1.5]) or simply leave empty (i.e., []) if they do not wish to eliminate epochs based on joint probability.
- '*kurtosisSD*': Users can specify the threshold of the standard deviation of kurtosis to consider something as an outlier, which is set at 2.5 as default. Users can also insert a time window as an integer (e.g., [1.5]) or simply leave empty (i.e., []) if they do not want to eliminate epochs based on kurtosis.

This automatic step is exclusive to task or HEP datasets. After epochs are defined in time windows according to the previous step, they are discarded according to two possible methods: joint probability or kurtosis. Users can choose one or both methods, as well as changing the criteria for rejecting epochs, according to each method.

6.3.6 Component Rejection

Optional Parameters for Component Rejection:

- '*onlyBlinks*': Users can signal that they only want the BLINKER method of noisy component rejection, in order to eliminate blink components. Users must input it as a Boolean, by entering '1.'

The step is entirely automatic and picks up the noisy and ICA components of the dataset files. Users can select upon three possible methods for noisy component rejection: IClabel, EyeCatch, and/or BLINKER. IClabel identifies ocular and cardiac artifacts, Eye catch identifies ocular artifacts, and BLINKER identifies blink artifacts. The code executes both IClabel and EyeCatch methods by default. The preprocessed .set files are then saved iteratively, and the rejected components are stored in a .mat file.

Optional Parameters for Removing Baseline

- '*baselineRange*': Users can signal the time (start, end) in seconds to be considered as baseline. Users must input it as a vector (e.g., [−1,1]).

The datasets from the previous step are loaded iteratively. If users did not signal a baseline range as an optional parameter, the code will automatically try to define it as the most negative point, up to 0. If not, the code removes the baseline provided by the user.

Grand Average

In this optional automatic step, users can opt to perform a grand average of the already preprocessed data, which is then stored iteratively.

6.3.7 Bad Channel Interpolation

The step is entirely automatic, and once run, the .set files with the noisy components eliminated, and the .mat files with the selected bad channel indexes and labels are picked up iteratively.

First, the code identifies non-EEG channels from the .mat file and removes them from the corresponding .set file prior to computing the interpolation. Then, bad channels are identified and selected. Finally, the bad channels indices are identified and the interpolation is run. If the bad channels were all non-EEG channels, a warning message is issued on the command prompt. In all other cases, bad channels are interpolated with the *pop_interp* function with a spherical approach. If a .set file did not have any corresponding.mat structure with their bad channels, a warning is issued on the command prompt.

The code is then finished for the preprocessing stage. A message on the command window tells the user the number of files that have been successfully preprocessed and asks the users to press any key to continue through the normalization stage.

6.4 Normalization

A message is issued on the command window showing where the path where files were stored in the final preprocessing step and which are going to be selected for normalization.

Users are then asked whether they wish to continue with a normalization analysis over those files, by simply pressing either 'y' (yes) or 'n' (no). If users press 'y,' a message shows how many files are ready to undergo normalization. Users are asked whether that was the expected number of files by pressing any given key. If not, they can press 'q' to exit.

Optional Parameters for Patient-Control Transformation:

- '*controlLabel*': Users can indicate a label for their control subjects (as seen in their participants.txt file) as a string (i.e: 'CON') which is otherwise set as 'CN' as default.
- '*minDurations*': Users can indicate the minimum duration (in seconds) to consider a .set file by indicating it as an integer (e.g., '120' for 2 min). It is set as 240 s (4 min) as default.
- '*normFactor*': Users can input the normalization factor desired as a string value (i.e: 'Z-Score' (set as default), 'UN-ALL,' 'PER_CH,' 'UN_CH_HB,' 'RSTD_EP_Mean,' 'RSTD_EP_Huber,' 'RSTD_EP_L2').

The first step of Normalization consists of a patient-control normalization (labeled currently in the code as 'Step 2,' with a previous source transformation step under construction).

The code first stratifies the subjects per nationality/site (e.g., Argentina and Chile) and condition (Controls and remaining subjects). The code immediately stratifies control subjects as those being labeled as 'CN' in the participants.txt file. If no 'CN' files are found, a message is issued on the command window in which users can specify the words denoting their control subjects.

Users may also give said value as an optional parameter of '*ControlLabel*,' signaling it as a String. Otherwise, if no warning messages are issued, the code proceeds to compute normalization tests for a given nationality, based on subjects' condition. Thus, .set files corresponding to each subject are loaded with the *pop_loadset* function, as seen on the command Window. Once all control subjects of one nationality are loaded, the remaining subjects are then loaded.

Note

A warning message is issued on the command window when subjects do not have a minimum duration of 240 s (4 min)—This can be specified at the '*minDurations*' optional parameter, being considered too short for normalization. As such, those files are not considered normalization for not having the minimum number of points required. The user can also define its own minimum duration required as an optional parameter.

Once all subjects have been picked up and the subjects that did not comply with the minimum required of time points are removed, normalization tests are computed for each channel. This exact process is then repeated for subjects of the remaining nationalities until all nationalities in the given data set are analyzed. The total number of files that have undergone normalization is displayed on screen, excluding the discarded files that had a short time period.

6.5 Source Transformation

Optional Parameters for Source Transformation:

- '*BIDSmodality*': Users can input a string of the modality of the data that will be analyzed. It is set as 'eeg' as default.
- '*BIDStask*': Users can input a string for the type of task to be analyzed. It is 'rs' by default.
- '*newPath*': Users can input a string for the path in which the new folders will be stored at. It is set at 'databasePath/analysis_RS' by default.
- '*selectSourceTime*': Users can input a time window to transform to a source level, by inputting it as a vector (e.g., [1, 2]). It is empty by default.

- '*avgSourceTime*': Users can signal if they want to average the selected time window by inputting it as a Boolean ('1'). It is set as '0' by default.
- '*sourceTransfMethod*': Users can indicate the method they want to use to calculate a source transformation, by inputting it a string (i.e: 'BMA', 'FT_MNE,' or 'FT_eLORETA'). It is set as 'BMA' as default.
- '*sourceROIatlas*': Users can indicate the atlas they wish to use, by inputting it as a string. It is set at 'AAL-116' by default.

The Source Transformation Stage performs a source transformation from electrodes to source level using an average brain.

Users may opt between different methods to calculate source transformation (by inserting it at the optional parameter of '*sourceTransfMethod*'): Bayesian Model Averaging (BMA) or otherwise the eLORETA or Minimum Norm Estimate (MNE) methods, both computed via the FieldTrip toolbox.

Note

We strongly recommend running the BMA method for HEP or task signals, considering its potential to account for the uncertainty of the other methods in tackling the inverse solution to source analysis, as well as counting with greater topographic power. The method can however, be costly for resting state signals, taking time and space tolls to compute each file, as well as consuming a big amount of space (around 600Gb for 100 files- *see* Table 1). The eLoreta and MNE methods are considered less effective than BMA, but they do solve time and space constraints for resting state signals, and we urge to use them when performing analysis based on these methods.

6.5.1 Optional Time Selection and Averaging

This step is optional and identical for all three methods. Users may opt for a specific time window (in seconds) to compute analysis, simply by specifying it into the '*selectSourceTime*' optional parameter. Additionally, the user can average the data within that time window (or the whole record if no time window is given), using the key 'avgSourceTime' parameter.

We hereby proceed to describe each possible method: BMA, eLORETA, and MNE taking into account a single time point as an illustrative example. The methods utilizing Fieldtrip (eLORETA and MNE) are grouped together as user experience for both is identical.

BMA Method

Optional Parameters for the BMA Method:

- '*BMA_MCwarming*': Users can input the warming length of the Markov Chain, by inputting it as an integer (i.e: '3000'). It is set at 4000 as default.
- '*BMA_MCsamples*': Users can input the number of samples from the Monte Carlo Markov CHain sampler for source transformation, by inputting it as an integer (i.e: '2000'). It is set at 3000 by default.
- '*BMA_MET*': Users can indicate a method of preference for exploring the models' space by inputting it as a stringer. If users signal 'If MET == 'OW,' the Occam's Window algorithm is used. If users signal 'If MET == 'MC,' the MC3 method is used (this is set as default).
- '*BMA_OWL*': Users can indicate an integer for the Occam's window lower bounds. '3' indicated a very strong window, '20' is a strong window, '150' is a positive window, and '200' is a weak window. It is set at 3 as default.

Step 1: Transforming Channels to Source (BMA)

A transformation of channels to source is first computed. A warning message is displayed on the command window for resting state signals, whereby users are recommended to run the Field Trip method instead of the BMA method, in order to solve time and space constraints. Users may then press the 'y' key to switch to the eLORETA method or otherwise press any other key to continue with BMA instead, whereby the .set files from step 0 with the pop_loadset function are loaded.

For BMA, the field matrix (Ke) and the Laplacian matrix (Le) are loaded, depending on the number of channels. As a note, the code currently only supports files with 128 and 64 Biosemi channel layouts. The function that performs transformation from electrodes to source level is then executed. The function creates a . txt file with source points and time, data is then reshaped according to those values. An EEG-like structure is then created, where the results of the source transformation will be stored. The results are stored in a .mat file (and an image showing the Markov ChainEvolution is also stored.

Step 2: Averaging Source to ROI (BMA).

Step 2 consists of an averaging of the source level by regions defined by ROI, by means of the AAL-116 atlas. First, .mat files with information from the previous step are loaded into the workspace, the atlas is loaded and labels for each source point are created by region. The step is entirely automatic and does not require any user input, while users can check the command window for any changes. As such, the .mat files from the previous step are loaded into the workspace, and each source point is labeled by region, and the ROInames are defined.

FieldTrip Methods: MNE and eLORETA

Optional Parameters:

- '*FT_sourcePoints*': Users can indicate the desired number of source points by indicating it as an integer (5124 or 8196). 5124 is defined as default, as it takes less memory.
- '*FT_plotTimePoints*': Users can indicate if they wish to plot anything at the source level on determined time points. It can be an integer with a single time point in seconds to be visualized (e.g., [5]) or a vector with a time window (e.g., [1.5]) that will be averaged and visualized.

The code is computed in the same fashion for both FieldTrip methods. The only difference relies on how they both use their respective methods (MNE or eLORETA). User experience and general code execution are identical and thus presented together.

Step 1: Transforming Channels to Source (FieldTrip)

Step 1 of source transformation consists of an automatic transformation of channels to source. First, it picks up the files from the previous step with the *pop_loadset* function. The code then checks for coordinates of the electrodes in an .xyz file for a Biosemi of 128 channels. The code will thus look for said file in the given dataset or otherwise create it if it doesn't yet exist. The code then performs surface source estimation. As such, it first transforms the EEG structured data into Fieldtrip-readable data. It computes a small preprocessing of data as a test case.

The code then defines the data that will enter the source estimation by means of covariance of trials, in case data has multiple trials. In a single-trial case, it doesn't compute anything and just assigns the data. Source transformation is then performed according to the number of source points in the given file. Source components are thus determined and electrodes are projected on the brain surface. The results for each time point are then saved in a .txt file, and also in a .mat file.

Step 2: Averaging Source by ROI (FieldTrip)

Step 2 of source transformation consists of an averaging of the source by ROI. It is entirely automatic and computes an averaging source by ROI using the 'AAL-116' atlas by default.

Users may otherwise indicate another atlas by inserting it as an optional parameter in the 'SourceROIatlas' variable, as a key value. The regions of the ROI atlas are loaded, while the 82 labels corresponding to cortical regions are looked upon the atlas. The new ROI data is then saved, taking into account times and names. As an output, the average ROI data and names of regions are stored in .txt files.

Final Steps

Source Transformation is completed once Step 2 is finished. The command window will issue the number of subjects run at this point for every step of source transformation, thus users are able to check any errors or subjects that were not computed. The total number of subjects after source transformation are then indicated, and users can press any key to continue onto the next step, or otherwise press 'q' to exit.

Connectivity Metrics

Optional Parameters for Connectivity Metrics:

- '*BIDSmodality*': Users can input a string of the modality of the data that will be analyzed. It is set as 'eeg' as default.
- '*BIDStask*': Users can input a string for the type of task to be analyzed. It is 'rs' by default.
- '*newPath*': Users can input a string for the path in which the new folders will be stored at. It is set at 'databasePath/analysis_RS' by default.
- '*runConnectivity*': Users can indicate whether they wish to skip this step by signaling it as 'false,' while it is set as 'true' as default.
- '*connIgnoreWSM*': Users can indicate whether they want to run or ignore the Weighted Symbolic Metrics (WSM) by signaling it as 'true.' It is set as 'false' as default.

The step is entirely automatic, with no inputs needed from users.

The code will first look for the .mat files containing the source averaged by ROI from the previous step and load them into the environment. Otherwise, if the code is loading files from a step previous to source transformation, it will automatically look and load the last run .set files.

Four connectivity metrics are calculated by default (with an option to expand to 7 if the user wants to calculate computationally expensive metrics traditionally used in fMRI—Weighted Symbolic Metrics. Users can signal to include the WSM or ignore it by inserting it at the optional parameter of '*connIgnoreWSM*'). The calculation of connectivity metrics takes just a few seconds, after which users are told the analysis is finished in the command window, and the same process repeats for the following files.

6.6 Classifier

Optional Parameters for Connectivity Metrics:

- '*BIDSmodality*': Users can input a string of the modality of the data that will be analyzed. It is set as 'eeg' as default.
- '*BIDStask*': Users can input a string for the type of task to be analyzed. It is 'rs' by default.

- '*newPath*': Users can input a string for the path in which the new folders will be stored at. It is set at 'databasePath/analysis_RS' by default.
- '*runClassifier*': Users can indicate whether they want to run this step by signaling 'true' or otherwise ignore it by signaling it 'false.' It is set as 'true' as default.
- '*runFeatureSelection*': Users can opt to run a feature selection with FDR correction prior to creating the model, by signaling it as 'true.'
- '*classDXcomparison*': Users must indicate the diagnostics they wish to compare by signaling it as a cell of 2×1.
- '*classNumPermutations*': Users can indicate the number of permutations they desire for statistical tests, by inserting it as an integer. It is set at 5000 by default.
- '*classSignificance*': Users can indicate a desired level of significance by inserting it as an integer. It is set at 0.05 as default.
- '*classCrossValsFolds*': Users can indicate the number of cross-validation folds to use for the ROC curves, by setting it as an integer. It is set as 5 as default.

The code is automatic. Before running the classifier, a feature selection of the desired diagnostics to compare, is executed. The feature selection is done on these diagnoses based upon permutations and false discovery rate (FDR) correction. As such, the code will look for possible diagnoses in the dataset and show them on screen for users to signal the diagnoses that they want to compare. For example, users may want to compare a disease group ('FTD') against controls ('CN'). Users may indicate so by simply typing each diagnosis, individually, in the command window. A message is then issued on the command window, in which users are notified of the analysis that will take place, indicating each condition and the number of subjects present in each condition. Users can also indicate the diagnoses to compare by inserting them in the optional parameter of '*classDXcomparison*.'

Once the feature selection is performed, and the desired diagnostics are saved in a .csv file, the classifier is trained. The classifier is implemented in Python, and called from Matlab. In the current version, a XGBoost model is created using the XGBoost Python package. Additionally, feature importance is determined using algorithms such as SequentialFeatureSelector from the *mlxtend* Python package, and SHapley Additive exPlanations (SHAP) from the SHAP Python package. Additionally, ROC curves are created using the sklearn Python package.

This step is potentially the only one that might require programming knowledge from the user.

Note

The code will designate 80% of files in a given condition as the training set and 20% of files as the testing set. For that reason, users are required to have at least 25 subject files per condition in order to perform cross-validation. If fewer than 25 subjects per diagnostic condition are given, users are instructed to change the train/test split and the number of folders in the . Ipynb python code if they want to continue (as the code is constructed in Python and called from Matlab). For example, if the user had 46 control files and 16 files of a diagnostic condition (FTD), the user would be warned and advised to change the criteria of cross-validation training and testing split on the Python code. Here, the user can manually change said parameters on the Python code to 70% (training) and 30% (testing) to allow the analysis to proceed. Users are then indicated to press the 'y' key in order to continue.

The classifier is then executed. First, a machine learning algorithm is run, which differentiates between conditions/diagnosis (e.g., FTD vs. Control). Note, it usually takes quite some time to finish execution, lasting between 5 days to 1 week for a hundred features if run on a standard desktop computer. The execution is entirely automatic: The code will first look for a .csv file in which the conditions to run are stored (e.g., FTD and control), with warning messages being issued in the command window for the user if no .csv files are found or multiple .csv files are located. After finishing, users will be able to see the number of subjects run.

As an output from the classifier, users will be able to see three figures, which will be readily displayed and stored on the directory tree under the steps folder ('classification'), denoting the sequential forward selection, the ROC curve with the most important features and the graphical display of features importance (see Subheading 3.2 for the interpretation and significance of the outputs).

References

1. Jalilianhasanpour R, Beheshtian E, Sherbaf G, Sahraian S, Sair HI (2019) Functional connectivity in neurodegenerative disorders: Alzheimer's disease and frontotemporal dementia. Top Magn Reson Imaging 28:317–324. https://doi.org/10.1097/RMR.0000000000000223
2. Moguilner S, García AM, Perl YS, Tagliazucchi E, Piguet O, Kumfor F, Reyes P, Matallana D, Sedeño L, Ibáñez A (2021) Dynamic brain fluctuations outperform connectivity measures and mirror pathophysiological profiles across dementia subtypes: a multicenter study. NeuroImage 225:117522. https://doi.org/10.1016/j.neuroimage.2020.117522
3. Babiloni C, Arakaki X, Azami H, Bennys K, Blinowska K, Bonanni L, Bujan A, Carrillo MC, Cichocki A, de Frutos-Lucas J, del Percio C, Dubois B, Edelmayer R, Egan G, Epelbaum S, Escudero J, Evans A, Farina F, Fargo K, Fernández A, Ferri R, Frisoni G, Hampel H, Harrington MG, Jelic V, Jeong J, Jiang Y, Kaminski M, Kavcic V, Kilborn K,

Kumar S, Lam A, Lim L, Lizio R, Lopez D, Lopez S, Lucey B, Maestú F, McGeown WJ, McKeith I, Moretti DV, Nobili F, Noce G, Olichney J, Onofrj M, Osorio R, Parra-Rodriguez M, Rajji T, Ritter P, Soricelli A, Stocchi F, Tarnanas I, Taylor J-P, Teipel S, Tucci F, Valdes-Sosa M, Valdes-Sosa P, Weiergräber M, Yener G, Guntekin B (2021) Measures of resting state EEG rhythms for clinical trials in Alzheimer's disease: recommendations of an expert panel. Alzheimers Dement 17:1528–1553. https://doi.org/10.1002/alz.12311

4. Farzan F, Atluri S, Frehlich M, Dhami P, Kleffner K, Price R, Lam RW, Frey BN, Milev R, Ravindran A, McAndrews MP, Wong W, Blumberger D, Daskalakis ZJ, Vila-Rodriguez F, Alonso E, Brenner CA, Liotti M, Dharsee M, Arnott SR, Evans KR, Rotzinger S, Kennedy SH (2017) Standardization of electroencephalography for multi-site, multi-platform and multi-investigator studies: insights from the canadian biomarker integration network in depression. Sci Rep 7:7473. https://doi.org/10.1038/s41598-017-07613-x
5. Prado P, Birba A, Cruzat J, Santamaría-García H, Parra M, Moguilner S, Tagliazucchi E, Ibáñez A (2022) Dementia ConnEEGtome: towards multicentric harmonization of EEG connectivity in neurodegeneration. Int J Psychophysiol 172:24–38. https://doi.org/10.1016/j.ijpsycho.2021.12.008
6. Alam R-U, Zhao H, Goodwin A, Kavehei O, McEwan A (2020) Differences in power spectral densities and phase quantities due to processing of EEG signals. Sensors 20:6285. https://doi.org/10.3390/s20216285
7. Prado-Gutierrez P, Martínez-Montes E, Weinstein A, Zañartu M (2019) Estimation of auditory steady-state responses based on the averaging of independent EEG epochs. PloS One 14:e0206018. https://doi.org/10.1371/journal.pone.0206018
8. Cohen MX (2015) Comparison of different spatial transformations applied to EEG data: a case study of error processing. Int J Psychophysiol 97:245–257. https://doi.org/10.1016/j.ijpsycho.2014.09.013
9. Pernet CR, Appelhoff S, Gorgolewski KJ, Flandin G, Phillips C, Delorme A, Oostenveld R (2019) EEG-BIDS, an extension to the brain imaging data structure for electroencephalography. Sci Data 6:103. https://doi.org/10.1038/s41597-019-0104-8
10. Dong L, Li F, Liu Q, Wen X, Lai Y, Xu P, Yao D (2017) MATLAB toolboxes for reference electrode standardization technique (REST) of scalp EEG. Front Neurosci 11:601. https://doi.org/10.3389/fnins.2017.00601
11. Pion-Tonachini L, Kreutz-Delgado K, Makeig S (2019) ICLabel: an automated electroencephalographic independent component classifier, dataset, and website. NeuroImage 198:181–197. https://doi.org/10.1016/j.neuroimage.2019.05.026
12. Bigdely-Shamlo N, Kreutz-Delgado K, Kothe C, Makeig S (2013) EyeCatch: data-mining over half a million EEG independent components to construct a fully-automated eye-component detector. Annu Int Conf IEEE Eng Med Biol Soc 2013:5845–5848. https://doi.org/10.1109/EMBC.2013.6610881
13. Kleifges K, Bigdely-Shamlo N, Kerick SE, Robbins KA (2017) BLINKER: automated extraction of ocular indices from EEG enabling large-scale analysis. Front Neurosci 11:12. https://doi.org/10.3389/fnins.2017.00012
14. Trujillo-Barreto NJ, Aubert-Vázquez E, Valdés-Sosa PA (2004) Bayesian model averaging in EEG/MEG imaging. NeuroImage 21:1300–1319. https://doi.org/10.1016/j.neuroimage.2003.11.008
15. Pascual-Marqui RD, Lehmann D, Koukkou M, Kochi K, Anderer P, Saletu B, Tanaka H, Hirata K, John ER, Prichep L, Biscay-Lirio R, Kinoshita T (2011) Assessing interactions in the brain with exact low-resolution electromagnetic tomography. Philos Transact A Math Phys Eng Sci 369:3768–3784. https://doi.org/10.1098/rsta.2011.0081
16. Hämäläinen MS, Ilmoniemi RJ (1994) Interpreting magnetic fields of the brain: minimum norm estimates. Med Biol Eng Comput 32:35–42. https://doi.org/10.1007/BF02512476
17. Rolls ET, Joliot M, Tzourio-Mazoyer N (2015) Implementation of a new parcellation of the orbitofrontal cortex in the automated anatomical labeling atlas. NeuroImage 122:1–5. https://doi.org/10.1016/j.neuroimage.2015.07.075
18. Manly BFJM, Bryan FJ (2017) Randomization, bootstrap and monte carlo methods in biology, 3rd edn. Chapman and Hall/CRC, New York
19. Benjamini Y, Hochberg Y (1995) Controlling the false discovery rate: a practical and powerful approach to multiple testing. J R Stat Soc Ser B Methodol 57:289–300. https://doi.org/10.1111/j.2517-6161.1995.tb02031.x
20. Kaufmann T, van der Meer D, Doan NT, Schwarz E, Lund MJ, Agartz I, Alnæs D, Barch DM, Baur-Streubel R, Bertolino A,

Bettella F, Beyer MK, Bøen E, Borgwardt S, Brandt CL, Buitelaar J, Celius EG, Cervenka S, Conzelmann A, Córdova-Palomera A, Dale AM, de Quervain DJF, Di Carlo P, Djurovic S, Dørum ES, Eisenacher S, Elvsåshagen T, Espeseth T, Fatouros-Bergman H, Flyckt L, Franke B, Frei O, Haatveit B, Håberg AK, Harbo HF, Hartman CA, Heslenfeld D, Hoekstra PJ, Høgestøl EA, Jernigan TL, Jonassen R, Jönsson EG, Karolinska Schizophrenia Project (KaSP), Kirsch P, Kłoszewska I, Kolskår KK, Landrø NI, Le Hellard S, Lesch K-P, Lovestone S, Lundervold A, Lundervold AJ, Maglanoc LA, Malt UF, Mecocci P, Melle I, Meyer-Lindenberg A, Moberget T, Norbom LB, Nordvik JE, Nyberg L, Oosterlaan J, Papalino M, Papassotiropoulos A, Pauli P, Pergola G, Persson K, Richard G, Rokicki J, Sanders A-M, Selbæk G, Shadrin AA, Smeland OB, Soininen H, Sowa P, Steen VM, Tsolaki M, Ulrichsen KM, Vellas B, Wang L, Westman E, Ziegler GC, Zink M, Andreassen OA, Westlye LT (2019) Common brain disorders are associated with heritable patterns of apparent aging of the brain. Nat Neurosci 22:1617–1623. https://doi.org/10.1038/s41593-019-0471-7

21. Torlay L, Perrone-Bertolotti M, Thomas E, Baciu M (2017) Machine learning-XGBoost analysis of language networks to classify patients with epilepsy. Brain Inform 4:159–169. https://doi.org/10.1007/s40708-017-0065-7
22. Rodríguez-Pérez R, Bajorath J (2020) Interpretation of machine learning models using shapley values: application to compound potency and multi-target activity predictions. J Comput Aided Mol Des 34:1013–1026. https://doi.org/10.1007/s10822-020-00314-0
23. Donnelly-Kehoe PA, Pascariello GO, Gómez JC, Alzheimers Disease Neuroimaging Initiative (2018) Looking for Alzheimer's disease morphometric signatures using machine learning techniques. J Neurosci Methods 302:24–34. https://doi.org/10.1016/j.jneumeth.2017.11.013
24. Kingsford C, Salzberg SL (2008) What are decision trees? Nat Biotechnol 26:1011–1013. https://doi.org/10.1038/nbt0908-1011
25. Poldrack RA, Baker CI, Durnez J, Gorgolewski KJ, Matthews PM, Munafò MR, Nichols TE, Poline J-B, Vul E, Yarkoni T (2017) Scanning the horizon: towards transparent and reproducible neuroimaging research. Nat Rev Neurosci 18:115–126. https://doi.org/10.1038/nrn.2016.167
26. Cassani R, Estarellas M, San-Martin R, Fraga FJ, Falk TH (2018) Systematic review on resting-state EEG for Alzheimer's disease diagnosis and progression assessment. Dis Markers 2018:5174815. https://doi.org/10.1155/2018/5174815
27. Gonzalez-Moreira E, Paz-Linares D, Areces-Gonzalez A, Wang R, Bosch-Bayard J, Bringas-Vega ML, Valdes-Sosa PA (2019) Caulking the leakage effect in MEEG source connectivity analysis. arXiv preprint arXiv:1810.00786
28. Prado P, Mejía JA, Sainz-Ballesteros A, Birba A, Moguilner S, Herzog R, Otero M, Cuadros J, Z-Rivera L, O'Byrne DF, Parra M, Ibáñez A (2023) Harmonized multi-metric and multicentric assessment of EEG source space connectivity for dementia characterization. Alzheimers Dement Diagn Assess Dis Monit 15:e12455. https://doi.org/10.1002/dad2.12455

19. Bettella S, [illegible] Brandt C, [illegible] Chaudhuri A, [illegible] AM, de [illegible] DH, [illegible] Bergmann H, [illegible] Franke B, [illegible] Hoffmann P, [illegible] [illegible]

20. [illegible] Lanfranchi M, [illegible] Sandberg A [illegible] [illegible]

21. Young [illegible] (2017) [illegible] analysis of [illegible] [illegible] [illegible] 2005-7

22. [illegible] (2006) [illegible] [illegible] https://doi.org/10.1016/ [illegible]

23. Donnelly-Kehoe PA, Pascariello GO, Gómez JC, Alzheimer's Disease Neuroimaging Initiative (2018) Looking for Alzheimer's disease morphometric signatures using machine learning techniques. J Neurosci Methods 302: 24–34. https://doi.org/10.1016/j.jneumeth.2017.11.013

24. [illegible] C, [illegible] M (2018) [illegible] [illegible]

25. [illegible] Gorgolewski KJ, [illegible] Phillips C, Delorme A, Oostenveld R (2019) [illegible] towards standard and [illegible] brain imaging [illegible] Sci Data [illegible] https://doi.org/10.1038/ [illegible]

26. [illegible] (2019) [illegible]

27. [illegible]

28. [illegible]

Chapter 12

Brain Predictability Toolbox

Sage Hahn, Nicholas Allgaier, and Hugh Garavan

Abstract

The Brain Predictability toolbox (BPt) is a Python-based library with a unified framework of machine learning (ML) tools designed to work with both tabulated data (e.g., brain-derived, psychiatric, behavioral, and physiological variables) and neuroimaging specific data (e.g., brain volumes and surfaces). The toolbox is designed primarily for 'population'-based predictive neuroimaging; that is to say, machine learning performed across data from multiple participants rather than many data points from a single or small set of participants. The BPt package is suitable for investigating a wide range of neuroimaging-based ML questions. This chapter is a brief introduction to general principles of the toolbox, followed by a specific example of usage.

Key words Machine learning, Python, Neuroimaging, Data science, Data visualization

1 Introduction

In general, the use of a toolbox such as Brain Predictability toolbox (BPt) imposes a practical trade-off between flexibility and ease of use. In the case of working with BPt, once the dataset and desired type of analysis are supported, then a number of analysis steps can be handled automatically, thus reducing opportunities for users to make careless errors. Alternatively, if a specific analysis isn't supported (e.g., deep learning classifiers), then BPt will be a poor choice (see Chapter 16 for an example of deep learning).

BPt is designed to be generalizable to different storage and computing requirements. In practice, data storage and computing requirements will depend on both the dataset of interest as well as predictive questions of interest. For example, performing machine learning on surface-projected data directly may require relatively large computational resources, but if the question or ML model of interest is simple, it could be run on a personal computer in a few hours. In general, BPt has been designed with single personal or workstation computing in mind and the vast majority of situations support this use case. However, this is not to say that BPt cannot be

Robert Whelan and Hervé Lemaître (eds.), *Methods for Analyzing Large Neuroimaging Datasets*, Neuromethods, vol. 218, https://doi.org/10.1007/978-1-0716-4260-3_12,

used by a more advanced user for more complex questions on large cloud-based computing clusters. Most functions within BPt allow for easy integration of multi-core processing to speed up potentially time-intensive ML modeling tasks, which tends to allow performing a greater range of analyses locally. Likewise, data storage requirements will obviously vary when dealing with a single csv file of a few hundred megabytes versus the raw fMRI files from a study with 10,000 participants (20 TB+).

2 Software and Coding

BPt is a Python 3.7+ based package that is tested regularly across all common operating systems (Windows, Mac, and Linux). Use of this package will therefore at the minimum require some proficiency and experience with Python and in setting up Python libraries. Prior experience with the standard data science Python libraries (e.g., pandas, numpy, scikit-learn) [see Resources] is encouraged but not strictly required. Likewise, some prior background knowledge on both neuroimaging and machine learning is expected as BPt tutorial material is not designed to be a user's first exposure to these topics. For new users, it is recommended that the library be used within a computation notebook (e.g., Jupyter notebook or Google Colab). These environments allow for an interactive and iterative approach to coding which is highly recommended when learning and exploring a new library or toolbox. Likewise, most available tutorial material is provided in this base format.

3 General Method

3.1 Inputs

Input data for the toolbox can take a wide range of forms, but generally speaking include outputs from a typical neuroimaging preprocessing pipeline (e.g., the example dataset used for this chapter), plus target and nuisance variables. The easiest data to work with are data already in tabular form (e.g., calculated mean values per region of interest). That said, the toolbox is capable of working with volumetric or surface projected structural or functional MRI (sMRI and fMRI, respectively) data as well. Other modalities, like EEG (electroencephalography), could also be analyzed using the toolbox, but in these cases, it may require additional formatting (as EEG requires quite different preprocessing steps).

There are no specific guidelines in terms of choice of preprocessing pipeline, or choice of parcellation size, atlas, voxel vs. vertex with respect to working with BPt. Instead, as BPt is a general utility toolbox, best practices with respect to all of these choices should be taken in consideration to the broader prediction-based neuroimaging literature. For the most part, these decisions will depend on the specific modalities employed as well as the predictive target(s) of interest. That said, there are a number of benchmark papers that address these questions empirically, including for surface-based sMRI [1] and functional connectome [2].

A good way of conceptualizing the 'readiness' level of data for machine learning is to consider any transformations that can be computed based solely on a single data point (e.g., a participant's data) versus transformations that utilize information across the entire dataset. That is, in most cases, any participant-level analysis or transformations should be already applied prior to machine learning. Importantly, the BPt toolbox provides support for some of these common data preparation/processing steps, which include: organization of the data, utilities for exploratory data visualization, common transformations such as k-binning and binarization, automatic outlier detection, information on missing data, and other summary measures. This chapter describes use of the interface to access common operations. Additional, more specific features exist as well (e.g., a built-in function to save a whole table of descriptive variables straight to a .docx file and built-in smart merging of index names; *see* Fig. 1).

3.1.1 Sample Size

No specific minimum number of participants are required, but when performing machine learning based experiments larger sample sizes are highly preferred (for a more detailed discussion on why [3]).

3.1.2 Missing Data

Missing data within predictive-based neuroimaging is a common occurrence given the 'messiness' of real-world data. BPt includes utilities both to identify (*see* Fig. 2) and purge existing datasets of missing data or alternatively if necessary to impute values properly within a machine learning pipeline (*see* Fig. 3). Supported strategies

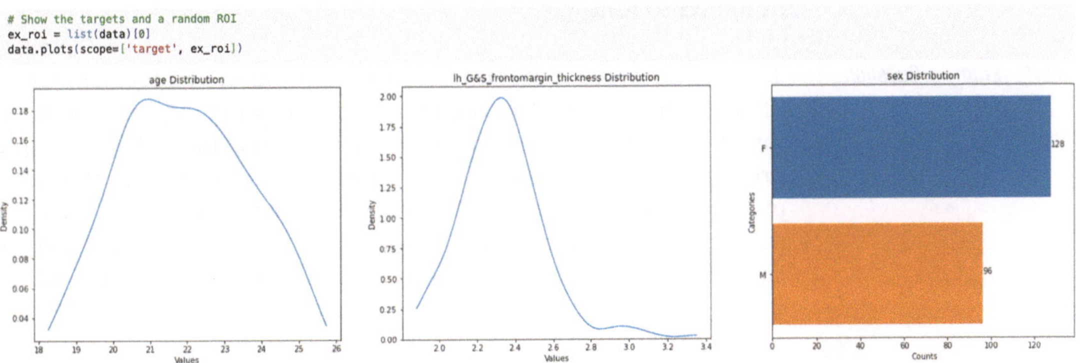

Fig. 1 Example showing build in dataset function for visualizing input data as a collage of plots

```
df.nan_info()
```

```
Loaded NaN Info:
There are: 23848 total missing values
24 columns found with 22 missing values (column name overlap: ['avg'])
23 columns found with 21 missing values (column name overlap: ['avg'])
20 columns found with 19 missing values (column name overlap: ['_surfavg'])
17 columns found with 20 missing values (column name overlap: ['avg'])
11 columns found with 23 missing values (column name overlap: ['avg', '_t'])
```

Fig. 2 Example of built-in function for providing information on patterns of loading missing data

BPt.Imputer

```
class BPt.Imputer(obj, params=0, scope='all', cache_loc=None,
base_model=None, base_model_type='default', **extra_params)
```
[source]

This input object is used to specify imputation steps for a `Pipeline`.

If there is any missing data (NaN's), then an imputation strategy is likely necessary (with some expections, i.e., a final model which can accept NaN values directly). This object allows for defining an imputation strategy. In general, you should need at most two Imputers, one for all *float* type data and one for all categorical data. If there is no missing data, this piece will be skipped.

Fig. 3 Screenshot of Imputer pipeline piece documentation

for imputation within a pipeline include mean and median imputation in addition to more complex strategies such as multiple rounds of iterative imputation.

3.2 Data Structure

BPt guides the user in how to structure and answer a question of interest within a predictive framework. Given the inherent vastness of this topic, it is important to note that there is no single 'right' way of doing things, and instead what we present here is a set of general recommendations in which the underlying library has been designed to follow.

3.2.1 Frame a Question

The very first step is to frame a research question of interest in terms of a prediction. For example, if our question of interest is to investigate age-related changes in cortical thickness, then a simple predictive re-framing could be: "how well can cortical thickness predict a participant's age?". What if we had longitudinal data per participant? Then maybe we could ask, how well does cortical thickness predict age at time point 1, what about time point 2, and so on?

```
import BPt as bp

# Load
data = bp.read_csv('quick_start.csv',
                   index_col='participant_id',
                   targets=['age', 'sex'])

# Set sex as a binary variable
data = data.to_binary('sex')
```

Fig. 4 Example of how data saved in a csv can be quickly loaded, and information around which columns are input data and which are target variables quickly set. Likewise, this example shows how columns can be easily transformed, in this case the variable 'sex' is binarized

Note

The key pieces of information to identify after composing a question of interest are: What are the input variables to the prediction? What variable(s) are being predicted? Furthermore, are there any other variables which might influence this prediction in an undesirable way (i.e., potential confounding variables).

3.2.2 Prepare Data in BPt

Once a question of interest has been identified, we load it into a Dataset object (*see* Fig. 4), which is a Python class based on the popular Pandas DataFrame. The key point here is that the Dataset object is inherently designed to enforce an explicit organization structure based on the question of interest. The idea is that each column of the Dataset class—where data points within the column are either single values or external references to (e.g., to a saved sMRI file)—are given a role: 'data', 'target' or 'non input'. These roles correspond to the variables used as input to a machine learning algorithm ('data'), the target variables that are predicted ('target') and everything else, including the potential confounding variables ('non input').

There are some pre-modelling steps that, depending on the dataset and the question, might also be explored at this stage, and can be performed using the Dataset object directly. For example, users may want to: generate exploratory plots of the different features in the dataset, remove any data based on status as an outlier, decide if missing data should be kept and imputed, or dropped, apply any pre-requisite transformations that should be applied to the data? (e.g., conversion from strings 'Male', 'Female' to 0 and 1's).

3.2.3 Define a ML Pipeline

A machine learning pipeline is not just the choice of ML model, it is the full set of transformations to the data prior to input to an ML algorithm. This is, in a lot of ways, the area with the most researcher degrees of freedom, as we can think of both the presence or absence of a transformation, as well as the choice of model and that model's parameters as all 'hyper-parameters' of the broader ML pipeline. These could be choices like what brain parcellation to use, to z-score each feature or not, which type of fMRI connectivity metric to use, the type of ML estimator, the parameters associated with that estimator, etc. The number of permutations grows quite rapidly, so in practice how should the researcher decide? We recommend treating each possible 'hyper-parameter' according to the following set of options.

Note

If a parameter is important to the research question, test and report the results by each possible value or a reasonable set of values of interest that this parameter might take. For example, let's say we want to know how our prediction varies by choice of parcellation, so we repeat our full ML experiment with three different parcellations, and report the results of each. Otherwise, if not directly important or related to the question of interest the researcher can either: (1) fix the value ahead of time based on a priori knowledge or best estimate or (2) assign the value through some nested validation strategy (e.g., train-validation/test split or nested K-fold). In general, option 1 is preferable, as it is simpler to both implement and conceptualize fixing a value ahead of time. That said, setting values through nested validation can be useful in certain cases, for example, it is often used for setting hyper-parameters specific to an ML estimator. In other words, option 2 is used as a way to try and improve down-stream performance, with an emphasis on 'try,' as it is difficult in practice to correctly identify the choices which will benefit from this approach.

While designing an ML pipeline can be daunting and introduce lots of researcher degrees of freedom, it is also the area most amenable to creativity. As long as proper validation, as discussed in the next section, is kept in mind, testing and trying new/different pipelines can be an important piece of ML modeling. This becomes especially important when the researcher starts to consider ML modeling in the context of potential confounds, where potential corrections for confounds are themselves steps within the pipeline. That said, especially as a newer researcher, it may be a good

```
results = bp.evaluate(pipeline='ridge_pipe', dataset=data, target='age')
```

Fig. 5 Screenshot showing how a default pipeline can be easily selected when defining a pipeline within an evaluation loop, in this case a default pipeline based on a regularized ridge regression

```
import BPt as bp

pipe = bp.Pipeline([bp.Scaler(obj='robust'),
                    bp.Model(obj='ridge',
                             params=1,
                             param_search=bp.ParamSearch(n_iter=60))])
```

Fig. 6 Customized creation of pipelines

idea to start by replicating previous strategies from the literature that have been found to work well. Default pipelines can be easily specified within BPt (*see* Fig. 5) or alternatively, we can easily customize the creation of pipelines (*see* Fig. 6).

3.2.4 Select and Evaluate According to a Validation Strategy

In order for the results from a ML-based predictive experiment to be valid, some sort of cross or external validation is essential. So how do we decide between say a training-test split between two matched samples and K-fold cross validation on the whole sample? In short, it depends. There is no silver bullet that works for every scenario, but the good news is that for the most part it really shouldn't matter! The most important element to properly using an external validation strategy isn't between threefolds versus tenfolds, but instead is in how the chosen strategy is used. That is to say, the validation data should only be used in answering the main predictive question of interest. If instead the current experiment isn't related to the primary research question, that is to say, the result will not be reported, then the validation data should not be used in any way. Let's consider an explicit example of what *not* to do: Let's say we decide to use a threefold cross validation strategy, predicting age from cortical thickness, and we start by evaluating a simple linear regression model, but it doesn't do very well. Next, we try a random forest model, which does a little better, but still not great, so we try changing a few of its parameters, run the threefold cross validation again, change a few more parameters, and after a little tweaking eventually get a score we are satisfied with. We then report just this result: "a random forest model predicted age, R2=XXX." The issue with the example above is, namely, one of over-using the validation data. By repeatedly testing different models with the same set of validation data, be it through K-fold or a left-aside testing set, we have increased our chances of obtaining an artificially high performance metric through chance alone (i.e., this is a phenomenon pretty similar in nature to p-hacking in classical statistics). Now in this example the fix is fairly

```
cv = bp.CV(splits=3, n_repeats=1, stratify='sex')
```

Fig. 7 Example of defining a cross-validation strategy where a three-fold validation is performed, and further the ratio of 'Males' and 'Females' as defined in variable 'sex' are preserved within every training and validation set

```
cv = bp.CV(splits='site')
```

Fig. 8 Example showing how a validation strategy for performing leave-site-out cross validation can be easily defined

easy. If we want to perform model selection and model hyper-parameter tuning, we can, but as long as both the model selection and hyper-parameter tuning are conducted with nested validation (e.g., on a set-aside training dataset). Fundamentally, it depends on what our ultimate question of interest is. For example, if we are explicitly interested in the difference in performance between different ML models, then it is reasonable to evaluate all of the different models of interest on the validation data, as long as all of their respective performances are reported.

There are of course other potential pitfalls in selecting and employing validation strategies that may vary depending on the underlying complexity of the problem of interest. For example, if using multi-site data, there is a difference between a model generalizing to other participants from the same site (random split k-fold validation) versus generalizing to new participants from unseen sites (group k-fold validation where site is preserved within fold). While choice of optimal strategy will vary, BPt provides an easy interface for employing varied and potentially complex validation strategies, such as internal cross-validation (*see* Fig. 7) or leave-site-out (*see* Fig. 8).

4 Interpreting and Reporting Results

Results from every machine learning based evaluation in BPt return a special results object called 'EvalResults' (*see* Fig. 9). This object stores by default key information related to the conducted experiment, which allows the user to then easily access or additionally compute a range of useful measures. Listed below are some of the available options.

Base common machine learning metrics are provided, across regression, binary, and multi-class predictions, for example R^2, negative mean squared error, ROC AUC (Receiver Operating Characteristic Area Under Curve) [see Glossary], balanced accuracy, and others. In the case of employing a cross-validation strategy like K-fold, these metrics can be accessed either per fold, or averaged across multiple folds (or even the weighted average across folds of different sizes).

```
EvalResults
------------
r2: 0.1027 ± 0.0454
neg_mean_squared_error: -2.83 ± 0.4594

Saved Attributes: ['estimators', 'preds', 'timing', 'estimator', 'train_subjects',
'val_subjects', 'feat_names', 'ps', 'mean_scores', 'std_scores', 'weighted_mean_scores',
'scores', 'fis_', 'coef_', 'cv']

Available Methods: ['to_pickle', 'compare', 'get_X_transform_df', 'get_inverse_fis',
'run_permutation_test', 'get_preds_dfs', 'subset_by', 'get_fis', 'get_coef_',
'permutation_importance']

Evaluated With:
target: age
problem_type: regression
scope: all
subjects: all
random_state: 1
```

Fig. 9 Example showing a string representation of an EvalResults object, with information on saved attributes and methods

```
results.get_preds_dfs()[0]
```

	predict	y_true
participant_id		
sub-0001	21.909845	25.50
sub-0005	22.126259	24.75
sub-0012	21.909031	22.75
sub-0017	21.716770	20.50
sub-0019	22.436428	21.25

Fig. 10 Example showing how predictions can be accessed from the results object

Raw predictions made per participant in the validation set (s) can be accessed in multiple formats (*see* Fig. 10 for an example) and can be useful in performing further analysis beyond those implemented in the base library (e.g., computing new metrics or feature importances).

In the case that the underlying machine learning model natively supports a measure of feature importance (e.g., beta weights in a linear model), then these importances can be directly accessed (*see* Fig. 11). Additionally, feature importances can be estimated regardless of underlying pipeline through a built-in permutation-based feature importance method. When working with neuroimaging objects directly (e.g., volumetric or surface representations of the data), users can back-project feature importances into their original space.

```
fis = results.get_fis().mean()
fis
```

```
lh_G&S_cingul-Ant_thickness            0.041987
lh_G&S_cingul-Mid-Ant_thickness       -0.107472
lh_G&S_cingul-Mid-Post_thickness      -0.115151
lh_G&S_frontomargin_thickness          0.022932
lh_G&S_occipital_inf_thickness        -0.003825
                                          ...
```

Fig. 11 Example showing how averaged feature importances can be quickly accessed

```
p_values, null_values = results.run_permutation_test(n_perm=10,
                                                     blocks=data['sex'], within_grp=True,
                                                     plot=True)
```

Fig. 12 Example showing how a constrained permutation test can be performed, where target labels are only permuted for participants with the same sex label

```
from neurotools.plotting import plot

plot(results.get_fis().mean())
```

Fig. 13 Example code, for plotting feature importances from ROIs automatically onto a set of brain surfaces

The results of a single evaluation, regardless of cross-validation method, can be investigated further in order to ask questions around the statistical significance of results and/or the potential influence of confounds on results. One of the most powerful tools for this type of analysis is a permutation test, wherein the analysis is repeated but with the target labels shuffled. An important extension to this base method is the ability to restrain the shuffling of target labels according to an underlying group or nested group structure (*see* Fig. 12 for an example).

Another available method related to probing the significance of results, is the ability to statistically compare between two and more similar results objects, that perhaps vary on choice of a meaningful hyper-parameter. It can also be useful in some instances to visualize the predictions made in other ways, for example, through ROC plots from a binary or multi-class analysis, or plots showing the residuals from regression prediction. Feature importances from BPt are further designed to be easily visualized through the related python package, from the same maintainers as BPt, bp-neurotools (*see* Fig. 13). This package contains one-line automatic plotting functions that handle a number of different cases (e.g., plotting ROIs, brain surfaces, brain volumes, or collages of different combinations).

When working with neuroimaging data files directly (e.g., performing machine learning on surfaces), BPt includes utilities that allow the user to back-project feature importances back into the original native space. This can be useful, along with the already mentioned neuroimaging specific plotting utilities, for visualizing results.

Note
When it comes to presenting a final set of results within a manuscript or project write up, there is no one-size fits all solution. Instead, how one reports results will depend fundamentally on the question(s) of interest. In practice, the typical advice is that all metrics from experiments related to questions should be reported. Likewise, all related experimental configurations tested should also be reported, the key point being that the user should do their best to accurately and fairly present their results. As tempting or desirable as publishing a very accurate classifier may be, authors should take care not to overstate their findings. This principle holds in the context of null findings as well, where it is valuable to highlight the areas where predictive models fail.

5 General Pitfalls

There are of course general pitfalls to be aware of when performing any type of analyses on observational data, which are not specific to this library itself but are prudent to keep in mind. Perhaps the most general is that despite machine learning, deep learning, or other variations, results will typically be correlational, not causal, in nature.

While machine learning can be a useful tool for identifying null findings, for example when a model is not predictive, the nature of predictive modeling means we cannot ever be fully confident. In other words, just because one model (or full pipeline/set of steps) isn't predictive does not mean that another one may produce a positive result. In practice, it is typically sufficient in the case of null findings to show that a representative range of pipelines all fail to predict the outcome.

The over-use of cross-validation or 'double-dipping' is a particularly insidious and sometimes hard to detect issue within machine learning and the broader literature [4]. These types of mistakes are often responsible for overly optimistic or inflated accuracy. Further, the conceptual difficulties with employing cross-validation correctly can multiply in the case of nested cross-validation, so potential users should be very careful if attempting to implement a custom-designed cross validation scheme. The cleverly titled "I tried a bunch of things: The dangers of unexpected overfitting in classification of brain data" provides a good, expanded description on this issue more broadly [5].

Be wary of results that look 'too' good. There are many different mistakes in machine learning which can lead to over-confident results, including problems with the data, mis-using cross-validation, and a whole host of other tricky issues. When encountering a situation like this, the best course of action is typically to perform small checks (building them into analysis code, whenever possible). These include little things such as printing the shape of a dataset, or maybe performing some assertion on the expected distribution of variables (e.g., confirm values for age in years are greater than 0 and less than 100). These types of checks are generally low effort and in the long run can be helpful in detecting small but disastrous bugs.

Interpreting, and in some cases overinterpreting, feature importances is a common problem. In practice, different measures of feature importance may have different drawbacks and constraints, and it is therefore a good idea to make sure one first understands a given importance's potential limitations. For example, in the case of multivariate linear models we refer readers to the excellent tutorial made available by scikit-learn. Hooker et al. [6] outline well other potential issues in interpreting some other common formulations of feature importance [6]. A more general discussion around interpretability in machine learning by Kaur et al. [7] may also be of interest [7].

6 Step-by-Step Example https:/colab.research.google.com/drive/1vFGw8HtpDeLCbDmYUiLKwa5JQwsiiUnF?usp=sharing

Intro/Setup

Within this example notebook, we will investigate as our test question of interest: Can cortical thickness ROIs predict participant age?

First though, let's download the brain predictability toolbox and additional neurotools package to this collab instance.

Note running this cell this first time will install and then crash the instance, after this we can just run it as normal.

```
try:
 import BPt as bp
except ImportError:
 !pip install brain-pred-toolbox
```

(continued)

```
!pip install bp-neurotools
# We need to force a restart of the runtime/kernel
# because of the version of matplotlib we are using
exit()
```

This example is designed to be run from the already prepared quick_start file, which went through a few simple steps to combine separately saved thickness ROIs into a single file, as well as to add age and sex columns. We will download this file to this instance directly, now.

```
!wget https://raw.githubusercontent.com/sahahn/methods_-
series/master/ds002790/quick_start.csv
--2022-10-31 17:01:10--https://raw.githubusercontent.
com/sahahn/methods_series/master/ds002790/quick_start.
csv
Resolving raw.githubusercontent.com (raw.githubusercon-
tent.com)... 185.199.108.133, 185.199.109.133,
185.199.110.133, ...
Connecting to raw.githubusercontent.com (raw.githubuser-
content.com)|185.199.108.133|:443... connected.
HTTP request sent, awaiting response... 200 OK
Length: 304024 (297K) [text/plain]
Saving to: 'quick_start.csv.1'

 quick_start.csv.1 0%[ ] 0 --.-KB/s quick_start.csv.1
100%[===================>] 296.90K --.-KB/s in
0.02s
2022-10-31 17:01:10 (13.8 MB/s) - 'quick_start.csv.1'
saved [304024/304024]
```

Preparing Our Dataset

As a first step, we will prepare our data into a BPt Dataset object. If you are already familiar with the python library pandas, then you might notice that this object looks awfully similar to the pandas DataFrame object – and you would be right! The Dataset class is built directly on top of the DataFrame class, just adding some extra functionality/special behavior for working with the BPt.

In this minimal example, this step will be rather simple. There are 3 different 'roles' that columns within our Dataset can take. The first is called 'data,' which by default every loaded column will be specified as, these are going to be our features that are used to predict some variable of interest (The X variable in a scikit-learn style setup). The second key role is

(continued)

'target,' which is going to be any of our feature(s) which we want to predict (using the columns as input variables). The last is 'non input' which as the name suggests are any variables which we do not want to ever use directly as an input/data variable. In this example, we will treat both age and sex as targets, and not use the 'non input' role.

Here we load data directly from a prepared csv, which gives us a BPt Dataset object. In loading the csv, we also specify a series of additional arguments, these are the file paths of the csv, which column we want to be treated as the index, then also which columns we want in different roles (where remember that by default every loaded variable is of role 'data' unless otherwise specified).

```
import BPt as bp

# Load
data = bp.read_csv('quick_start.csv',
                   index_col='participant_id',
                   targets=['age', 'sex'])


# Set sex as a binary variable
data = data.to_binary('sex')

# Show the first five rows of our Dataset

data.head()
  lh_G&S_frontomargin_thickness  lh_G&S_occipital_inf_th-
ickness \
participant_id
sub-0001      1.925    2.517
sub-0002      2.405    2.340
sub-0003      2.477    2.041
sub-0004      2.179    2.137
sub-0005      2.483    2.438


  lh_G&S_paracentral_thickness  lh_G&S_subcentral_thick-
ness \
participant_id
sub-0001      2.266    2.636
sub-0002      2.400    2.849
sub-0003      2.255    2.648
sub-0004      2.366    2.885
sub-0005      2.219    2.832
```

(continued)

```
 lh_G&S_transv_frontopol_thickness \
participant_id
sub-0001 2.600
sub-0002 2.724
sub-0003 2.616
sub-0004 2.736
sub-0005 2.686


 lh_G&S_cingul-Ant_thickness lh_G&S_cingul-Mid-An-
t_thickness \
participant_id
sub-0001 2.777 2.606
sub-0002 2.888 2.658
sub-0003 2.855 2.924
sub-0004 2.968 2.576
sub-0005 3.397 2.985


 lh_G&S_cingul-Mid-Post_thickness \
participant_id
sub-0001 2.736
sub-0002 2.493
sub-0003 2.632
sub-0004 2.593
sub-0005 2.585


 lh_G_cingul-Post-dorsal_thickness \
participant_id
sub-0001 2.956
sub-0002 3.202
sub-0003 2.984
sub-0004 3.211
sub-0005 3.028


 lh_G_cingul-Post-ventral_thickness ... \
participant_id ...
sub-0001 2.925 ...
sub-0002 2.868 ...
sub-0003 2.972 ...
sub-0004 2.428 ...
sub-0005 3.361 ...
```

(continued)

```
 rh_S_postcentral_thickness \
participant_id
sub-0001 2.038
sub-0002 1.882
sub-0003 2.066
sub-0004 1.930
sub-0005 1.938


 rh_S_precentral-inf-part_thickness \
participant_id
sub-0001 2.425
sub-0002 2.513
sub-0003 2.410
sub-0004 2.241
sub-0005 2.445


 rh_S_precentral-sup-part_thickness rh_S_suborbi-
tal_thickness \
participant_id
sub-0001 2.324 2.273
sub-0002 2.429 2.664
sub-0003 2.579 3.494
sub-0004 2.296 3.092
sub-0005 2.218 3.712


 rh_S_subparietal_thickness rh_S_temporal_inf_thickness \
participant_id
sub-0001 2.588 2.548
sub-0002 2.676 2.220
sub-0003 2.375 2.625
sub-0004 2.641 2.622
sub-0005 2.360 2.402


 rh_S_temporal_sup_thickness \
participant_id
sub-0001 2.465
sub-0002 2.291
sub-0003 2.497
sub-0004 2.487
sub-0005 2.442
```

(continued)

```
rh_S_temporal_transverse_thickness age sex
participant_id
sub-0001 2.675 25.50 1
sub-0002 2.714 23.25 0
sub-0003 2.674 25.00 0
sub-0004 2.556 20.00 0
sub-0005 1.864 24.75 1
```

[5 rows × 150 columns]

There are other steps we could potentially perform here as well, e.g., let's look to see if there are any extreme outliers. Like you can find other available options for different common encodings or filtering.

```
data = data.filter_outliers_by_std(scope='float',
n_std=10)
data.shape
(224, 150)
```

We can see that no data was dropped with a strict filter of 10 standard deviations. We can also confirm before moving on that there is no missing data in this prepared dataset.

If there were any missing data, this would print something

data.nan_info()

Visualize Features

BPt includes a few different utilities for easy visualizations, in particular, our Dataset class has some built in plotting functions. In the example below, we specify what variables/ columns we want to plot with a special scope argument. We can see the distributions of variables with plotting methods plot and plots.

Show the targets and a random ROI

```
ex_roi = list(data)[0]
data.plots(scope=['target', ex_roi])
```

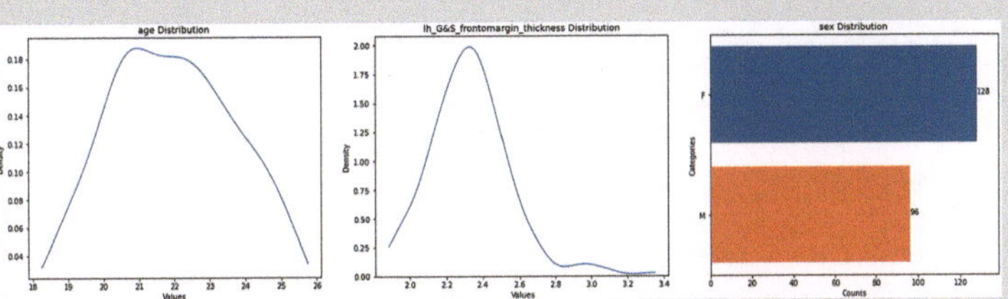

(continued)

Another built in plotting option we can use is for visualizing bi-variate relationships between variables, with plot_bivar. Note that internally these plotting functions make use the library seaborn which is addition to matplotlib for creating more complex out of the box plots.

```
# Plot age vs sex and the roi
data.plot_bivar(scope1='age', scope2=['sex', ex_roi])
```

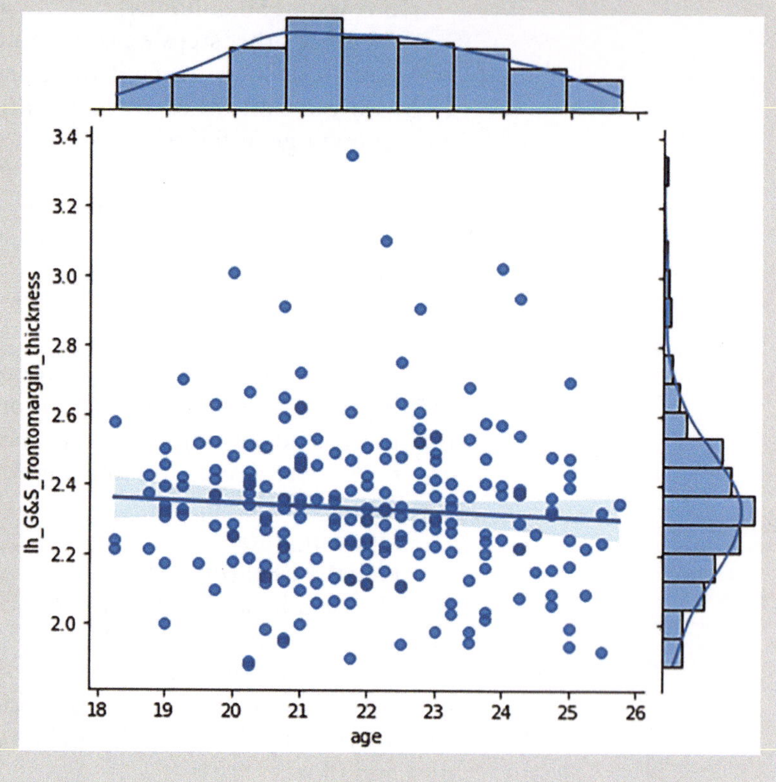

(continued)

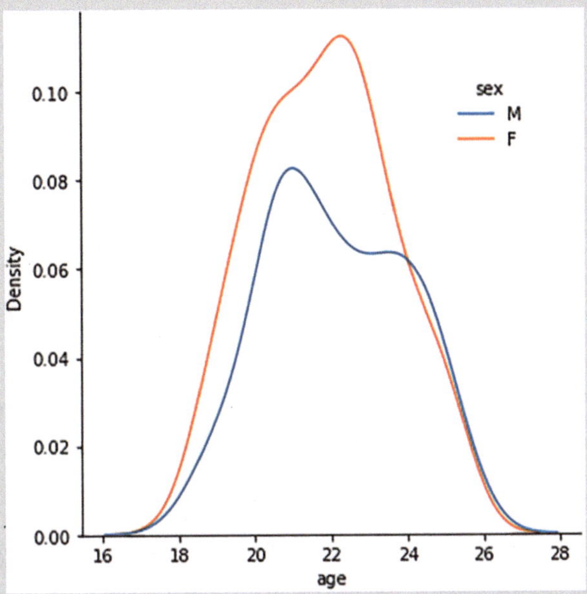

Machine Learning

Next, we will jump in directly to a minimal machine learning example. In this case, our question of interest is can our thickness ROIs predict age, which we have already setup within our dataset by virtue of specifying all cortical thickness ROI's as role='data' and age with role='target,' though since we have two loaded targets, we could also optionally make sure that age is predicted by passing target='age.'

To run the experiment itself, we are going to use the evaluate method from BPt. This method allows us to perform a number of different ML evaluations and can be customized according to a large number of different parameters. In this case though, we are going to provide the bare minimum input needed, and let the default settings take care of everything else.

In particular, we specify that we want to use a default ML pipeline from BPt called 'ridge_pipe' which is a pre-defined pipeline based on a regularized ridge regressor w/nested hyper-parameter search, and also we let the function know our dataset.

```
results = bp.evaluate(pipeline='ridge_pipe', dataset=da-
ta, target='age')
Predicting target = age
Using problem_type = regression
```

(continued)

```
Using scope = all (defining a total of 148 features).
Evaluating 224 total data points.
{"version_major":2,"version_minor":0,"model_i-
d":"66a13cc8677948f583ebc060aec166c4"}


Training Set: (179, 148)
Validation Set: (45, 148)
Fit fold in 1.8 seconds.
r2: 0.1416
neg_mean_squared_error: -3.27


Training Set: (179, 148)
Validation Set: (45, 148)
Fit fold in 1.8 seconds.
r2: 0.1444
neg_mean_squared_error: -2.19


Training Set: (179, 148)
Validation Set: (45, 148)
Fit fold in 1.7 seconds.
r2: 0.1131
neg_mean_squared_error: -3.44


Training Set: (179, 148)
Validation Set: (45, 148)
Fit fold in 1.8 seconds.
r2: 0.0944
neg_mean_squared_error: -2.66


Training Set: (180, 148)
Validation Set: (44, 148)
Fit fold in 5.6 seconds.
r2: 0.0198
neg_mean_squared_error: -2.59
```

We can see from the verbose output above that five different training and validation sets was evaluated. This is because the default cross-validation behavior is to run a K-Fold cross validation with 5 folds.

(continued)

In essence that brief example was designed to be as implicit as possible or mostly rely on default values. That said, we could run the same exact evaluation, but this time explicitly providing a number of the default arguments as:

```
bp.evaluate(pipeline='ridge_pipe', dataset=data
 target='age', scorer=['r2', 'neg_mean_squared_error'],
 scope='all', subjects='all',
 problem_type='regression',
 n_jobs=1, random_state=1, cv=5,
 progress_bar=True, eval_verbose=1)
```

Next, let's look at the returned results object, an instance of EvalResults which we saved in variable results.

results

EvalResults

----

r2: 0.1027 ± 0.0454

neg_mean_squared_error: -2.83 ± 0.4594

Saved Attributes: ['estimators', 'preds', 'timing', 'estimator', 'train_subjects', 'val_subjects', 'feat_names', 'ps', 'mean_scores', 'std_scores', 'weighted_mean_scores', 'scores', 'fis_', 'coef_', 'cv']

Available Methods: ['to_pickle', 'compare', 'get_X_transform_df', 'get_inverse_fis', 'run_permutation_test', 'get_preds_dfs', 'subset_by', 'get_fis', 'get_coef_', 'permutation_importance']

Evaluated With:

target: age

problem_type: regression

scope: all

subjects: all

random_state: 1

This object saves by default a large amount of potentially useful information from the experiment. This includes the mean evaluation metrics, actual estimator objects, predictions made, information on feature importance, and more.

For example, we can look at the more 'raw' object of what exactly we just ran.

results.estimator

```
BPtPipeline(steps=[('mean float',
 ScopeTransformer(estimator=SimpleImputer(), inds=Ellip-
sis)),
 ('median category',
```

(continued)

```
ScopeTransformer(estimator=SimpleImputer(strategy='me-
dian'), inds=[])),
 ('robust float',
 ScopeTransformer(estimator=RobustScaler(quantile_range=
(5, 95)), inds=Ellipsis)),
 ('one hot encoder category',
 BPtTransformer(estimator=OneHotEncoder(handle_unknow...
```

BPtModel(estimator=NevergradSearchCV(estimator=–Ridge(max_iter=100, random_state=1, solver='lsqr'), param_distributions={'alpha': Log(lower=0.001, upper=100000.0)}, ps={'cv': BPtCV(cv_strategy=CVStrategy(), n_repeats=1, splits=3, splits_vals=None), 'cv__cv_strategy': CVStrategy(), 'cv__cv_strategy__groups': None, 'cv__cv_strategy__stratify': None, 'cv__cv_strategy__-train_only_subjects': None, 'cv__n_repeats': 1, 'cv__only_-fold': None, 'cv__random_state': 'context', 'cv__splits': 3, 'dask_ip': None, 'memmap_X': False, 'mp_context': 'loky', 'n_iter': 60, 'n_jobs': 1, 'progress_loc': None, 'random_state': 1, 'scorer': make_scorer(r2_score), 'search_only_params': {}, 'search_type': 'RandomSearch', 'verbose': 0, 'weight_scorer': False}, random_state=1), inds=Ellipsis))])

Yikes, that's a handful... another way of looking at the pipeline we ran is to look at it in the BPt pipeline object syntax rather than the raw scikit-learn style.

Essentially when we pass 'ridge_pipe' as our pipeline, it will grab some default code from BPt.default.pipelines, we can also import it directly.

from BPt.default.pipelines import ridge_pipe

```
ridge_pipe
```

```
Pipeline(steps=[Imputer(obj='mean', scope='float'),
 Imputer(obj='median', scope='category'), Scaler(obj='-
robust'),
 Transformer(obj='one hot encoder', scope='category'),
 Model(obj='ridge',
 param_search=ParamSearch(cv=CV(cv_strategy=CVStrategy
()),
 n_iter=60),
 params=1)])
```

As a careful user might note that in this case a number of these steps are actually redundant, e.g., we have no missing data, so no need for imputation and we have no categorical data, so no need for one hot encoding. The beauty here is that

(continued)

if not needed, or if out of scope given a certain input, these pipeline steps are just skipped. This is helpful for designing re-usable pipelines, that are robust to different types of inputs (e.g., includes categorical variables or not).

We can also look at some other options available when working with a result's object, for example, viewing the raw predictions.

```
results.get_preds_dfs()[0]
 predict y_true
participant_id
sub-0001 21.909845 25.50
sub-0005 22.126259 24.75
sub-0012 21.909031 22.75
sub-0017 21.716770 20.50
sub-0019 22.436428 21.25
sub-0020 22.145191 20.00
sub-0029 22.262978 21.75
sub-0030 21.805216 20.50
sub-0032 22.008379 20.50
sub-0034 22.045389 24.75
sub-0035 22.154486 23.25
sub-0036 22.001040 24.00
sub-0039 22.097433 22.25
sub-0040 21.570295 19.75
sub-0045 22.572697 25.00
sub-0052 21.528109 19.00
sub-0059 21.977768 23.25
sub-0063 22.722687 25.25
sub-0068 21.829220 22.25
sub-0070 22.035213 18.75
sub-0074 21.457249 20.25
sub-0079 22.089703 20.25
sub-0085 21.702505 23.00
sub-0086 21.735174 21.00
sub-0092 22.166224 21.50
sub-0100 22.488585 23.75
sub-0103 21.614632 21.50
sub-0107 21.745026 19.25
sub-0108 21.288305 21.25
sub-0120 22.015606 22.75
sub-0121 21.495790 21.75
sub-0125 22.300081 23.00
sub-0152 22.030699 23.75
sub-0153 21.711060 19.00
sub-0156 21.993868 20.75
```

(continued)

```
sub-0161 21.311602 23.50
sub-0163 22.200386 22.25
sub-0167 21.918657 24.75
sub-0169 21.757229 25.00
sub-0188 22.429842 23.75
sub-0189 21.426121 20.50
sub-0192 21.896299 22.25
sub-0196 21.228672 19.25
sub-0198 22.514311 25.00
sub-0204 21.584047 19.75
```

Visualizing Results

Let's say we want to look at feature importances. Because the model we ran was a regularized ridge regression, the importances we are going to be looking at are the beta weights from the model. Further, because we ran a 5-fold CV, we want to look at the beta weights as averaged across each of our 5 models:

```
fis = results.get_fis().mean()
fis
lh_G&S_cingul-Ant_thickness 0.041987
lh_G&S_cingul-Mid-Ant_thickness -0.107472
lh_G&S_cingul-Mid-Post_thickness -0.115151
lh_G&S_frontomargin_thickness 0.022932
lh_G&S_occipital_inf_thickness -0.003825
 ...
rh_S_suborbital_thickness -0.033135
rh_S_subparietal_thickness -0.026897
rh_S_temporal_inf_thickness 0.008185
rh_S_temporal_sup_thickness -0.004152
rh_S_temporal_transverse_thickness -0.080884
```

Length: 148, dtype: float32

Next, let's plot our results. For this, we will use an 'automagical' plotting function, plot, from the library neurotools (a library designed to complement BPt, but with a less ML focus).

from neurotools.plotting import plot

```
plot(fis)
Downloading latest neurotools_data to/root/neurotools_da-
ta
Downloaded data version = 1.2.5 complete!
Current version saved at: /root/neurotools_data/neuro-
tools_data-1.2.5/data
```

(continued)

```
If you move this directory, make sure to update saved
location in data ref at /usr/local/lib/python3.7/dist-
packages/neurotools/data_ref.txt.
```

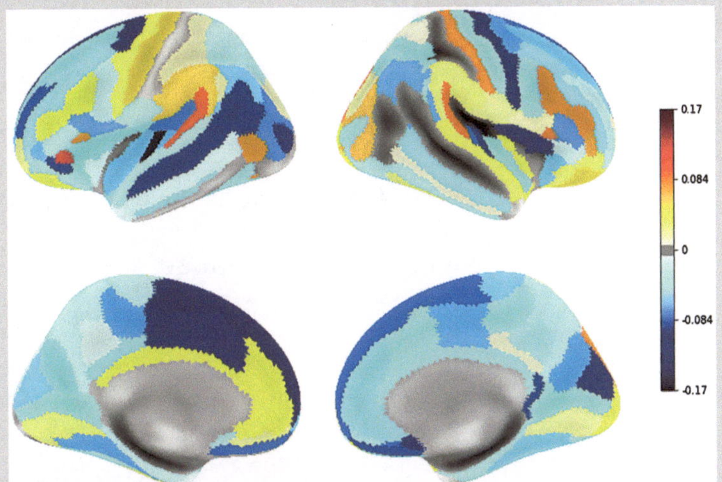

This plotting function tries to basically automate everything. This includes, as an important caveat, an automatic conversion from ROI names to their associated parcellation – which for now only supports a small number of underlying parcellations, including of course our current freesurfer based destr. parcellation.

Next, we will plot the same feature importances again, but this time go a little further to add some customization. We will customize here by adding a user-defined threshold in which to not show results under and also go a little deeper and add it as a part of a collage of plots. We will also save the figure with matplotlib.

```
import matplotlib.pyplot as plt
from neurotools.plotting import plot_bars

# Initialize two subplots on the same row
fig, axes = plt.subplots(nrows=1, ncols=2,
 figsize=(18, 6),
 gridspec_kw={'wspace': 0})

# Share threshold
threshold = .1
# Make a bar plot, with fi values before each fold
plot_bars(results.get_fis(), threshold=threshold, ax=axes
[0])
```

(continued)

Use the same plot as before, but with some extra arguments

```
plot(fis,
 threshold=.1,
 space='fsaverage5', # Note could change to fsaverage for
higher resolution
 ax=axes[1]
 )
```

Add a title to the whole figure

```
plt.suptitle('Predict Age - Avg. Beta Weights', font-
size=22)


plt.savefig('example.png', dpi=100)
```

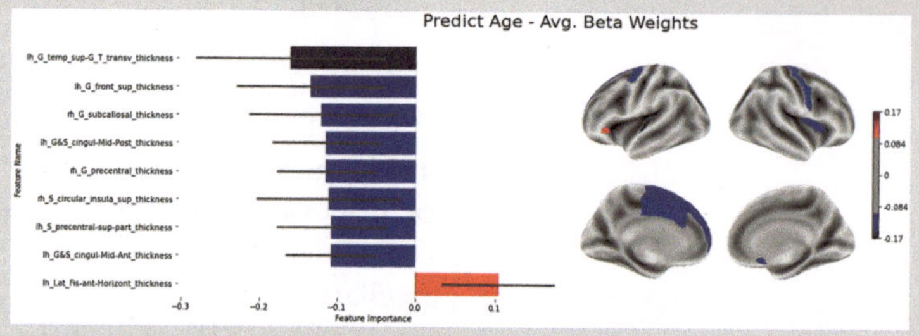

Permutation Test

There are of course other useful things we can do with this result's object. One useful one is to be able to easily run permutation tests as a way of estimating the significance of our results. In the context of results generated from cross-validation and if significance tests are desired, than permutation based methods are preferred.

While this is useful, we can also easily extend this idea with a powerful extension, that is, constraining the permutations in a meaningful way. For example, we will run 10 permutations, but with the added specification that values within the target only be allowed to be swapped with other participants of the same sex. Now what we are testing is a more specific null model, one where any potential sex-age effects will be preserved within our null distribution. In this version, we are essentially testing to see if sex effects are driving our observed R2. If the null dist mean is still the same as before, it is likely not, but if it is higher, than to some degree it might be. This

(continued)

type of constrained permutation test is especially useful with multi-site data, where a variable representing site is passed.

```
p_values, null_values = results.run_permutation_test
(n_perm=10,
 blocks=data['sex'], within_grp=True,
 plot=True)
```

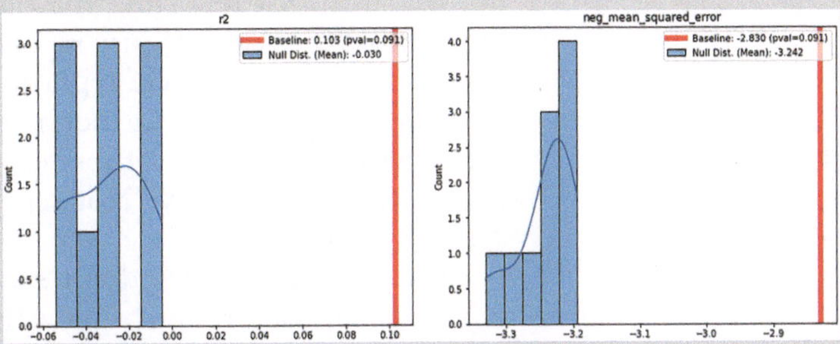

Other ML Models

There are plenty of other choices for ML models we could have used besides the regularized linear ridge regression, even while staying within the default pipelines made available by BPt. That said ...

Testing a bunch of different models is a common area where essentially too many researcher degrees of freedom are often introduced. In order for the results from a ML-based predictive experiment to be valid, some sort of cross or external validation is essential. In this example, we have used a 5-fold cross-validation on the entire available dataset. What that means in practice is that the only ML experiments we want to run this full 5-fold cross validation on, are those which are directly related to our main question of interest – or in other words – we should be reporting any results which we test using this full 5-fold cross validation.

So let's say we do test multiple models using this full 5-fold CV. Let's test additionally the default elastic-net based pipeline and the default gradient boosting based pipeline. To do this, we will use an object from BPt called Compare which allows the same evaluate function from earlier to perform a comparison between a few different options.

```
compare_pipes = bp.Compare(['ridge_pipe', 'elastic_-
pipe', 'gb_pipe'])
```

(continued)

```
all_results = bp.evaluate(pipeline=compare_pipes,
 dataset=data, eval_verbose=-2, n_jobs=1)
```

```
all_results.summary()
```

In this case, now that we have run and tested three different models, we have a perfect example of what not to do, which is, just reporting the ridge regression results. Or more broadly speaking, just reporting the best performing results. Instead, if we were reporting this result in a paper or write-up, we NEED to include information related to all of the experiments we ran, if those experiments made use of our main validation strategy! Not reporting this information is similar to p-hacking.

Let's say though we don't want to have choice of ML model fill up our results, but we still want to find a high performing model. One common strategy for this is done with nested cross-validation, the easiest being a front-end global train-test split. I'll show a quick example of that below (again under the assumption that this was done INSTEAD of what we already did in this notebook).

```
# Make a copy of our dataset
data_tr_test = data.copy()
```

```
# Split it into train and test sets
data_tr_test = data_tr_test.set_test_split(size=.2, ran-
dom_state=4)
data_tr_test
```

Re-run the same compare style analysis from before, but now with subjects = 'train.' So, explicitly, in this alternate analysis we are performing 5-fold CV, but only on the subset of subjects we labelled as part of the training set.

```
all_results  =  bp.evaluate(compare_pipes,  dataset=da-
ta_tr_test,
 subjects='train', eval_verbose=-2)
all_results.summary()
```

Then, the next step of this work-flow would be once the correct pipeline is identified, our new validation test is training on the full training set, and testing on the set of subjects that

(continued)

we said was the test set. And we do it with just our best identified 'ridge pipe.'

```
results = bp.evaluate('ridge_pipe', dataset=data_tr_t-
est, subjects='all', cv='test')
results
```

Now, we could report just these ridge regression results, with our checks between different pipelines conducted safely on a training set. That said, in this particular example – given the small sample sizes involved, a train-test approach, where our test set only has 45 subjects is likely a pretty bad idea... or rather, we should expect large error bars. We can formalize this intuition by returning to working with the full dataset and essentially as our CV strategy is simulating a repeated train and test split of 20%.

This is a custom CV object that sats, use a random test split of 20% 10 times.

cv = bp.CV(splits=.2, n_repeats=10)

results = bp.evaluate('ridge_pipe', dataset=data, cv=cv, eval_verbose=0)

Plot the distribution of scores with library seaborn

import seaborn as sns

sns.displot(results.scores['r2'], kde=True)

Seriously don't double dip

I want to make sure this point is very clear, so next we will look at an ever more extreme example than just trying a few models and just reporting the best one. In this case we will be repeating a 3-fold CV, but with a bunch of different random splits.

for random_state in range(10):

```
 score = bp.evaluate('ridge_pipe', data, cv=3, random_-
state=random_state,
 eval_verbose=0, progress_bar=False).score
 print(random_state, score)
```

First thing to notice is that cross-validation, especially with small-ish sample sizes is not perfect. Depending on just luck of different random split, we can observe some variability in mean explained variance.

The second, is that a dishonest person could use a strategy like the one above and run and just report the following:

bp.evaluate('ridge_pipe', data, cv=3, random_state=0,

```
 eval_verbose=0, progress_bar=False)
```

(continued)

The 'cheating' part here, is that we essentially tried 10 different things using our full dataset, then chose a result from those 10, didn't report the others. That is to say, running 10 repeats of different 3-fold CV is actually not a bad thing – as long as you report all of them (well, the average).

Well, this example is obvious, it can be easy to accidently end up doing something similar. Say for example you try one model, and then another, and then with slightly different features, then maybe you go back to a different model, etc...

Predict Sex

The other thing to note is that all of the BPt style objects when possible are generic to problem type also. So that means, even if we switch to a binary prediction, we can still use the same code from the ridge pipe. Let's try that now:

from BPt.default.pipelines import ridge_pipe

We can pass either the default str, or the object itself for pipeline

Also note that now that we have two targets, we need to specify which one we want to predict

Let's also add one more parameter, n_jobs, which let's use multi-process our evaluation

```
results = bp.evaluate(pipeline=ridge_pipe,
 dataset=data,
 target='sex',
 n_jobs=2)
results
```

Now if we look at the fully composed scikit-learn style estimator again:

results.estimator

We see that there are some changes from before, e.g., now our base model is a LogisticRegression and the set of parameters it searches over are different

```
param_distributions={'C': Log(lower=1e-05,
upper=1000.0), 'class_weight': TransitionChoice([None,
'balanced'])}
```

Where before the default hyper-parameter search parameters were:

```
param_distributions={'alpha': Log(lower=0.001,
upper=100000.0)}
```

(continued)

On the end of the user, specifying these hyper-parameter distributions is done when building the model as just: Model (obj='ridge,' params=1, ...)

Where 1 refers which default distribution to select (see all choices for all supported models here: https://sahahn.github.io/BPt/options/pipeline_options/models.html).

A more advanced used could also manually specify this choice as well, for example:

```
# Make a copy of the default ridge pipeline
our_ridge_pipe = ridge_pipe.copy()


# Look in our pipeline at where the param distribution is
saved
our_ridge_pipe.steps[-1].params
# Replace it with one of our choosing
our_ridge_pipe.steps[-1].params = {'C': bp.p.Log(low-
er=1, upper=1000)}


# Then re-run
results = bp.evaluate(pipeline=our_ridge_pipe,
 dataset=data,
 target='sex',
 n_jobs=2)
results
```

Something to keep in mind of course when performing custom hyper-parameter tuning like this is not to abuse it, re-running different choices until by chance you get a high performing results. This of course is another area like with what we discussed before, and is why a lot of time in practice it can be helpful to just use default settings, either defaults from BPt, or defaults from other libraries as we will see in the next example:

We can also just as easily use custom sklearn objects as either a step in the pipeline – or instead of the pipeline.

from sklearn.linear_model import LogisticRegressionCV
# We just need to wrap it in a Model object, so BPt
# knows how to handle it correctly
sk_model = bp.Model(LogisticRegressionCV())

```
results = bp.evaluate(pipeline=sk_model,
 dataset=data,
 target='sex',
```

(continued)

```
  mute_warnings=True # Mute ConvergenceWarning's
  )
results
```

Conclusion

BPt as a library is still a work in progress, and may not be as polished as some other available libraries, but I hope it still may be useful for some people.

If you are interested, I encourage you to check out the User Guide on the documentation page, which includes a number of other Full Examples. Likewise, the github repository method_series contains a work in progress larger tutorial, which will feature the use of BPt and neurotools in analyzing the three AOIMIC datasets, across a number of different analysis.

I also encourage anyone interested in contributing code, ideas, bugs, etc... to check out the project github and/or open an issue with your question or comment.

7 Sustainability

The software is currently hosted on github at https://github.com/sahahn/BPt as well as through the python PIP repository under 'brain-pred-toolbox.' Users are welcome to submit any code/improvements to the library. Users are also welcome to comment with any suggestions or features they would like to see implemented.

8 Conclusions

Frameworks like BPt can be helpful for some projects, but require a tradeoff between flexibility to usefulness. That said, we hope this library can be as useful a tool as possible moving forward and welcome any suggestions, feedback, or bug reports on the library's github page (https://github.com/sahahn/BPt).

References

1. Hahn S, Owens MM, Yuan D, Juliano AC, Potter A, Garavan H, Allgaier N (2022) Performance scaling for structural MRI surface parcellations: a machine learning analysis in the ABCD Study. Cereb Cortex 33:176–194. https://doi.org/10.1093/cercor/bhac060
2. Dadi K, Rahim M, Abraham A, Chyzhyk D, Milham M, Thirion B, Varoquaux G,

Alzheimer's Disease Neuroimaging Initiative (2019) Benchmarking functional connectome-based predictive models for resting-state fMRI. NeuroImage 192:115–134. https://doi.org/10.1016/j.neuroimage.2019.02.062

3. Varoquaux G (2018) Cross-validation failure: small sample sizes lead to large error bars. NeuroImage 180:68–77. https://doi.org/10.1016/j.neuroimage.2017.06.061
4. Button KS (2019) Double-dipping revisited. Nat Neurosci 22:688–690. https://doi.org/10.1038/s41593-019-0398-z
5. Hosseini M, Powell M, Collins J, Callahan-Flintoft C, Jones W, Bowman H, Wyble B (2020) I tried a bunch of things: the dangers of unexpected overfitting in classification of brain data. Neurosci Biobehav Rev 119:456–467. https://doi.org/10.1016/j.neubiorev.2020.09.036
6. Hooker G, Mentch L, Zhou S (2021) Unrestricted permutation forces extrapolation: variable importance requires at least one more model, or there is no free variable importance. Stat Comput 31:82. https://doi.org/10.1007/s11222-021-10057-z
7. Kaur H, Nori H, Jenkins S, Caruana R, Wallach H, Wortman Vaughan J (2020) Interpreting interpretability: understanding data Scientists' use of interpretability tools for machine learning. In: Proceedings of the 2020 CHI conference on human factors in computing systems. Association for Computing Machinery, New York, pp 1–14

Chapter 13

NBS-Predict: An Easy-to-Use Toolbox for Connectome-Based Machine Learning

Emin Serin, Nilakshi Vaidya, Henrik Walter, and Johann D. Kruschwitz

Abstract

NBS-Predict is a prediction-based extension of the Network-based Statistic (NBS) approach, which aims to alleviate the curse of dimensionality, lack of interpretability, and problem of generalizability when analyzing brain connectivity. NBS-Predict provides an easy and quick way to identify highly generalizable neuroimaging-based biomarkers by combining machine learning (ML) with NBS in a cross-validation structure. Compared with generic ML algorithms (e.g., support vector machines, elastic net, etc.), the results from NBS-Predict are more straightforward to interpret. Additionally, NBS-Predict does not require any expertise in programming as it comes with a well-organized graphical user interface (GUI) with a good selection of ML algorithms and additional functionalities. The toolbox also provides an interactive viewer to visualize the results. This chapter gives a practical overview of the NBS-Predict's core concepts with regard to building and evaluating connectome-based predictive models with two real-world examples using publicly available neuroimaging data. We showed that, using resting-state functional connectomes, NBS-Predict: (i) predicted fluid intelligence scores with a prediction performance of $r = 0.243$; (ii) distinguished subjects' biological sexes with an average accuracy of 65.9%, as well as identified large-scale brain networks associated with fluid intelligence and biological sex.

Key words Network-based statistic, Machine learning, Graph theory, Biomarkers, Connectome-based prediction, Sex prediction, Fluid intelligence, Tutorial

1 Introduction

The increasing interest in structural and functional connectivity networks of the human brain has given rise to a need for models that unveil their secrets [1, 2]. Graph models [*see* Glossary] that define brain networks as a set of nodes (i.e., brain regions) and edges (i.e., connections between brain regions) are among the most powerful representations of brain networks that can be used to investigate effect-associated connections and sub-networks [1, 3]. In this context, mass-univariate [*see* Glossary] testing of hypotheses is a popular method for unraveling sub-networks of interest, in which a statistical model is fit at each edge in the graph, and a corresponding p-value is then computed. However,

Robert Whelan and Hervé Lemaître (eds.), *Methods for Analyzing Large Neuroimaging Datasets*, Neuromethods, vol. 218, https://doi.org/10.1007/978-1-0716-4260-3_13,

this standard approach is limited due to the multiple comparisons problem [4]. Specifically, the high number of simultaneous statistical tests across all edges in the graph leads to an accumulation of a high statistical error rate ("alpha inflation"). Correction methods for alpha inflation such as Bonferroni correction [*see* Glossary] or the False Discovery Rate [*see* Glossary] correction (FDR) are common practices. However, they have been repeatedly criticized as being too conservative, particularly when a great number of statistical tests are performed.

One well-known method to mitigate the issue of multiple comparisons is Network-based Statistic (NBS) [*see* Glossary] [3]. NBS is a statistical method that controls the family-wise error rate, in a weak sense, by combining the concept of connected components and cluster-based thresholding. Although NBS provides more power than traditional methods used to correct for multiple comparisons (e.g., Bonferroni or FDR) by resulting in lower false-negative errors [3, 5], it has limitations related to the traditional statistical method that is utilized in its framework: the general linear model (GLM). Statistical inference methods such as the GLM have repeatedly faced criticism with respect to reproducibility and generalizability [6, 7]. While generalizability is a hard-to-meet criterion with traditional statistical approaches, it is essential in developing neuroimaging-based biomarkers, which are critical in the realm of precision medicine, where erroneous results can lead to misdiagnosis. Generalizability can be quantified by applying out-of-sample estimation techniques (e.g., cross-validation (CV) [*see* Glossary] [8].

Machine learning (ML) has become extensively popular among neuroimaging researchers since computing power, and the availability of open-access large-scale neuroimaging datasets has exponentially increased over the last decade. A critical advantage of ML algorithms over classical statistical inference models is that they aim to extract latent factors as functions of observed data, thereby attempting to reverse the data generating process [9], thereby harnessing the multivariate nature of input data. As such, ML models are a great tool for predicting behavioral or clinical variables, especially when working with enormous feature sets such as those present in brain networks, because underlying associations between dependent and independent variables are not obvious. Unfortunately, ML models commonly suffer from two methodological drawbacks when applied to brain networks: the "curse of dimensionality" and the "lack of interpretability." The curse of dimensionality refers to the fact that ML models tend to overfit [*see* Glossary] when the ratio of dimensions (i.e., the number of features) is high, relative to the sample size. In such cases, overly complex models are formed in the training data that do not generalize to the test data [10]. To avoid overfitting, several feature selection [*see* Glossary] methods have been proposed, including filter-based, wrapper-based, and embedded methods [11]. The second issue is the lack of interpretability when one is interested in explaining which features contributed to the prediction. This is because interpretation of the coefficients derived from machine learning models (even from the linear ones) is not straightforward [12, 13].

To alleviate the lack of generalizability, the curse of dimensionality, and the lack of interpretability, Serin et al. (2021) proposed *NBS-Predict*, a novel connectome-based prediction method [14]. This new approach provides a fast way to identify highly generalizable neuroimaging-based biomarkers by combining ML with NBS in a cross-validation structure. NBS-Predict holds several advantages over existing ML methods. Compared with generic ML algorithms (e.g., support vector machines, elastic net, etc.), the results from NBS-Predict are more straightforward to interpret: NBS-Predict outputs a weighted network [*see* Glossary] indicating the extent to which input features contributed to the model. On the contrary, results derived from other generic ML algorithms are often difficult to interpret, regardless of the linearity of the algorithms [12, 13].

A ML method that is closely comparable to NBS-Predict is the Connectome-based Prediction (CPM) [15, 16]. CPM is a connectome-based machine learning method for structural or functional brain networks to predict neurobiological-related individual differences in behavior. Although NBS-Predict and CPM are both connectome-based machine learning methods, they are notably different in many ways. The most fundamental difference is that CPM was designed for regression problems (but cf. [*see* Chapter 15] for an extension of CPM to classification), which means it can only predict continuous values such as individuals' scale scores. In contrast, NBS-Predict can handle both discrete and continuous variables. Also, while CPM selects individual and spatially dispersed features based on a linear association to the target variable, NBS-Predict takes the topological structure of features into account by selecting a subnetwork of suprathreshold features (i.e., NBS-Predict uses connected graph components as features). Furthermore, unlike CPM, NBS-Predict does not require any expertise in programming as it comes with a well-organized graphical user interface (GUI) with a good selection of machine learning algorithms and additional functionalities.

This chapter serves as a practical guide for applying NBS-Predict on real-world functional MRI (fMRI) data to predict clinical and cognitive outcome variables. Specifically, this chapter briefly introduces the rationale underpinning the NBS-Predict methodology, which has been presented in detail in Serin et al. (2021) [14]. This chapter provides a practical overview of the NBS-Predict core concepts with regard to building and evaluating connectome-based predictive models with two real-world examples using publicly available neuroimaging data [17]. We demonstrate the usage of NBS-Predict, for (1) predicting fluid intelligence and (2) classifying sex from task-based functional connectivity matrices. The version of the NBS-Predict toolbox used in both applications is v1.0.0-beta.9.

2 NBS-Predict

2.1 General Algorithm

NBS-Predict operates in a repeated cross-validation (CV) structure (nested if hyperparameter optimization is desired), where the algorithm comprises several stages, in which suprathreshold edge selection (i.e., feature selection), model training, hyperparameter optimization (optional), ML algorithm optimization (optional), and model evaluation are employed. A full description of the NBS-Predict algorithm is given in Serin et al. (2021) [14].The general workflow of the NBS-Predict algorithm is depicted in Fig. 1.

2.1.1 Suprathreshold Edge Selection

To select relevant edges, NBS-Predict performs suprathreshold edge selection, which is an in-house developed feature selection algorithm inside the CV procedure (and also in the inner-most CV loop if hyperparameter optimization is desired). This feature selection algorithm combines univariate feature selection algorithms with the graph-theoretical concept of connected components. Briefly, a GLM model based on a given contrast is fitted to each edge (i.e., feature) in the brain connectome, and the corresponding p-value is then computed. The largest connected component, which is formed by a subset of connected edges that are determined as significant based on their GLM derived p-values (e.g., lower than 0.01), is then selected as a multivariate feature for subsequent training of the ML model. Figure 2 demonstrates the workflow of the suprathreshold edge selection algorithm.

2.1.2 Model Evaluation

After selecting the component of relevant edges, model evaluation (training and testing) is performed. Trained and tested prediction performance (e.g., balanced accuracy) of the applied ML model is then assigned to the edges present in the selected component

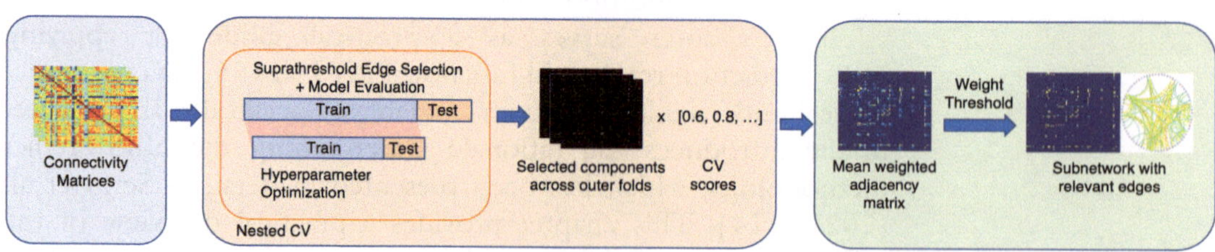

Fig. 1 Workflow of NBS-Predict [14]

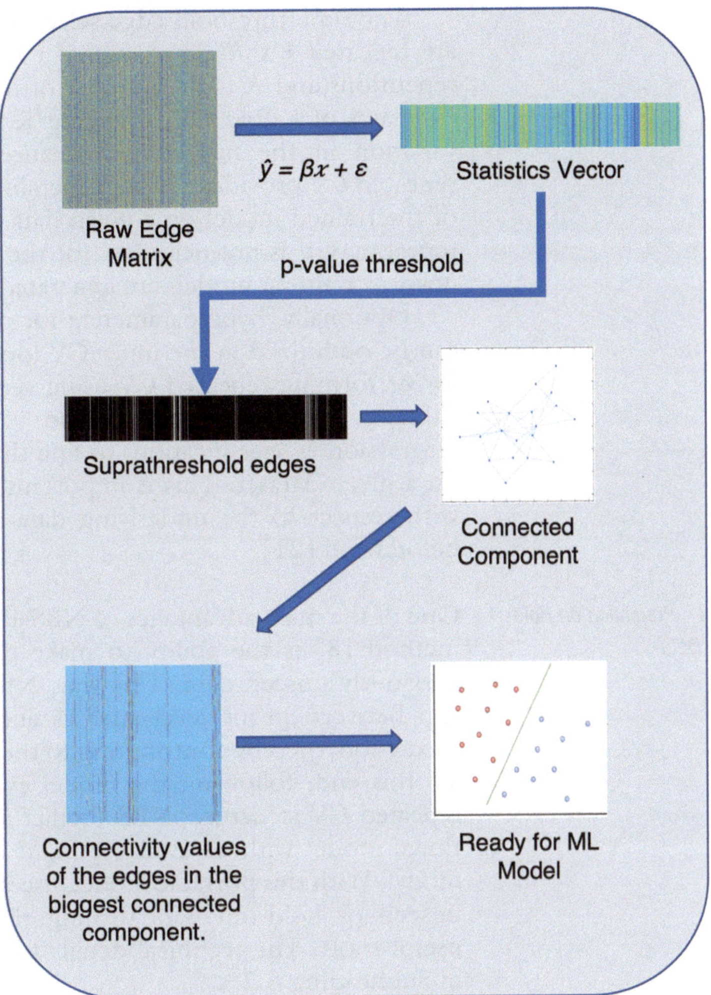

Fig. 2 Suprathreshold edge selection workflow

(0, otherwise, is set to unselected edges). This component-wise assignment is performed to quantify the predictive power of component-specific edges across CV folds. For instance, in a particular CV fold, it could be the case that the connected component of selected features predicts the target variable with relatively low prediction performance, whereas another connected component may achieve a high prediction performance in the subsequent CV iteration. To take the varying contribution of these two sets of edges to the overall model into account, they are assigned with the prediction performance of the corresponding CV fold. In this way, the weight for each edge in the weighted output matrix does not only represent the frequency of being present in a selected connected component but also the selected components' out-of-sample performance, thereby providing an easy way to assess the contribution of each edge to the overall model.

The suprathreshold edge selection and model evaluation steps are repeated r x K times, where r represents the number of CV repetitions and K is the number of folds. By generating averaged estimates of K-fold CV, repeating K-fold CV r times reduces the variation in the model performance estimates [18–20]. Thus, repeated CV provides a more generalizable and accurate estimation of the trained model on a given data set. It should be noted that repeating CV is not necessary for the leave-one-out method as all possible training models are generated in each run.

Optionally, hyperparameters for machine learning algorithms can be optimized in the inner CV loop. Also, NBS-Predict allows for performing repeated CV using various machine learning algorithms suitable for the given ML problem (regression vs. classification) to find the best performing algorithm on a given dataset. This is important because the optimal model with respect to the underlying data structure cannot be known beforehand [21].

2.1.3 Prediction on Hold-Out Data

One of the main advantages of NBS-Predict over the original NBS method [3] is the ability to make predictions about novel and previously unseen data. This way, NBS-Predict aims to close the gap between group-level analysis and interpretable subject-level prediction, thereby contributing to the field of precision psychiatry. To this end, following the model evaluation process within the repeated CV structure, NBS-Predict automatically trains an additional model with all available input data to generate a portable model. With this portable model, users can make predictions about unseen or hold-out data turning their analyses into practically useful tools. The technical details for model exporting are given in Subheading 8.2.

2.2 Software and Coding

An important advantage of NBS-Predict is its user-friendly GUI allowing users to perform complex connectome-based ML tasks. Therefore, coding knowledge is not required to run the toolbox. Since NBS-Predict is an extension of the NBS toolbox [3] (https://www.nitrc.org/projects/nbs), it has a similar interface design with addition of several advanced functionalities. The GUI allows users to easily make predictions for a given set of connectome data by providing nothing more than individual connectivity matrices, a spreadsheet with brain regions of the network, a design matrix, and a contrast vector. A specific machine learning algorithm can simply be selected by a drop-down window. Additionally, the GUI allows users to perform hyperparameter optimization and permutation testing. Further, parameters such as the number of CV folds or p-value thresholds can be defined via text input boxes. Following the analysis, users may visualize obtained results in a heatmap as a weighted network, on a circular network, or as a 3D brain surface

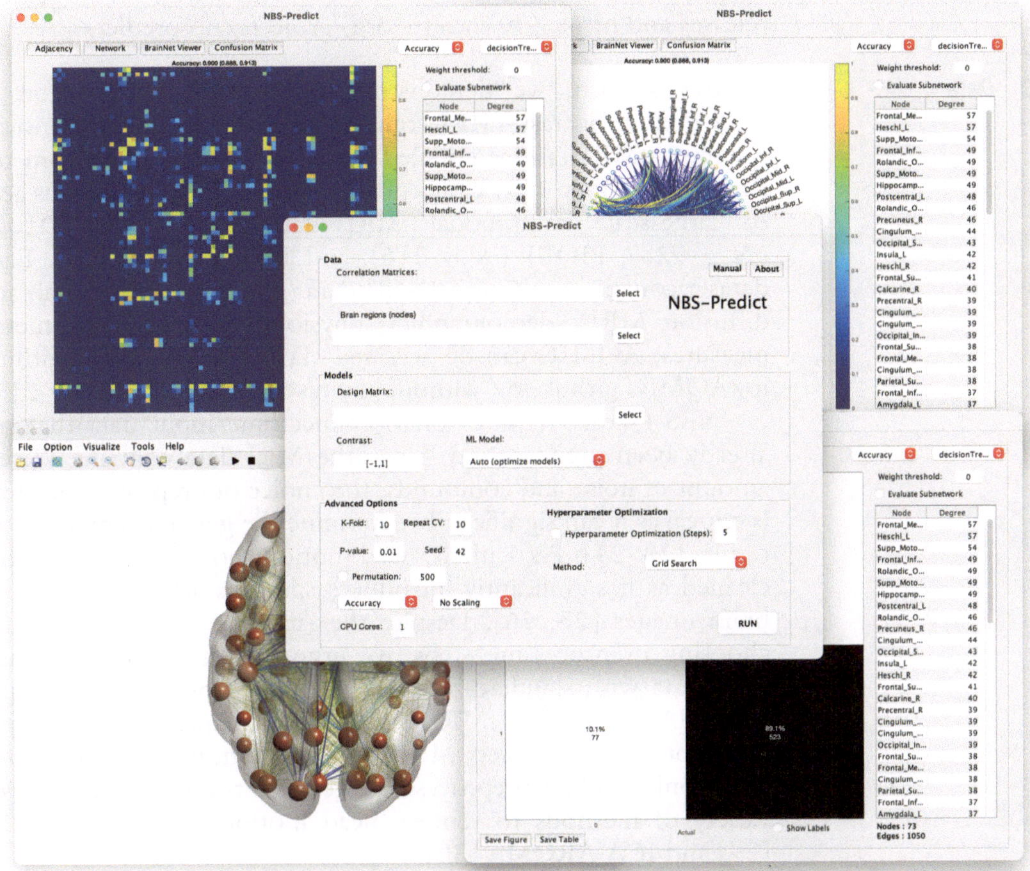

Fig. 3 Screenshots of the NBS-Predict GUI

generated by the BrainNet Viewer [22]. The GUI of the NBS-Predict toolbox is shown in Fig. 3.

3 Setup and Data

3.1 NBS-Predict Installation

NBS-Predict requires MATLAB (The MathWorks, Inc.) version r2016b or newer and the Statistics and Machine Learning Toolbox to run properly. Additionally, the Parallel Computing Toolbox is required if parallel processing is desired to quicken the analyses—this is recommended for the analysis of big data sets. NBS-Predict, unfortunately, does not work on Octave (GNU) due to incompatibilities in the functions and GUI libraries. We strongly recommend users to install the latest version of NBS-Predict since the toolbox is regularly updated and expanded with new features and more intuitive GUI elements. The most recent version of the toolbox can be downloaded from https://github.com/eminSerin/NBS-Predict/releases and https://www.nitrc.org/projects/nbspredict/.

3.2 Data

In this section, we demonstrate the practical application of NBS-Predict on task-fMRI connectivity data. We use the movie-watching fMRI task as entailed in the ID1000 dataset from the Amsterdam Open MRI Collection (AOMIC) [17]. AOMIC is a set of large-scale multimodal MRI dataset including ID1000 ($N = 928$), PIOP1 ($N = 216$), and PIOP2 ($N = 226$). Each dataset contains multimodal MRI data (structural, functional, and diffusion MRI), demographics, physiological and psychometric measures. AOMIC can be accessed via https://nilab-uva.github.io/AOMIC.github.io/ without any restrictions [*see* Chapter 2].

NBS-Predict requires brain connectome input data that has already been preprocessed. Since the MRI data contains a great amount of noise and confounds, the choice of preprocessing steps is critical as it can significantly determine the generalizability of the results [23, 24]. Particularly, head motion should be meticulously cleaned as it significantly introduces spurious activity patterns in brain images [25, 26]. Despite the lack of a gold standard for choosing preprocessing steps, we urge users to preprocess their MRI data with standardized and established preprocessing pipelines: HALFpipe [27], fMRIPrep [28] [*see* Chapter 8]. These pipelines not only preprocess MRI data automatically using the most commonly applied preprocessing steps, but also provide a wide variety of methods to remove head motion such as CompCor [29] and ICA-AROMA [30].

One advantage of using the AOMIC data set is that it provides preprocessed fMRI data (with fMRIPrep), with details of the scanner, scanning protocol, and preprocessing given in Snoek et al. (2021) [17]. For further use of this data, we applied the following two noise correction procedures on the preprocessed fMRI data: (1) removing tCompcor components [29] which explain 50% of the variance in the data and 24 motion parameters [31], and (2) high-pass filter of 1/128 Hz. Temporal filtering and removal of motion-related covariates were performed simultaneously using a single linear regression since the application of these preprocessing steps in isolation can be the cause for undesired and problematic artifacts [32]. Furthermore, we performed spatial smoothing using a Gaussian kernel of 6 mm FWHM. Brain images were then parcellated into 268 functionally coherent regions using the Shen atlas [15, 33] and functional connectivity between each pair of brain regions was computed using Pearson's correlation coefficient, yielding a 268×268 correlation matrix for each subject. Procedures as given above were performed with Nilearn [34]. Further, subjects with excessive head motion (i.e., average frame-wise displacement rate above 0.5 mm) were subsequently excluded, leaving a total of 861 subjects. The final dataset can be found in https://github.com/eminSerin/NBSPredict_SpringerNature.

4 Model Construction and Validation

The subsequent Subheadings (4.1, 4.2, 4.3, 4.4, 4.5, 4.7, 4.8, 4.9, 4.10, 4.11, 5.1, 5.2, 6.1, and 6.2) provide a practical overview of the application of the NBS-Predict GUI to predict individuals' fluid IQ scores from ID1000 fMRI connectome data [17] and highlight associated analysis choices. In these sections, we discuss general concepts (e.g., design matrix, contrast, etc.). Wherever suitable, we discuss implications for regression and classification problems in conjunction. Starting Subheading 7, we specifically demonstrate the usage of NBS-Predict for the classification of sex and discuss related analysis decisions.

4.1 Input Data

NBS-Predict requires the following input data to perform ML analysis (Fig. 4): (1) individual connectivity matrices, (2) a spreadsheet with brain regions of the network, (3) a design matrix, and (4) a contrast vector.

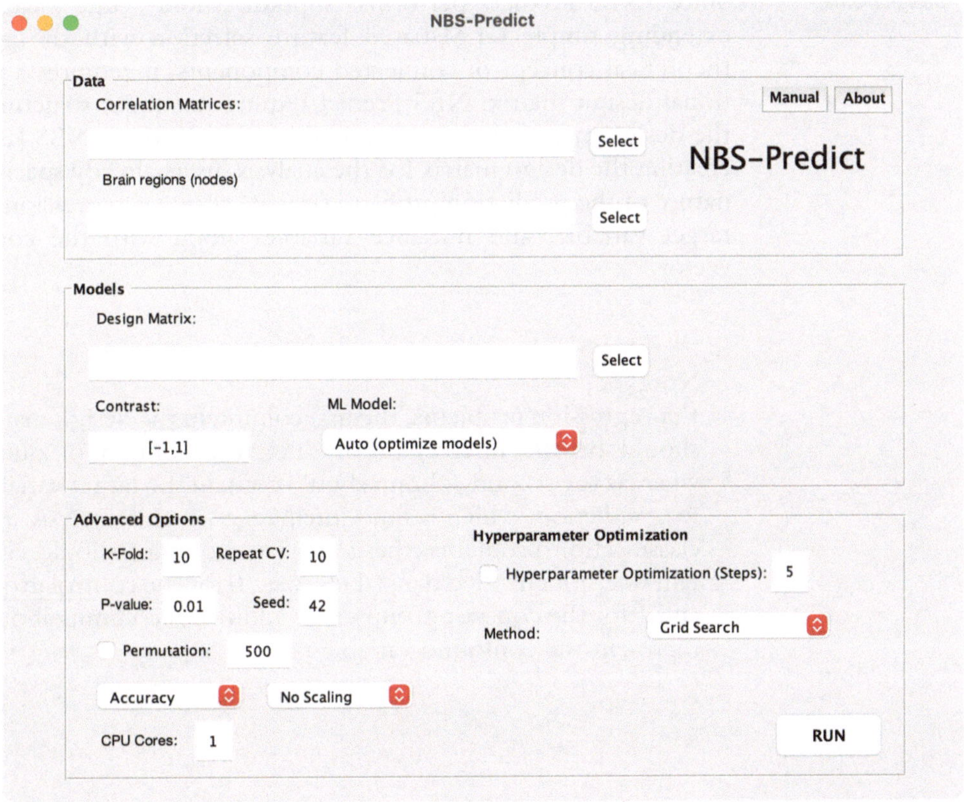

Fig. 4 Screenshot the analysis setup window of NBS-Predict

4.1.1 Connectivity Matrices

Similar to NBS [3], users must provide symmetrical $N \times N$ connectivity matrices as feature sets (one matrix per subject), where N is the total number of nodes (i.e., brain regions). In symmetrical fully connected connectivity matrices (i.e., where all possible pairs of nodes are connected), the total number of edges between nodes is $N \times (N\text{-}1)/2$. The number of edges is determined by the parcellation atlas used to define brain regions. For example, as the Shen atlas [15, 33] consists of 268 brain regions, one could have a maximum number of 35,778 pairwise edges. In contrast, the Harvard-Oxford atlas [35] provides 69 brain regions (69 × 69 connectivity matrix) with up to 2346 edges. Therefore, we recommend to choose a parcellation that is consistent with prior hypotheses of brain function since various atlases provide different levels of resolution [36]. Notably, although NBS-Predict was designed and developed for analyzing functional connectomes, connectome data from other imaging modalities such as DTI derived white-mater networks can also be used.

4.1.2 Brain Parcellation Spreadsheet

NBS-Predict requires a brain parcellation file containing x, y, z coordinates for brain regions, and labels. This file must contain only four columns (x, y, z coordinates and node labels) without any column names (i.e., entries start at the first row).

4.1.3 Design Matrix

Since NBS-Predict performs suprathreshold edge selection, extending simple GLM-based feature selection with the graph-theoretical concept of connected components, it requires a traditional design matrix. NBS-Predict requires a similar structure for the design matrix as those required by the GLMs in NBS [3]. By creating the design matrix for the analysis, users already specify the nature of the prediction problem (classification vs. regression), the target variable, and nuisance variables along with the contrast vector.

Note

For regression problems, the first column in the design matrix should be the intercept term (i.e., the column of ones), whereas the second column should contain the target variable for prediction, which is fluid intelligence in this analysis. For classification problems, these first two columns should contain one-hot encoded data labels (e.g., 0 for the control group and 1 for the contrast group). The additional columns should represent the confound variables.

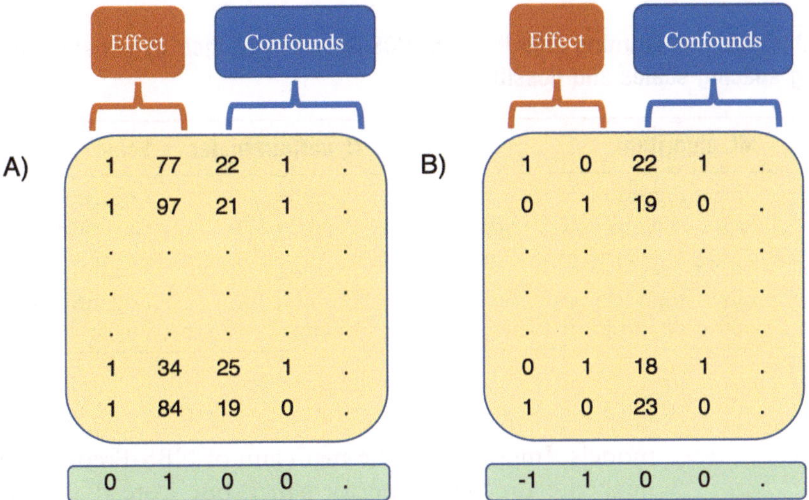

Fig. 5 Example design matrices (upper) and contrast vectors (lower) for regression (A) and classification (B) problems

It should be noted that multilabel classification (i.e., distinguishing more than three groups) is not supported in the current version of NBS-Predict. In this application, we will regress out age and sex, which are the most common confounding variables that need to be controlled within neuroimaging studies. Similar to the brain parcellation spreadsheet, the design matrix should not contain any column names. Examples of the design matrices for regression and classification problems are given in Fig. 5.

4.1.4 Contrast

Along with the design matrix, users must specify a contrast vector (Fig. 5) for the GLM used in the suprathreshold edge selection algorithm. NBS-Predict uses the standard structure for the contrast vector as implemented in the FSL guideline for GLM (https://fsl.fmrib.ox.ac.uk/fsl/fslwiki/GLM; [37]). In our analysis, we specify the contrast vector as [0, 1, 0, 0], representing fluid IQ scores as the effect of interest (i.e., target variable) and age and sex variables as nuisance covariates. We strongly recommend users carefully create and double-check the desired contrast vectors as ill-constructed contrast vectors would return incorrect results.

4.2 Machine Learning Models

As shown in Table 1, NBS-Predict can apply several popular ML algorithms. The user can select between classification or regression ML algorithms, which of course, depends on the target variable as defined in the design matrix (i.e., only if the second column is binary, classification algorithms may be selected). Note that we refrained from offering more complex ML models such as ensemble algorithms or artificial neural networks. Such models have lower interpretability and are more vulnerable to overfitting, thereby requiring significantly more observations than relatively simpler

Table 1
List of available machine learning algorithms in NBS-Predict and their hyperparameters with corresponding spacing scales and possible ranges

	ML algorithm	Hyperparameter	Scale	Range
Regression	Linear regression	Lambda (L_2)	Logarithmic	10^{-2}—10^{3}
	Linear support vector regression	Lambda (L_2)	Logarithmic	10^{-2}—10^{3}
Classification	Logistic regression	Lambda (L_2)	Logarithmic	10^{-2}—10^{3}
	Linear support vector classification	Lambda (L_2)	Logarithmic	10^{-2}—10^{3}
	Linear discriminant analysis (LDA)	Gamma	Linear	0–1

models. Importantly, the main aim of NBS-Predict is not to maximize the prediction performance but instead to provide highly generalizable and interpretable models with an overall good prediction performance. If only maximizing prediction performance is the sole aim of the analysis and not interpretability, we suggest running pipelines with more complex ML algorithms.

Notably, we removed the decision tree algorithm from the current version of the toolbox due to incompatibility issues with confound regression. Specifically, as discussed in Subheading 4.4, the cross-validated confound regression used in the toolbox only removes the linear association between the feature set and confounding variables. However, as the decision tree algorithm is a non-linear ML algorithm, it still may pick up the non-linear effects of confounding variables, complicating the interpretation of the results.

As mentioned in Subheading 2.1, NBS-Predict also allows users to automatically consecutively run all applicable machine learning algorithms for given input data (by selecting the "Auto (optimize models)" option). We recommend this option in the exploratory stage of analyses if prior knowledge about the most suitable ML algorithm for a given dataset is not present. In our application, we left NBS-Predict to run all the ML algorithms suitable for our dataset to detect the best-performing algorithm.

Note

In general, it is good practice to hold out a subset of the data (i.e., not to load this data into NBS-predict) and subsequently apply the final NBS-predict model to this never-touched data (*see* Subheading 8.2 for how to apply NBS-predict models to previously unseen data).

All implemented ML algorithms available in NBS-Predict were initially developed for the Statistics and Machine Learning Toolbox (The MathWorks, Inc.).

4.3 Cross-Validation

To test prediction performance and generalizability, the trained model needs to be evaluated on novel data. One common way to do this is to separate the input dataset into two subsets: a training set and a test set. This way, the model can be trained on the training set and evaluated on the test set, where the latter was not used in any training processes. However, as the prediction performance may strongly depend on the pseudo-random choice for dividing training and test sets, cross-validation is commonly used to overcome this potential drawback. Briefly, in cross-validation, the input data (e.g., features and target) is divided into K subsets, and in K times, one subset is selected as a test set, and other K-1 subsets are used as a training set. Importantly, it has been shown that optimizing and evaluating a learning model on the same CV-splits can result in an overly-optimistic estimate of model performance [38, 39]. Therefore, to overcome this limitation of the traditional "flat CV design" [*see* Glossary], Cawley and Talbot, 2010 [38] suggested use of a "nested cross-validation" approach, in which model tuning and model evaluation are performed independently by generating a series of train/test splits within the same original CV-split. Specifically, the nested-CV structure consists of at least two folds: outer and inner folds. In the outer fold, the whole dataset is divided into train and test sets as discussed for "flat CV design." In the inner fold, the train set from the outer fold is then further divided into another combination of train and test sets used to tune the hyperparameters of the model. Once the model is tuned inside the inner fold, it is carried over to the outer fold to be evaluated. In this way, the nested CV structure can prevent an information "leak" and thus overfitting. Further, as the estimate of the trained model can slightly differ across various K-fold CV repetitions due to the stochastic nature of randomly creating train and test splits, it has been suggested to repeat the entire procedure multiple times (e.g., 100 times) and to report the average of its repetitions. Such a procedure, called "repeated cross-validation" has been suggested to reduce the variability of model performance estimation [18–20].

As indicated in Subheading 2.1 and Fig. 1, NBS-Predict employs repeated cross-validation (nested if hyperparameter tuning is desired) to estimate the out-of-sample prediction performance of the trained model. Through the GUI, the user can easily define the number of CV folds and repetitions. The choice of the number of CV-folds is quite arbitrary, however, five-fold CV or ten-fold CV are the most commonly used procedures. Although five-fold CV and ten-fold CV hold different advantages over each other, in practice, both structures yield similar model estimates if the dataset is large enough (e.g., ID1000). With respect to the number of repetitions of the entire CV procedure, it can be said that higher numbers of repetitions provide lower variance in model estimates, especially if the sample size is small. We suggest users to repeat the CV as many times as possible, considering their computation power. Furthermore, the distribution of the edge weights in the outcome weighted network is determined by the number of CV folds and repetitions [14]. That is, the weight of each edge in the weighted output matrix represents not only the prediction performance of these selected components but also how frequently the edge is present across selected components. Therefore, higher number of CV repetitions will result in more fine-grained results (refer to Discussion in [14] for more detail). Alternative and more specific CV structures such as group-based K-fold CV will be implemented in the following versions of NBS-Predict. In this chapter, we use 50-repeated five-fold cross-validation to evaluate the prediction performance of the trained models.

4.4 Confound Regression

The relationship between target and features may be confounded with several subject-related variables such as age, sex, comorbid pathology, and head movement. Confound regression is the most common way to control confounds in the data [40]. This usually involves regressing out the variance in the target that can be described by the confounds.

Note

Confound regression should be performed in a fold-wise manner (i.e., separately in each CV fold). If confounds are regressed out from the entire data set at once, it might create dependence between subsets, thereby violating the CV assumption of independent subsets [41].

It is important to understand that the effect of cross-validated confound regression depends on the method used to regress out influencing variables. If linear models such as the GLM are utilized in confound regression, only linear associations between confounding variables and feature sets are eliminated. Thus, potential non-linear relationships between these variables might remain after deconfounding [41, 42]. In NBS-Predict, confound regression is performed with a GLM before feature selection along with scaling (if desired, *see* Subheading 4.9 for feature scaling). In our example, we regressed out age and sex confound using the confound regression in our analysis.

4.5 Feature Selection Parameter

As described under Subheading 2.1, for suprathreshold edge selection, in NBS-Predict, a GLM model is fitted to each edge (i.e., feature) in the brain connectome, and a corresponding p-value is computed. The largest connected component formed by a subset of connected edges that are determined as significant based on their GLM derived p-values (e.g., lower than 0.01) is then used as a feature set for model training. Based on this procedure, the p-value parameter can be tuned manually to define the conservativeness of the suprathreshold edge selection process. The number of edges selected in each CV-fold, and thus the weighted outcome network, is directly determined by this p-value. Therefore, users should consider the size of the input network when defining this parameter. For instance, if the input connectome network is very sparse, using very conservative p-value thresholds can be a bad option as no adjacent edges might survive and no connected components can be created. In such a case, the ML algorithm would fail to converge. In our experience and most cases, the default parameter of 0.01 is a sufficient choice. In this application, we used the default p-value suggestion.

4.6 Hyperparameter Optimization

Hyperparameters are parameters that control the learning process of ML algorithms. They are predefined and can be tuned using hyperparameter optimization. Hyperparameter optimization is a technique to determine the optimal hyperparameters for the corresponding ML algorithm on a specific dataset using different searching algorithms such as grid search, random search, and other algorithms [43]. Table 1 depicts the hyperparameters for the available ML algorithms in NBS-Predict and their corresponding spacing scales and possible ranges. Notably, NBS-Predict only optimizes the regularization term strength (i.e., lambda) for the L2 regularization method (not L1) for SVM and regression algorithms. The reason for using L2 regularization instead of L1 is that the latter inherently promotes sparsity and performs feature selection by potentially shrinking less relevant features' coefficients to zero. This might complicate the interpretation of the weighted outcome network since features present in the selected connected component may not be fully utilized by the ML algorithm but instead removed due to the L1 penalty.

NBS-Predict allows users to perform hyperparameter optimization using grid search, random search, and Bayesian optimization algorithms [*see* Glossary] [44, 45].

Note

The number of optimization steps directly determines the exhaustiveness of the optimization algorithms such that a

(continued)

higher number of steps results in higher dimensions of the parameter space. Therefore, increasing the number of optimization steps is more computationally expensive. Thus, we recommend random search or Bayesian optimization over simple grid search in such cases as they provide similar or better model performance with a significantly smaller number of iterations.

It should be noted that the choice of the number of optimization steps itself is arbitrary and that higher numbers may not always yield better performance. Therefore, users should decide on the number of optimizations by mainly considering their computation power. Note that the current version of NBS-Predict does not allow manual set up of hyperparameters to be optimized. In this chapter, we set the optimization steps to 10 and the searching algorithm to "grid search."

4.7 Permutation Testing

Permutation testing [*see* Glossary] is used to evaluate the statistical validity of the model's prediction performance and is generally required when reporting the final model. Permutation testing procedure generates an input data-specific empirical null distribution that is used to estimate how likely the model's prediction performance could be achieved by chance [46]. To establish the empirical null distribution, labels (i.e., target variable) are permuted n times, and significance is computed as the fraction of iterations where the model trained on the permuted dataset but performed similar or better than the ground model (i.e., model trained using the original data). In NBS-Predict, the number of permutations is customizable and should be defined based on sample size and computation power. Notably, the statistical power of permutation testing increases with sample size, yet an even lower number of permutations (e.g., ~100) might be enough for datasets with a moderate-large sample size. The reader should keep in mind that the sample size for lower numbers of permutations may also be influenced by factors such as the true effect size in the data or the number of features and other determinants [46]. Consequently, although we set the default number of permutations to 500 in NBS-Predict, users may choose to perform fewer or more permutations based on their sample size and desired significance level. Importantly, running repeated CV and permutation testing multiple times on the same data, one might observe results with slightly different estimates of model performance. This is because the ground model is evaluated only once in the permutation testing (i.e., 1-repeated CV), while performance estimates obtained from the repeated CV represent the average of a given number of CV estimates (e.g.,

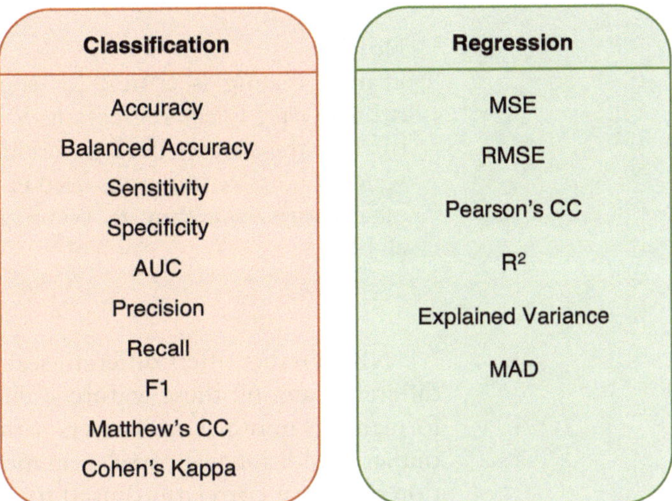

Fig. 6 Available model performance metrics in NBS-Predict

50-repeated CV). Taking computation time into account, here we used 1000 permutations in the default model and 100 permutations in the optimized model (i.e., with hyperparameter optimization), to evaluate model significance respectively.

4.8 Performance Metrics

NBS-Predict offers a wide variety of predictive performance metrics (Fig. 6) suitable for different prediction problems (classification vs. regression) and datasets (e.g., balanced vs. imbalanced). Although the GUI helps by presenting suitable performance metrics for a given target, a further in-depth decision should be made considering the nature of the input data. For instance, despite its popularity in machine learning, accuracy often fails to estimate the true performance of the trained model if group labels are imbalanced [47]. In such a case, it has been suggested to use balanced accuracy and F1 scores instead of simple accuracy alone [48]. Also, mean squared error (MSE) is a widespread performance metric for regression problems, but it does not allow for a straightforward interpretation. In contrast, Pearson's coefficient or R^2 scores can be more straightforwardly interpreted. Therefore, we strongly suggest that researchers may select performance metrics based on the needs of their data and report various performance metrics in their manuscripts. In this application, we used Pearson's correlation as the primary performance metric, where R^2 is also reported.

4.9 Feature Scaling

Feature scaling is a technique to scale the range of features to ensure each feature is distributed within the same range.

Note

Feature scaling is critical for some machine learning algorithms (e.g., linear regression, SVM), particularly if the features in the dataset have different ranges. If features are not scaled, objective functions used in these ML algorithms might take a very long time to converge or may not converge at all [49].

NBS-Predict offers different scaling options. In most cases, the difference among these feature scaling methods in prediction performance is minimal on datasets without a considerable number of outliers. Additionally, correlation methods (e.g., Pearson's correlation or partial correlation) used to construct connectivity matrices initially scale the data into the range between −1 and 1. Thus, the choice of feature scaling methods will not meaningfully change results. Therefore, we recommend users to select one of the commonly used scaling methods implemented in the toolbox. We used "MinMaxScaler" in this analysis, which scales the features between 0 and 1.

4.10 Computational Expense

We ran the analyses on a computer with 32-core Xeon CPU (E5–2630 v3) and 128GB RAM. However, depending on the size of the dataset, researchers can analyze their connectome data also on a standard desktop computer (i.e., with an 8-core CPU and 16GB or 32GB RAM). Running our analysis pipeline without hyperparameter optimization (1000 permutations, 50-repeated five-fold CV, Fig. 7) and with hyperparameter optimization (100 permutations, 50-repeated five-fold CV, Fig. 7) took approximately 2 hours and 5 hours respectively. Since running permutations and CV procedures are linear processes, on a personal computer with an 8-core CPU and at least 16GB RAM, these pipelines would take around 7–9 hours and 19–21 hours. The RAM usage is strongly dependent on the size of the dataset. For datasets with a similar sample size as the ID1000, we recommend at least 16GB RAM.

4.11 Data Analysis Practices

An important goal in machine learning is to train a generalizable and interpretable ML model, ideally without consuming large amounts of time and energy. To ensure that the trained ML model is generalizable, researchers must avoid actions that can result in overfitting as much as possible. For example, in the current version of the NBS-Predict toolbox, the p-value threshold is a hyperparameter that is not tuned in a CV procedure. To prevent overfitting, users should only set a statistically sound value (e.g., 0.01, 0.05, 0.001) and not optimize this value based on the final

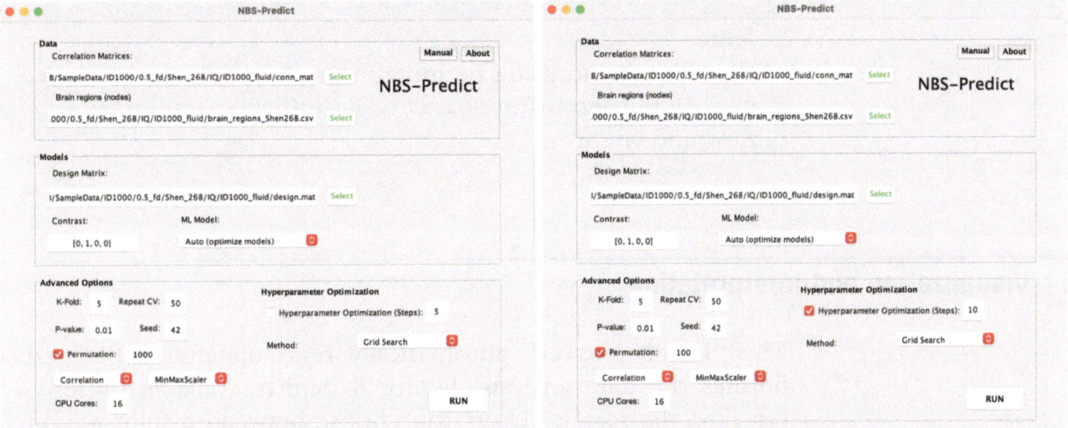

Fig. 7 The analysis setup window of NBS-Predict for the two presented prediction procedures after loading data and setting parameters (left: pipeline without hyperparameter optimization, 1000 permutations, 50-repeated five-fold CV; right: with hyperparameter optimization, 100 permutations, 50-repeated five-fold CV)

overall model performance. As mentioned in Subheading 4.5, the p-value threshold of 0.01 is a good choice in most cases, but it can still be tweaked based on the density of the input network. NBS-Predict offers various feature scaling methods. As mentioned in Subheading 4.9, the selection of feature scaling methods is mostly trivial. However, due to several reasons (e.g., the topology of input network, distribution of connectivity values, the scale of confounding variables, etc.), the selected scaling method might yield unexpected problematic results, especially in the suprathreshold edge selection algorithm (e.g., almost all—or zero—edges are selected). In such cases, users should switch to other feature scaling methods or consider not scaling the features at all. Of note, the scaling methods should not be optimized based on the model performance as that could lead to overfitting.

Machine learning algorithms require a significant amount of CPU power, further, machine learning-based analyses are commonly exploratory in nature, requiring exploring different setups and tweaks, thereby making ML even more computationally costly. Since, in most cases, computation power is limited and the available power might not be sufficient for all the exploration possible, researchers should determine an analysis plan taking their available computation power into account to finish their analyses as quickly as possible. One strategy might be taking a stepwise approach for running ML methods instead of running all at once.

> **Note**
> We recommend using permutation testing only for the best-performing algorithm or a set of algorithms of interest to save time and energy.

5 Visualization and Interpretation

The "Result Viewer" automatically pops up after NBS-Predict finishes the data analysis. Figure 8 depicts available options to visualize the results. Users can choose between a fully weighted network and thresholded subnetworks in the form of a heatmap, or visualization as a circular network, or rendering results on a 3D brain surface generated by the BrainNet Viewer [22]. Detailed information regarding nodes and edges is present on these plots, such as node label and nodal degree. Edge weights are shown by clicking on the corresponding node or edge.

In addition to these plots, for classification, the user can visualize a confusion matrix, summarizing the model's predictive performance over all CV folds. Specifically, true negatives, false negatives, true positives, and false positives are displayed in the confusion matrix along with percentage values, allowing users to explore model +-performance in detail.

A table depicting node labels and the corresponding nodal degrees is displayed on the right side of the "Result Viewer" of NBS-Predict. In this table, the nodes are sorted based on their degrees. That is because the degree of nodes (i.e., brain region) in the weighted network represents the number of connections between the corresponding node and the rest of the nodes in the network, which are associated with the target variable

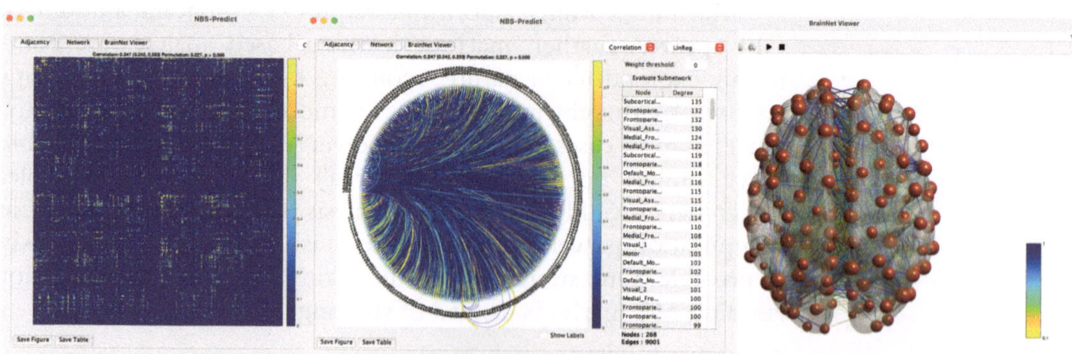

Fig. 8 The screenshots of the three main visualization methods (the heatmap, circular graph, and the 3D brain generated by the BrainNet Viewer [22]) available in the NBS-Predict's "Results Viewer" window

[50]. Therefore, the degree of a node can directly reflect its relevance since it has a considerable number of informative connections with other brain regions, thereby being the central region in the network. Hence, we recommend users to report the nodes with high nodal degrees. Also, if functional network atlases are used [33, 51], researchers can sum the nodal degree of brain regions from each brain network to evaluate the overall importance of the brain network to the trained model (*see* Subheading 6.1 for an example).

5.1 Weight Threshold

An important criterion that can guide visualization is to consider the weight threshold of the network. Although the main outcome of NBS-Predict is the weighted network and corresponding prediction performance of the model, users may visualize the most relevant set of features by weight-thresholding the network. The weight threshold is a cut-off for the contribution of edges to the performance of the overall model. Thresholding the weighted network allows for a straightforward interpretation of the results and provides information on the contribution of the edges to the overall model. Thus, by thresholding the network, researchers can evaluate the robustness and importance of an edge or subnetwork (i.e., biomarker) in a neuroimaging-based prediction. Users should keep in mind that this weight threshold is arbitrary by its nature, and various strategies can be used to identify a plausible threshold [14, 16]. Although a very conservative approach, users may set the weight threshold to 1 to visualize the subnetwork comprising only the most relevant edges. If this approach results in too few or no nodes, more lenient thresholds (e.g., 0.9–0.8) might be used.

Note

Connectome density and the p-value used in the suprathreshold edge selection should be considered when setting the weight threshold. For instance, smaller weight thresholds might be used if the p-value is very strict (e.g., 0.0001) or the input connectome is sparse. We suggest that users report subnetworks obtained from conservative and more lenient thresholding along with the weighted outcome network to provide more detailed information on the distribution of edge weights and the underlying structure of the predictive brain network.

In this chapter, we used threshold values of 1 and 0.9 to visualize the subnetwork of features that highly contributed to the final model performance.

5.2 Subnetwork Evaluation

The prediction performance of subnetworks obtained after thresholding the weighted outcome network will be different from the prediction performance of the overall model comprising all data. That is, weight thresholding necessarily leads to a smaller set of features in comparison to those used for the model's overall evaluation. Therefore, NBS-Predict optionally allows users to evaluate the out-of-sample prediction performance of the suprathreshold subnetwork. The main aim of this feature is to provide a general idea of how the subnetwork would perform independently. To this end, the current version of NBS-Predict optionally evaluates the subnetworks' prediction performance within a 10-repeated ten--fold CV scheme (without hyperparameter optimization). Of note, this feature should not be used to choose an optimal weight threshold as this could result in an overfitting issue.

6 Application: Fluid Intelligence Prediction

6.1 Results

Table 2 indicates the prediction performance of the two presented pipelines in NBS-Predict (i.e., with and without hyperparameter optimization) to predict fluid intelligence. Specifically, NBS-Predict with hyperparameter optimization yielded a prediction performance with a Pearson's coefficient of $r = 0.243$ (95% CI: 0.238–0.249; permutation: 0.268, $p < 0.01$; $R^2 = 0.012$). The prediction performance of the pipeline without hyperparameter optimization was slightly lower with $r = 0.241$ (95% CI: 235–247, permutation: 0.246, $p < 0.001$, $R^2 = -0.041$) than the pipeline with model tuning. In both pipelines, linear regression outperformed the support vector regressor.

Figure 9 depicts the weighted outcome network for the prediction of fluid intelligence scores in a heatmap and a circular graph. Out of 35,778 possible connections present in the input network,

Table 2
Prediction performance of ML algorithms used to predict fluid intelligence

		Repeated CV		Permutation test	
Pipeline	**Algorithms**	μ_r	σ_r	r	p
With model tuning	Linear regression	0.243	0.020	0.268	0.00**
	Linear support vector Regressor	0.182	0.021	0.201	0.00**
Without model tuning	Linear regression	0.241	0.022	0.246	0.000*
	Linear support vector Regressor	0.194	0.019	0.197	0.000*

Note: Pearson's correlation coefficient was used as a performance metric. The statistical validity of the pipeline with and without model tuning (i.e., hyperparameter optimization) was evaluated using the total number of 100 and 1000 permutations. * $p < 0.001$, ** $p < 0.01$

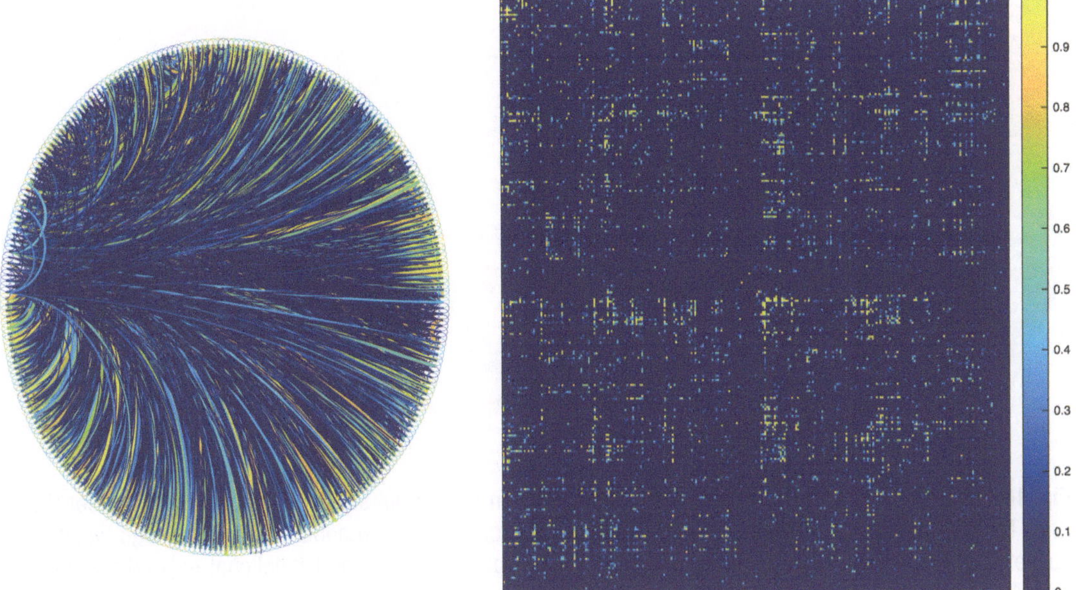

Fig. 9 Main result of NBS-Predict: a weighted network (no threshold applied) plotted on a circular graph and in a heatmap showing network connections associated with fluid intelligence. Edge and node colors depict weights and nodal degrees

only 9072 edges (i.e., 25.36%) connecting 268 regions were selected at least once in all CV iterations. With a weight threshold of 1, the subnetwork of the most relevant features was reduced to 54 edges connecting 44 nodes ($r = 0.299$) from several functional networks such as frontal network, default mode network, visual network as well as subcortical-cerebellar networks (Fig. 10). In this subnetwork, regions from the frontoparietal network were found to be associated with fluid intelligence to the greatest extent (nodal degree = 39), followed by areas in medial-frontal (degree = 35) and default mode networks (degree = 14).

A more lenient threshold of 0.9 yielded a larger-scale subnetwork comprising a total number of 245 connections between 124 brain regions ($r = 0.397$) from a wide range of brain networks (e.g., frontal, DMN, visual, motor, and subcortical-cerebellar; Fig. 11). In this network, again, the fronto-parietal network (degree = 144) was found to be the most contributing network for the overall model performance, followed by the medial-frontal (degree = 101) network and DMN (degree = 71).

6.2 Discussion of Fluid Intelligence Prediction

In this application, we aimed to predict fluid intelligence scores based on functional brain networks as measured during movie watching. To this end, we ran NBS-Predict on task-based (movie watching) fMRI data from the ID1000 dataset [17]. NBS-Predict yielded a prediction performance of $r = 0.243$, which is broadly

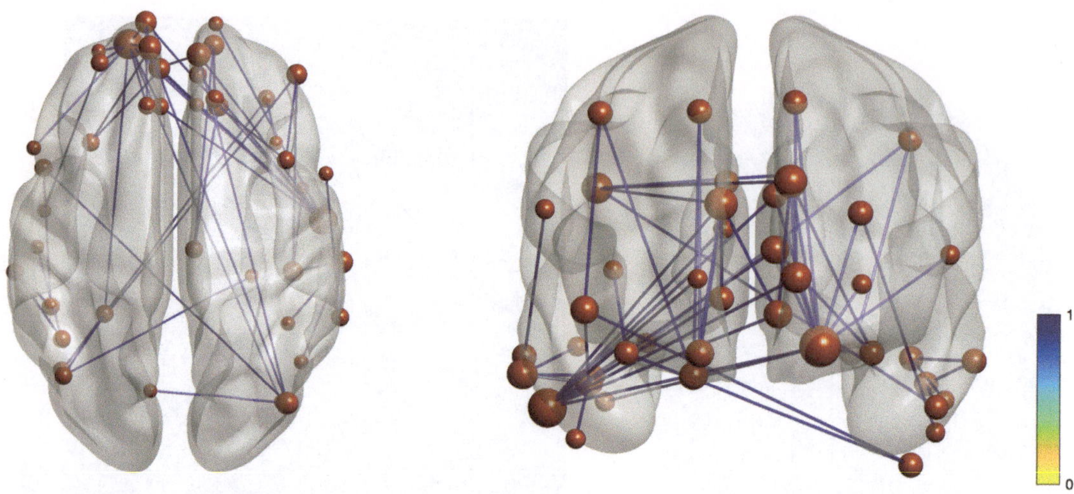

Fig. 10 Subnetwork (weight threshold = 1.0) visualized on a 3D brain surface generated by BrainNet Viewer [22]. The size and color of the nodes and edges are depicted based on nodal degree and edge weight. The subnetwork comprises 54 edges connecting 44 brain regions from several functional networks associated with fluid intelligence

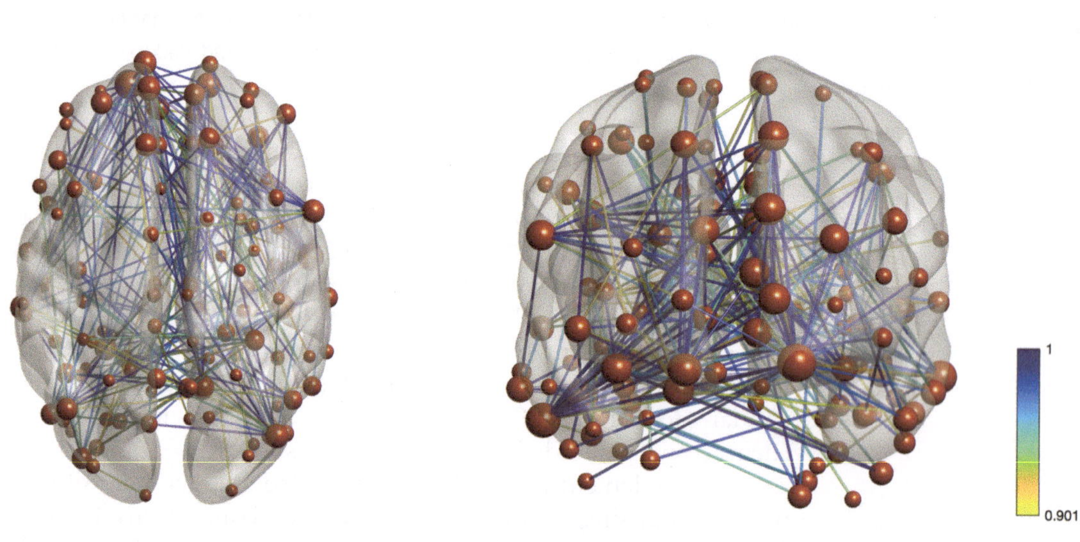

Fig. 11 Subnetwork (weight threshold = 0.9) visualized on a 3D brain surface generated by BrainNet Viewer [22]. The size and color of the nodes and edges are depicted based on nodal degree and edge weight. The subnetwork consists of 245 edges between 124 brain regions associated with fluid intelligence

comparable with previously reported results despite few performance dissimilarities due to differences in some of the data analyses settings (e.g., different datasets, MRI modalities, parcellation atlases, ML models, etc.) [52–54]. In line with previous work, we observed a variety of brain regions from many different brain networks contributing to the prediction of intelligence when applying the presented post-hoc thresholding of the weighted outcome network. Specifically, the thresholded subnetwork comprised regions of the default mode network, fronto-parietal, and medial-frontal networks, which were also previously found to exhibit the strongest association with fluid intelligence [15, 55]. Overall, our results support the idea that human intelligence is realized through the large-scale interaction of many brain regions [56–58]. Of note, in Serin et al. (2021) [14], we also employed NBS-Predict to predict intelligence from data of the Human Connectome Project 1200-subject release [59] but achieved a relatively poor prediction performance. This performance difference could be attributed to various non-trivial factors. For example, in our previous study, we used resting-state (not task fMRI connectome data) as feature set and predicted general intelligence [60] instead of fluid intelligence measured with the Intelligence Structure Test [61]. Further, we used the Power atlas [51] to parcellate the brain images and not the Shen atlas [33]. As such, the reader should be aware that these can be critical factors contributing to variations in prediction performance in real-life data sets.

7 Application: Sex Classification

NBS-Predict can also be applied to classification problems. This is one of the main advantages of NBS-Predict over the most comparable method, the Connectome-based Predictive Modeling [16], which can only perform regression (cf. [*see* Chapter 15]). In this section, we demonstrate an application of NBS-Predict in a generic classification setting in neuroscience, which is the prediction of sex from brain readouts.

7.1 Data

In this analysis, we used the same task-based functional connectomes of the ID1000 dataset that were used in the regression setting. Detailed information about preprocessing and data cleaning is given in Subheading 3.2.

7.2 Data Analysis

7.2.1 Target and Confounds

We predict sex from task-based fMRI connectomes in this application (here, we consider just two categories, male and female). As it has been previously shown that brain size is significantly associated with sex [62, 63], we regressed out this confound along with the individuals' age using cross-validated confound regression implemented in NBS-Predict. Brain size was computed using the total intracranial volume (TIV) estimated with FreeSurfer (v6.0.1) using the Destrieux2009 atlas [64]. TIV information is provided within the ID1000 dataset [17].

7.2.2 Design Matrix and Contrast Vector

Sex was one-hot encoded (i.e., 0: male, 1: female). As shown in the example depicted in Fig. 5, the design matrix needs to be constructed differently than the one depicted in the application for regression problems. Specifically, the first two columns should contain binary-coded vectors representing female and male subjects, while the subsequent two columns represent the subjects' TIV and age values. Related to the design matrix, the contrast vector is also specified differently. Of note, the construction of a contrast vector in classification problems is quite important because features are selected based on the contrast vector in the suprathreshold edge selection. For example, in this application, we run NBS-Predict twice using two sets of contrast vectors: [−1, 1, 0, 0] and [1, −1, 0, 0]. The first contrast vector tells NBS-Predict to select a set of edges whose mean values (i.e., task functional connectivity in this application) are higher in males than females while controlling the confounding effect of TIV and age, whereas the second vector tells NBS-Predict the exact opposite. Thus, analyses using these two sets of contrast vectors return completely different weighted output networks representing the contribution of edges, which are greater in males or females, to the overall model distinguishing sex. Thus, we suggest that users run their analyses in both directions (i.e., G1 > G2, and G1 < G2) unless they have a particular hypothesis regarding the direction of the effect.

7.2.3 Scaling

Unlike the first application in the chapter, here we scaled the features using "StandardScaler," which is a z-score transformation, transforming the distribution of features to the Gaussian distribution of $\mu = 0$ and $\sigma = 1$. This is because we identified that the suprathreshold feature selection did not run properly (i.e., *all* the features were selected in most CV folds) with the "MinMaxScaler" method for this particular problem, so we decided to use "StandardScaler" instead.

Besides the change of settings mentioned above, the remaining analysis settings are the same as in the first regression application (i.e., the number of CV folds, the number of permutations, the hyperparameter tuning steps, etc.). The analysis setup window of NBS-Predict after loading data and setting parameters for sex classification is depicted in Fig. 12.

7.3 Results

Here, we report results from the application of NBS-Predict for sex classification using two different contrast vectors (*see* Subheading 7.2 for details).

7.3.1 Females > Males

NBS-Predict with and without hyperparameter optimization distinguished sex with the classification accuracy of 0.637 (CI: 0.634–0.640, Permutation: 0.631, $p < 0.01$, AUC: 0.637) and 0.613 (CI: 0.608–0.618, Permutation: 0.638, $p < 0.001$, AUC: 0.613), respectively. Linear Discriminant Analysis (LDA) was found to be the best performing algorithm when hyperparameters were tuned. In contrast, the logistic regression outperformed the

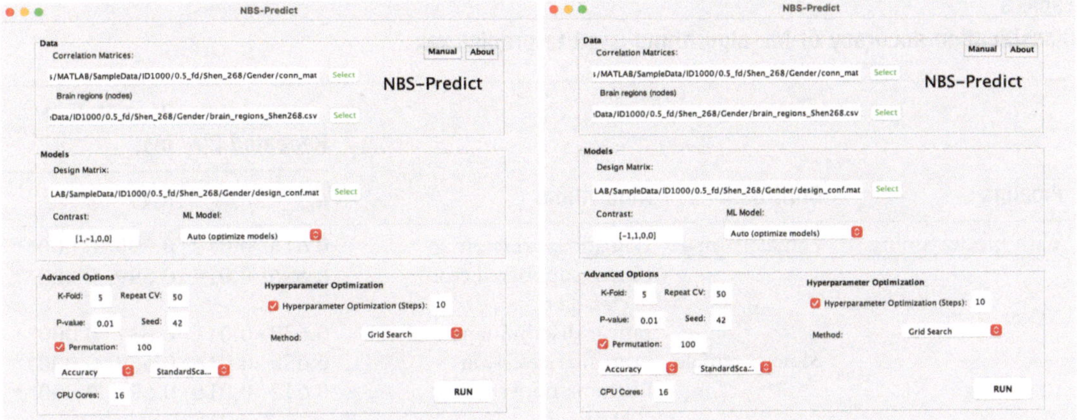

Fig. 12 The analysis setup window of NBS-Predict after loading data and setting parameters for sex classification. Two sets of contrast vectors were used to select sets of edges that showed stronger connectivity in the group of males as compared to females or vice versa

LDA in the pipeline without hyperparameter optimization while performing similarly to the SVM (Table 3).

The weighted networks comprising a set of selected edges with greater values in females than males are depicted in Fig. 13. The most conservative threshold of 1 yielded a subnetwork comprising 49 brain regions, connected by 70 connections (accuracy = 0.617, Fig. 14). The subcortical-cerebellum had the highest nodal degree, followed by the motor, fronto-parietal, visual association networks, medial-frontal, and visual networks.

The threshold of 0.9 returned a large-scale subnetwork consisting of 111 nodes and 282 edges (accuracy = 0.611, Fig. 15). This subnetwork associated with sex comprised, again, brain regions mostly from the subcortical-cerebellum, motor, visual association, and fronto-parietal areas as well as regions from the visual, medial frontal, and default mode networks.

7.3.2 Males > Females

By using the edges that were greater in males than females, NBS-Predict with and without hyperparameter optimization distinguished sex with a classification accuracy of 0.680 (CI: 0.677–0.683, Permutation: 0.688, $p < 0.01$, AUC: 0.680) and 0.647 (CI: 0.643–0.651, Permutation: 0.646, $p < 0.001$, AUC: 0.646), respectively. Like the female > male condition, LDA and logistic regression (slightly better than the SVM) were found to be the best performing ML algorithms in the pipeline with and without hyperparameter tuning (Table 3).

The size of the weighted network and threshold subnetworks were significantly greater in the male > female condition than the female > male condition. The most conservative threshold of 1 yielded a large-scale subnetwork comprising 161 brain regions

Table 3
Classification accuracy of ML algorithms used to predict sex

Pipeline	Contrast	Algorithms	Repeated CV		Permutation test	
			$\mu_{acc.}$	$\sigma_{acc.}$	Acc.	*p*
With model tuning	Female > male	Logistic regression	0.615	0.012	0.599	0.000**
		Linear support vector classifier	0.606	0.014	0.599	0.000**
		Linear discriminant analysis	0.637	0.011	0.631	0.000**
	Male > female	Logistic regression	0.650	0.014	0.659	0.000**
		Linear support vector classifier	0.647	0.016	0.650	0.000**
		Linear discriminant analysis	0.680	0.011	0.688	0.000**
Without model tuning	Female > male	Logistic regression	0.613	0.016	0.638	0.00*
		Linear support vector classifier	0.606	0.017	0.611	0.00*
		Linear discriminant analysis	0.583	0.006	0.571	0.00*
	Male > female	Logistic regression	0.647	0.014	0.646	0.00*
		Linear support vector classifier	0.640	0.018	0.654	0.00*
		Linear discriminant analysis	0.591	0.006	0.590	0.00*

Note: Classification accuracy was used as a performance metric. The statistical validity of the pipeline with and without hyperparameter optimization was evaluated using the total number of 100 and 1000 permutations

* $p < 0.001$, ** $p < 0.01$

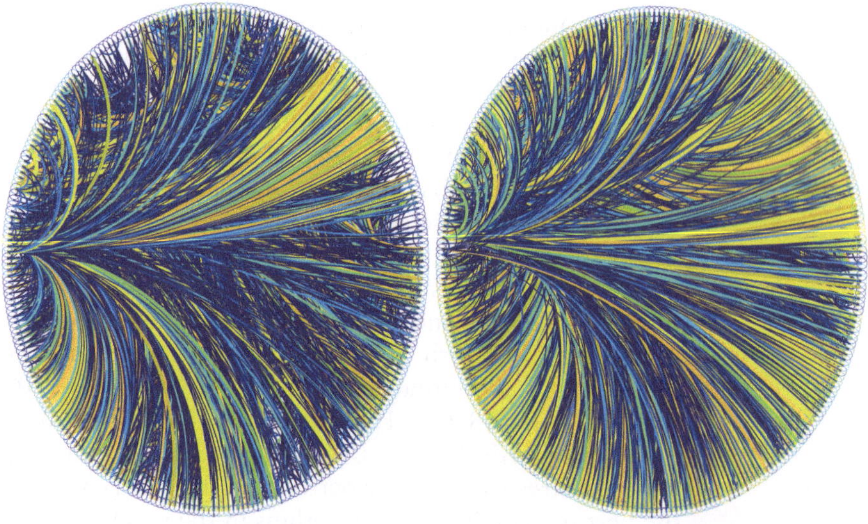

Fig. 13 Weighted network (no threshold applied) on a circular graph associated with sex. The color of edges and nodes depicts weights and nodal degrees. Females > Males on the left and Males > Females on the right

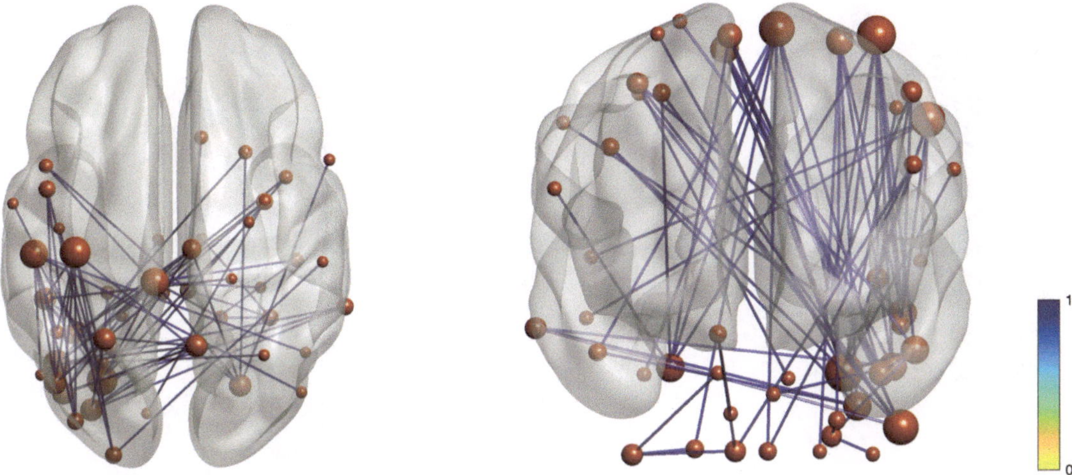

Fig. 14 Females > Males subnetwork (weight threshold = 1.0) visualized on a 3D brain surface generated by BrainNet Viewer [22]. The size and color of the nodes and edges are depicted based on nodal degree and edge weight. The subnetwork consists of 49 brain regions connected by 70 edges whose values were greater in females than males

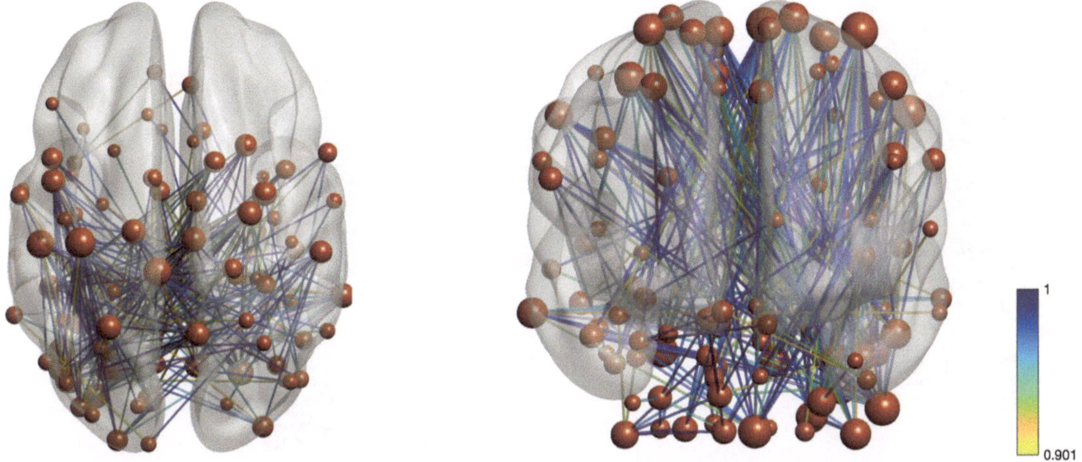

Fig. 15 Females > Males subnetwork (weight threshold = 0.9) visualized on a 3D brain surface generated by BrainNet Viewer [22]. The size and color of the nodes and edges are depicted based on nodal degree and edge weight. The subnetwork consists of 111 brain regions connected by 282 edges (female > male)

connected by 425 connections (accuracy = 0.630, Fig. 16). Likewise, the subcortical-cerebellum network had the highest sum of nodal degree among functional networks. This network was followed by the motor, medial-frontal, fronto-parietal, default mode, visual, and visual association networks.

The threshold of 0.9, however, returned a significantly denser subnetwork consisting of 225 nodes and 1219 edges (accuracy = 0.624, Fig. 17). The subnetwork relevant to sex classification was comprised of similar functional networks as identified above.

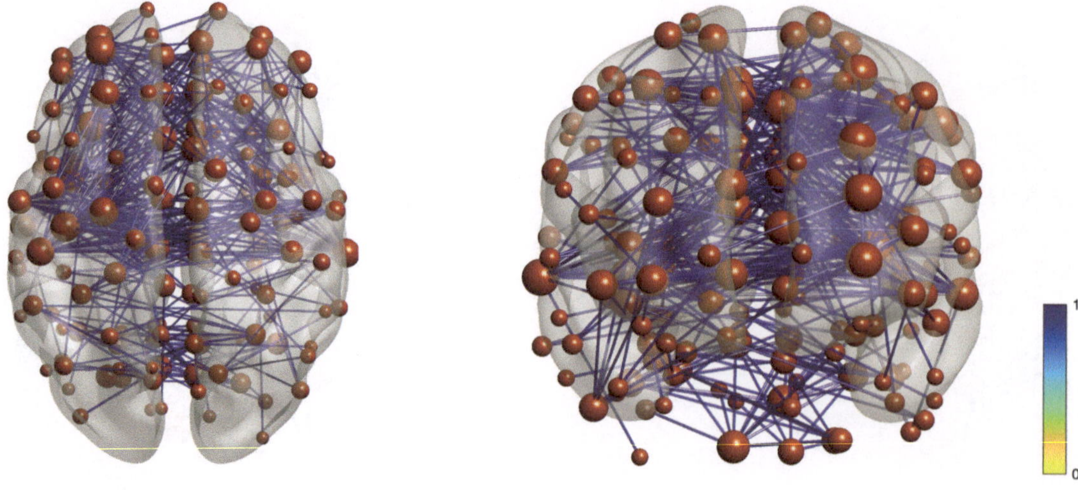

Fig. 16 Males > Females subnetwork (weight threshold = 1.0) visualized on a 3D brain surface generated by BrainNet Viewer [22]. The size and color of the nodes and edges are depicted based on nodal degree and edge weight. The subnetwork consists of 161 brain regions connected by 425 edges with greater connectivity in males than females

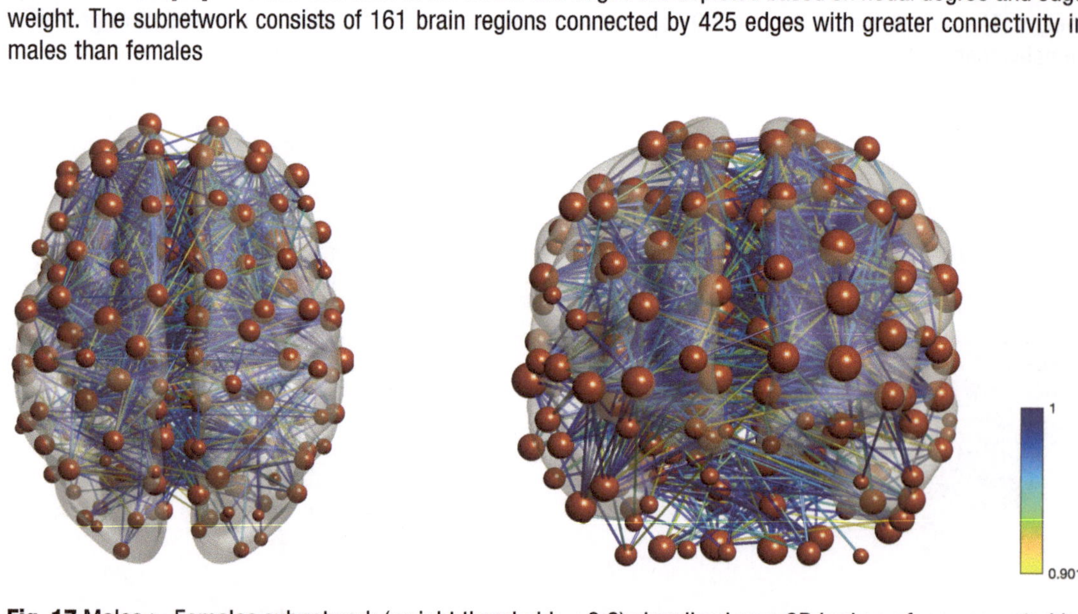

Fig. 17 Males > Females subnetwork (weight threshold = 0.9) visualized on a 3D brain surface generated by BrainNet Viewer [22]. The size and color of the nodes and edges are depicted based on nodal degree and edge weight. The subnetwork consists of 225 brain regions connected by 1219 edges (male > female)

7.4 Discussion of Sex Difference Classification

In this application, we aimed to classify sex from functional brain networks. To this end, similar to the first application, we employed NBS-Predict using the task-based (movie watching) fMRI data from the ID1000 dataset [17]. Using the connected component of edges selected in the female > male and male > female conditions, NBS-Predict distinguished sex groups with a classification accuracy of 63.7% and 68%, respectively.

The magnitude of the classification accuracy is significantly smaller than some previous studies, which reported very high classification accuracies (>85%) using structural [65, 66] and functional [67, 68] multimodal MRI data [69]. The primary reason for the significant discrepancy between previous results and our findings is that the confounding effect of brain size was not controlled for in these prior studies. As mentioned in Subheading 7.2, brain size has a substantial confounding effect when sex-related neurobiological differences are investigated. Various studies have shown that males have higher brain volumes (cortical and subcortical) than females [62, 63, 70]. Given the strong correlation between brain size and sex, disregarding this dimorphic difference could thus jeopardize the statistical validity and the generalizability of the findings. Therefore, without adequately controlling the confounding effects of brain size, true contribution of brain measures to sex classification can never be fully understood. In line with our results, studies that controlled for brain size [68, 71–74] yielded significantly lower classification accuracies than those without deconfounding (with deconfounding: 50% to ~76%; without: >85%). Therefore, as we considered these confounding effects and accordingly regressed out age and brain size, the classification performance of NBS-Predict is clearly comparable.

In both female > male and male > female conditions, regions from the subcortical-cerebellum network had the highest nodal degree. As regions with a high degree have various connections with other brain regions within or between networks, our results corroborate with previous findings that suggested that sex differences are related to cerebellar hemispheric asymmetry [75], morphology [76], and functional connectivity density [77]. Recent studies have also highlighted the importance of the cerebellum in sex classification [68, 78]. Importantly, Tiemeier et al. (2010) [76] showed that the morphological difference between females and males remained even after covarying total brain volume, which might also affect the functional connectivity of the cerebellum. As such, the cerebellum appears to be a significant feature for sex classification. Some subcortical areas (e.g., thalamus) have also been reported to be associated with sex [68, 73]. The importance of other functional brain networks in sex classification has been reported elsewhere [68, 69, 74, 79].

8 Advanced Features

A detailed description of advanced features in NBS-Predict is given in the user manual document provided within the toolbox.

Table 4
MATLAB structure to store several important information regarding analysis including parameters, input data, and the results

Structure	Substructure 1	Substructure 2	Description
NBS-predict	Parameter		Analysis parameters (e.g., K-fold, p-value, etc.)
	Info		General information about the toolbox and the analysis (e.g., toolbox version, analysis date, etc.)
	Data		Input data and directories for it
	searchHandle		Function handler for the searching algorithm (e.g., grid search)
	Results	LinReg	Results derived from linear regression
		svmR	Results derived from SVM regressor
		bestEstimator	The best performing estimator (if ML algorithm optimization selected)

8.1 Command Line

Although the NBS-Predict toolbox was primarily designed for GUI usage, it partly allows researchers to run their analyses through the MATLAB command window. Specifically, the toolbox requires a MATLAB structure (i.e., NBSPredict, Table 4) to save and load parameters and input data to run the analysis properly. For running analyses through the command line, we suggest that users load an "NBSPredict.mat" file saved immediately after starting the analysis and use this file as a template. Users can change the parameters or directories for the input file in this template. This edited structure will then serve as an input structure for the subsequent analyses. Users may load this edited structure file to MATLAB and run the "run_NBSPredict(NBSPredict)" command to perform the analysis. Alternatively, the edited and saved structure can directly be run using the "run_NBSPredict('filename.mat')" command. Users can also access input and output data stored in the "NBSPredict" structure and experimental metrics can be accessed inside the structure. For instance, within-sample feature stability [80] can be accessed via "NBSPredict.results.model_name.stability," where "model_name" represents the name of the ML model (e.g., LogReg, svmC). Of note, as these experimental metrics are not yet made available in the GUI, they might be subject to change, and some may not be fully validated within the NBS-Predict framework.

8.2 Making Predictions Using Novel Data

As mentioned in Subheading 2.1, users can further use their trained models to predict outcomes using novel data. This is quite important and one of the most useful advantages of machine learning over traditional inference statistics methods.

The *portable* model (i.e., the model trained using the whole input set) is automatically trained and saved under the subdirectory of "NBSPredict.results.ML_name.model," where "ML_name" is the name of the ML algorithm employed. To make predictions

```
holdoutDir = '/Users/user/Documents/MATLAB/SampleData/holdoutPrediction';

fprintf('Started predicting holdout subjects...\n')

% Look for subject files in the holdout directory.
subFiles = dir([holdoutDir, filesep, 'conn_mat', filesep, 'sub-*']);
nSubs = numel({subFiles.name}); % number of subjects.

% Load confound matrix
confoundMat = csvread('confoundMat.csv');

% Preallocate yhat vector
yPred = zeros(nSubs,1);

% Predict subjects' labels using trained lda model.
for s = 1: nSubs
    cSubDir = [subFiles(s).folder, filesep, subFiles(s).name]; % Current subject
    yPred(s) = NBSPredict_predict(NBSPredict.results.lda.model,...
        'connectome',cSubDir, 'confMat', confoundMat(s,:)); % Predict
end

% Load true labels
labels = csvread('labels.csv');

% Compute explained variance between yTrue and yPred
fprintf('Holdout performance: %.3f\n', ...
    compute_modelMetrics(labels, yPred, 'explained_variance'));
```

Fig. 18 An exemplary code for estimating the holdout set performance of the trained model. The image for the exemplary code was generated on carbon.now.sh

about unseen individuals, users should employ the "NBSPredict_-predict" function available within the toolbox. The function only requires connectome data of unseen individuals (e.g., holdout dataset), the portable model, and the confound matrix if confound regression had been performed. A key feature of this function is that it estimates the performance of the trained portable model on a "never seen" holdout dataset. As the holdout set has never been used in any model training, the models' performance on the holdout set represents its generalizability in the "real world." A sample code to estimate the models' holdout performance is depicted in Fig. 18.

9 Limitations and Caveats

A detailed discussion about the limitations of the NBS-Predict method is given elsewhere [14]. Briefly, NBS-Predict requires thresholds in the suprathreshold edge selection algorithms and for visualization purposes. As the choice of these thresholds can be arbitrary, some best practice strategies for thresholding are provided in Subheading 4.11 and Serin et al. (2021) [14]. Further, the distribution of weights in the weighted outcome network is strictly determined by the number of CV folds and repetitions. A higher number of CV folds or repetitions provides a more fine-grained edge weight distribution. In contrast, a lower number would result in coarse-grained weight distributions. Therefore, we strongly suggest that users perform more repetitions if the input connectome is dense. Critically, since connected components (i.e., features for the ML models) are based on significant edges selected by the suprathreshold edge selection algorithm, the method of NBS-Predict is only beneficial if the selected edges form a network (i.e., are connected). In cases of very sparse connectome data, we suggest using alternative methods such as CPM (Shen et al., 2017). Of note, NBS-Predict does not aim to necessarily maximize the prediction performance but instead strives to provide a clearly interpretable model that can predict the target of interest with a good predictive performance. Therefore, NBS-Predict relies on linear techniques. However, on some datasets, non-linear solutions such as ensemble algorithms (e.g., Gradient Boosting, Random Forest) or advanced deep neural network architectures might perform significantly better than NBS-Predict in exchange for interpretability. Therefore, if maximizing the predictive power is the sole aim of the analysis, we recommend using more complex non-linear ML algorithms.

Additionally, there are several limitations of the current version of NBS-Predict, which will be mitigated with continuous updates. First, since the toolbox is mainly designed and developed for GUI usage, it is not fully available for command-line use (*see* Subheading 8.1). Second, the hyperparameter optimization is limited in the current version of the toolbox (v1.0.0-beta.9), such that only one hyperparameter is tuned for each ML algorithm (Table 1). Thus, other critical hyperparameters such as "solvers" will be implemented in the following versions. Furthermore, hyperparameters of underlying ML algorithms as implemented in NBS-Predict are predefined. Therefore, in the current version of the toolbox, users cannot enter manual hyperparameters to tune. Third, the model performance is evaluated using only K-fold CV and leave-one-out CV (LOOCV) procedures. More advanced CV procedures (e.g., "Group K-Fold" that takes a group membership such as family membership into account) and other out-of-sample evaluation techniques such as the 0.632 bootstrapping technique [81] will be provided with future updates. Although NBS-Predict is an open-source toolbox, it requires the paid MATLAB platform (The MathWorks, Inc.), thereby making it not completely free-to-use. A completely free-to-use version of NBS-Predict developed in Python will be released in the future.

10 Development and Contribution

NBS-Predict is an open-source toolbox mainly stored in GitHub (https://github.com/eminSerin/NBS-Predict). Released versions of the toolbox can be downloaded from GitHub and NITRC (https://www.nitrc.org/projects/nbspredict/). Pre-release versions of the toolbox comprising experimental features can be downloaded from its GitHub repository.

We highly appreciate any form of contribution, such as bringing new features to NBS-Predict, reporting bugs, and improving readability of the code and the documentation. In case of any bugs, users should create an issue in the GitHub repository (https://github.com/eminSerin/NBS-Predict/issues) with a clear description. Developers should use the "dev" branch to pull the most recent version of the toolbox and push their new features or bug fixes.

11 Conclusion

This chapter provides an example-based walkthrough for the novel connectome-based predictive method of NBS-Predict. We present two application scenarios with freely available data covering two main prediction problems in machine learning: regression and classification. For regression, we employed NBS-Predict to infer individuals' fluid intelligence scores from their brain connectomes generated using task-based fMRI data from the ID1000 dataset. In the classification application, we inferred individuals' sex using the same data as in the regression scenario. NBS-Predict predicted fluid intelligence scores with $r = 0.243$ and identified a contributing subnetwork that spanned regions from various large-scale brain networks. With respect to sex classification, NBS-Predict achieved a classification accuracy of 63.7% (female > male) and 68% (male > female) and identified connectome wide coupled cerebellar and frontal regions as a driving factor for classification success.

We anticipate that this novel toolbox will make machine-learning approaches more accessible to a broader audience of researchers thanks to its easy-to-use GUI and thus encourage the exploration of highly generalizable neuroimaging-based biomarkers.

Acknowledgments

The development of NBS-Predict was partially funded by a grant from the Melbourne/Berlin Research Partnership through the Berlin University Alliance.

References

1. Bullmore E, Sporns O (2009) Complex brain networks: graph theoretical analysis of structural and functional systems. Nat Rev Neurosci 10:186–198. https://doi.org/10.1038/nrn2575
2. Poldrack RA, Baker CI, Durnez J, Gorgolewski KJ, Matthews PM, Munafò MR, Nichols TE, Poline J-B, Vul E, Yarkoni T (2017) Scanning the horizon: towards transparent and reproducible neuroimaging research. Nat Rev Neurosci 18:115–126. https://doi.org/10.1038/nrn.2016.167
3. Zalesky A, Fornito A, Bullmore ET (2010) Network-based statistic: identifying differences in brain networks. NeuroImage 53:1197–1207. https://doi.org/10.1016/j.neuroimage.2010.06.041
4. Benjamini Y (2010) Simultaneous and selective inference: current successes and future challenges. Biom J Biom Z 52:708–721. https://doi.org/10.1002/bimj.200900299
5. Meskaldji DE, Fischi-Gomez E, Griffa A, Hagmann P, Morgenthaler S, Thiran J-P (2013) Comparing connectomes across subjects and populations at different scales. NeuroImage 80:416–425. https://doi.org/10.1016/j.neuroimage.2013.04.084
6. Bennett CM, Miller MB (2010) How reliable are the results from functional magnetic resonance imaging? Ann N Y Acad Sci 1191:133–155. https://doi.org/10.1111/j.1749-6632.2010.05446.x
7. Pashler H, Wagenmakers E (2012) Editors' introduction to the special section on replicability in psychological science: a crisis of confidence? Perspect Psychol Sci 7:528–530. https://doi.org/10.1177/1745691612465253
8. Tabe-Bordbar S, Emad A, Zhao SD, Sinha S (2018) A closer look at cross-validation for assessing the accuracy of gene regulatory networks and models. Sci Rep 8:6620. https://doi.org/10.1038/s41598-018-24937-4
9. Waller L, Brovkin A, Dorfschmidt L, Bzdok D, Walter H, Kruschwitz JD (2018) GraphVar 2.0: a user-friendly toolbox for machine learning on functional connectivity measures. J Neurosci Methods 308:21–33. https://doi.org/10.1016/j.jneumeth.2018.07.001
10. Whelan R, Garavan H (2014) When optimism hurts: inflated predictions in psychiatric neuroimaging. Biol Psychiatry 75:746–748. https://doi.org/10.1016/j.biopsych.2013.05.014
11. Mwangi B, Tian TS, Soares JC (2014) A review of feature reduction techniques in neuroimaging. Neuroinformatics 12:229–244. https://doi.org/10.1007/s12021-013-9204-3
12. Haufe S, Meinecke F, Görgen K, Dähne S, Haynes J-D, Blankertz B, Bießmann F (2014) On the interpretation of weight vectors of linear models in multivariate neuroimaging. NeuroImage 87:96–110. https://doi.org/10.1016/j.neuroimage.2013.10.067
13. Hebart MN, Baker CI (2018) Deconstructing multivariate decoding for the study of brain function. NeuroImage 180:4–18. https://doi.org/10.1016/j.neuroimage.2017.08.005
14. Serin E, Zalesky A, Matory A, Walter H, Kruschwitz JD (2021) NBS-predict: a prediction-based extension of the network-based statistic. NeuroImage 244:118625. https://doi.org/10.1016/j.neuroimage.2021.118625
15. Finn ES, Shen X, Scheinost D, Rosenberg MD, Huang J, Chun MM, Papademetris X, Constable RT (2015) Functional connectome fingerprinting: identifying individuals based on patterns of brain connectivity. Nat Neurosci 18:1664–1671. https://doi.org/10.1038/nn.4135
16. Shen X, Finn ES, Scheinost D, Rosenberg MD, Chun MM, Papademetris X, Constable RT (2017) Using connectome-based predictive modeling to predict individual behavior from brain connectivity. Nat Protoc 12:506–518. https://doi.org/10.1038/nprot.2016.178
17. Snoek L, van der Miesen MM, Beemsterboer T, van der Leij A, Eigenhuis A, Steven Scholte H (2021) The Amsterdam Open MRI Collection, a set of multimodal MRI datasets for individual difference analyses. Sci Data 8:85. https://doi.org/10.1038/s41597-021-00870-6
18. Braga-Neto UM, Dougherty ER (2004) Is cross-validation valid for small-sample microarray classification? Bioinformatic 20:374–380. https://doi.org/10.1093/bioinformatics/btg419
19. Kim J-H (2009) Estimating classification error rate: repeated cross-validation, repeated hold-out and bootstrap. Comput Stat Data Anal 53:3735–3745. https://doi.org/10.1016/j.csda.2009.04.009
20. Krstajic D, Buturovic LJ, Leahy DE, Thomas S (2014) Cross-validation pitfalls when selecting and assessing regression and classification models. J Cheminformatics 6:10. https://doi.org/10.1186/1758-2946-6-10

21. Jollans L, Boyle R, Artiges E, Banaschewski T, Desrivières S, Grigis A, Martinot J-L, Paus T, Smolka MN, Walter H, Schumann G, Garavan H, Whelan R (2019) Quantifying performance of machine learning methods for neuroimaging data. NeuroImage 199:351–365. https://doi.org/10.1016/j.neuroimage.2019.05.082
22. Xia M, Wang J, He Y (2013) BrainNet viewer: a network visualization tool for human brain connectomics. PLoS One 8:e68910. https://doi.org/10.1371/journal.pone.0068910
23. Bright MG, Murphy K (2015) Is fMRI "noise" really noise? Resting state nuisance regressors remove variance with network structure. NeuroImage 114:158–169. https://doi.org/10.1016/j.neuroimage.2015.03.070
24. Parkes L, Fulcher B, Yücel M, Fornito A (2018) An evaluation of the efficacy, reliability, and sensitivity of motion correction strategies for resting-state functional MRI. NeuroImage 171:415–436. https://doi.org/10.1016/j.neuroimage.2017.12.073
25. Van Dijk KRA, Sabuncu MR, Buckner RL (2012) The influence of head motion on intrinsic functional connectivity MRI. NeuroImage 59:431–438. https://doi.org/10.1016/j.neuroimage.2011.07.044
26. Yendiki A, Koldewyn K, Kakunoori S, Kanwisher N, Fischl B (2014) Spurious group differences due to head motion in a diffusion MRI study. NeuroImage 88:79–90. https://doi.org/10.1016/j.neuroimage.2013.11.027
27. Waller L, Erk S, Pozzi E, Toenders YJ, Haswell CC, Büttner M, Thompson PM, Schmaal L, Morey RA, Walter H, Veer IM (2022) ENIGMA HALFpipe: interactive, reproducible, and efficient analysis for resting-state and task-based fMRI data. Hum Brain Mapp 43:2727–2742. https://doi.org/10.1002/hbm.25829
28. Esteban O, Markiewicz CJ, Blair RW, Moodie CA, Isik AI, Erramuzpe A, Kent JD, Goncalves M, DuPre E, Snyder M, Oya H, Ghosh SS, Wright J, Durnez J, Poldrack RA, Gorgolewski KJ (2019) fMRIPrep: a robust preprocessing pipeline for functional MRI. Nat Methods 16:111–116. https://doi.org/10.1038/s41592-018-0235-4
29. Behzadi Y, Restom K, Liau J, Liu TT (2007) A component based noise correction method (CompCor) for BOLD and perfusion based fMRI. NeuroImage 37:90–101. https://doi.org/10.1016/j.neuroimage.2007.04.042
30. Pruim RHR, Mennes M, van Rooij D, Llera A, Buitelaar JK, Beckmann CF (2015) ICA-AROMA: a robust ICA-based strategy for removing motion artifacts from fMRI data. NeuroImage 112:267–277. https://doi.org/10.1016/j.neuroimage.2015.02.064
31. Friston KJ, Williams S, Howard R, Frackowiak RS, Turner R (1996) Movement-related effects in fMRI time-series. Magn Reson Med 35:346–355. https://doi.org/10.1002/mrm.1910350312
32. Lindquist MA, Geuter S, Wager TD, Caffo BS (2019) Modular preprocessing pipelines can reintroduce artifacts into fMRI data. Hum Brain Mapp 40:2358–2376. https://doi.org/10.1002/hbm.24528
33. Shen X, Tokoglu F, Papademetris X, Constable RT (2013) Groupwise whole-brain parcellation from resting-state fMRI data for network node identification. NeuroImage 82:403–415. https://doi.org/10.1016/j.neuroimage.2013.05.081
34. Abraham A, Pedregosa F, Eickenberg M, Gervais P, Mueller A, Kossaifi J, Gramfort A, Thirion B, Varoquaux G (2014) Machine learning for neuroimaging with scikit-learn. Front Neuroinform 8:14. https://doi.org/10.3389/fninf.2014.00014
35. Diedrichsen J, Balsters JH, Flavell J, Cussans E, Ramnani N (2009) A probabilistic MR atlas of the human cerebellum. NeuroImage 46:39–46. https://doi.org/10.1016/j.neuroimage.2009.01.045
36. Pervaiz U, Vidaurre D, Woolrich MW, Smith SM (2020) Optimising network modelling methods for fMRI. NeuroImage 211:116604. https://doi.org/10.1016/j.neuroimage.2020.116604
37. Jenkinson M, Beckmann CF, Behrens TEJ, Woolrich MW, Smith SM (2012) FSL. NeuroImage 62:782–790. https://doi.org/10.1016/j.neuroimage.2011.09.015
38. Cawley GC, Talbot NLC (2010) On over-fitting in model selection and subsequent selection bias in performance evaluation. J Mach Learn Res 11:2079–2107
39. Varma S, Simon R (2006) Bias in error estimation when using cross-validation for model selection. BMC Bioinformatics 7:91. https://doi.org/10.1186/1471-2105-7-91
40. Rao A, Monteiro JM, Mourao-Miranda J, Alzheimer's Disease Initiative (2017) Predictive modelling using neuroimaging data in the presence of confounds. NeuroImage 150:23–49. https://doi.org/10.1016/j.neuroimage.2017.01.066
41. Snoek L, Miletić S, Scholte HS (2019) How to control for confounds in decoding analyses of neuroimaging data. NeuroImage 184:741–760. https://doi.org/10.1016/j.neuroimage.2018.09.074

42. Dinga R, Schmaal L, Penninx BWJH, Veltman DJ, Marquand AF (2020) Controlling for effects of confounding variables on machine learning predictions. bioRxiv
43. Hastie T, Tibshirani R, Friedman J (2009) The elements of statistical learning. Springer, New York, NY
44. Bergstra J, Bardenet R, Bengio Y, Kégl B (2011) Algorithms for hyper-parameter optimization. In: Advances in neural information processing systems. Curran Associates, Inc
45. Hutter F, Hoos HH, Leyton-Brown K (2011) Sequential model-based optimization for general algorithm configuration. In: Coello CAC (ed) Learning and intelligent optimization. Springer, Berlin, Heidelberg, pp 507–523
46. Ojala M, Garriga G (2010) Permutation tests for studying classifier performance. J Mach Learn Res 11:1833–1863
47. Branco P, Torgo L, Ribeiro RP (2016) A survey of predictive modeling on imbalanced domains. ACM Comput Surv 49:31:1–31:50. https://doi.org/10.1145/2907070
48. Bej S, Galow A-M, David R, Wolfien M, Wolkenhauer O (2021) Automated annotation of rare-cell types from single-cell RNA-sequencing data through synthetic oversampling. BMC Bioinformatics 22:557. https://doi.org/10.1186/s12859-021-04469-x
49. Ioffe S, Szegedy C (2015) Batch normalization: accelerating deep network training by reducing internal covariate shift. arXiv
50. Bullmore ET, Bassett DS (2011) Brain graphs: graphical models of the human brain connectome. Annu Rev Clin Psychol 7:113–140. https://doi.org/10.1146/annurev-clinpsy-040510-143934
51. Power JD, Cohen AL, Nelson SM, Wig GS, Barnes KA, Church JA, Vogel AC, Laumann TO, Miezin FM, Schlaggar BL, Petersen SE (2011) Functional network organization of the human brain. Neuron 72:665–678. https://doi.org/10.1016/j.neuron.2011.09.006
52. Dubois J, Galdi P, Han Y, Paul LK, Adolphs R (2018) Resting-state functional brain connectivity best predicts the personality dimension of openness to experience. Personal Neurosci 1: e6. https://doi.org/10.1017/pen.2018.8
53. He T, Kong R, Holmes AJ, Sabuncu MR, Eickhoff SB, Bzdok D, Feng J, Yeo BTT (2018) Is deep learning better than kernel regression for functional connectivity prediction of fluid intelligence? In: 2018 international workshop on pattern recognition in neuroimaging, PRNI 2018. Institute of Electrical and Electronics Engineers Inc., p 8423958
54. Noble S, Spann MN, Tokoglu F, Shen X, Constable RT, Scheinost D (2017) Influences on the test-retest reliability of functional connectivity MRI and its relationship with Behavioral utility. Cereb Cortex 27:5415–5429. https://doi.org/10.1093/cercor/bhx230
55. Mantwill M, Gell M, Krohn S, Finke C (2022) Fingerprinting and behavioural prediction rest on distinct functional systems of the human connectome. Commun Biol 5(1):261
56. Hearne LJ, Mattingley JB, Cocchi L (2016) Functional brain networks related to individual differences in human intelligence at rest. Sci Rep 6:32328. https://doi.org/10.1038/srep32328
57. Song M, Zhou Y, Li J, Liu Y, Tian L, Yu C, Jiang T (2008) Brain spontaneous functional connectivity and intelligence. NeuroImage 41: 1168–1176. https://doi.org/10.1016/j.neuroimage.2008.02.036
58. van den Heuvel MP, Stam CJ, Kahn RS, Hulshoff Pol HE (2009) Efficiency of functional brain networks and intellectual performance. J Neurosci 29:7619–7624. https://doi.org/10.1523/JNEUROSCI.1443-09.2009
59. Van Essen DC, Smith SM, Barch DM, TEJ B, Yacoub E, Ugurbil K, WU-Minn HCP Consortium (2013) The WU-Minn human connectome project: an overview. NeuroImage 80:62–79. https://doi.org/10.1016/j.neuroimage.2013.05.041
60. Cattell RB (1974) Raymond B. Cattell. In: A history of psychology in autobiography, vol VI. Prentice-Hall, Inc, Englewood Cliffs, NJ, US, pp 61–100
61. Schmidt-Atzert L (2002) Tests und tools. Z Für Pers 1:50–56. https://doi.org/10.1026/1617-6391.1.1.50
62. Barnes J, Ridgway GR, Bartlett J, Henley SMD, Lehmann M, Hobbs N, Clarkson MJ, MacManus DG, Ourselin S, Fox NC (2010) Head size, age and gender adjustment in MRI studies: a necessary nuisance? NeuroImage 53: 1244–1255. https://doi.org/10.1016/j.neuroimage.2010.06.025
63. Ritchie SJ, Cox SR, Shen X, Lombardo MV, Reus LM, Alloza C, Harris MA, Alderson HL, Hunter S, Neilson E, Liewald DCM, Auyeung B, Whalley HC, Lawrie SM, Gale CR, Bastin ME, McIntosh AM, Deary IJ (2018) Sex differences in the adult human brain: evidence from 5216 UK biobank participants. Cereb Cortex 28:2959–2975. https://doi.org/10.1093/cercor/bhy109

64. Destrieux C, Fischl B, Dale A, Halgren E (2010) Automatic parcellation of human cortical gyri and sulci using standard anatomical nomenclature. NeuroImage 53:1–15. https://doi.org/10.1016/j.neuroimage.2010.06.010
65. Nieuwenhuis M, Schnack HG, van Haren NE, Lappin J, Morgan C, Reinders AA, Gutierrez-Tordesillas D, Roiz-Santiañez R, Schaufelberger MS, Rosa PG, Zanetti MV, Busatto GF, Crespo-Facorro B, McGorry PD, Velakoulis D, Pantelis C, Wood SJ, Kahn RS, Mourao-Miranda J, Dazzan P (2017) Multicenter MRI prediction models: predicting sex and illness course in first episode psychosis patients. NeuroImage 145:246–253. https://doi.org/10.1016/j.neuroimage.2016.07.027
66. Peng H, Gong W, Beckmann CF, Vedaldi A, Smith SM (2021) Accurate brain age prediction with lightweight deep neural networks. Med Image Anal 68:101871. https://doi.org/10.1016/j.media.2020.101871
67. Al Zoubi O, Misaki M, Tsuchiyagaito A, Zotev V, White E, Paulus M, Bodurka J (2022) Machine learning evidence for sex differences consistently influences resting-state functional magnetic resonance imaging fluctuations across multiple independently acquired data sets. Brain Connect 12:348–361. https://doi.org/10.1089/brain.2020.0878
68. Zhang C, Dougherty CC, Baum SA, White T, Michael AM (2018) Functional connectivity predicts gender: evidence for gender differences in resting brain connectivity. Hum Brain Mapp 39:1765–1776. https://doi.org/10.1002/hbm.23950
69. Zhang X, Liang M, Qin W, Wan B, Yu C, Ming D (2020) Gender differences are encoded differently in the structure and function of the human brain revealed by multimodal MRI. Front Hum Neurosci 14:244. https://doi.org/10.3389/fnhum.2020.00244
70. Ruigrok ANV, Salimi-Khorshidi G, Lai M-C, Baron-Cohen S, Lombardo MV, Tait RJ, Suckling J (2014) A meta-analysis of sex differences in human brain structure. Neurosci Biobehav Rev 39:34–50. https://doi.org/10.1016/j.neubiorev.2013.12.004
71. Chekroud AM, Ward EJ, Rosenberg MD, Holmes AJ (2016) Patterns in the human brain mosaic discriminate males from females. Proc Natl Acad Sci USA 113:E1968. https://doi.org/10.1073/pnas.1523888113
72. More S, Eickhoff SB, Caspers J, Patil KR (2021) Confound removal and normalization in practice: a neuroimaging based sex prediction case study. In: Dong Y, Ifrim G, Mladenić D, Saunders C, Van Hoecke S (eds) Machine learning and knowledge discovery in databases. Applied data science and demo track. Springer, Cham, pp 3–18
73. Sanchis-Segura C, Ibañez-Gual MV, Aguirre N, Cruz-Gómez ÁJ, Forn C (2020) Effects of different intracranial volume correction methods on univariate sex differences in grey matter volume and multivariate sex prediction. Sci Rep 10:12953. https://doi.org/10.1038/s41598-020-69361-9
74. Weis S, Patil KR, Hoffstaedter F, Nostro A, Yeo BTT, Eickhoff SB (2020) Sex classification by resting state brain connectivity. Cereb Cortex 30:824–835. https://doi.org/10.1093/cercor/bhz129
75. Fan L, Tang Y, Sun B, Gong G, Chen ZJ, Lin X, Yu T, Li Z, Evans AC, Liu S (2010) Sexual dimorphism and asymmetry in human cerebellum: an MRI-based morphometric study. Brain Res 1353:60–73. https://doi.org/10.1016/j.brainres.2010.07.031
76. Tiemeier H, Lenroot RK, Greenstein DK, Tran L, Pierson R, Giedd JN (2010) Cerebellum development during childhood and adolescence: a longitudinal morphometric MRI study. NeuroImage 49:63–70. https://doi.org/10.1016/j.neuroimage.2009.08.016
77. Tomasi D, Volkow ND (2012) Gender differences in brain functional connectivity density. Hum Brain Mapp 33:849–860. https://doi.org/10.1002/hbm.21252
78. Xin J, Zhang Y, Tang Y, Yang Y (2019) Brain differences between men and women: evidence from deep learning. Front Neurosci 13:185. https://doi.org/10.3389/fnins.2019.00185
79. Fan L, Su J, Qin J, Hu D, Shen H (2020) A deep network model on dynamic functional connectivity with applications to gender classification and intelligence prediction. Front Neurosci 14:881. https://doi.org/10.3389/fnins.2020.00881
80. Nogueira S, Sechidis K, Brown G (2018) On the stability of feature selection algorithms. J Mach Learn Res 18:1–54
81. Efron B, Tibshirani R (1997) Improvements on cross-validation: the 632+ bootstrap method. J Am Stat Assoc 92:548–560. https://doi.org/10.1080/01621459.1997.10474007

BY

Chapter 14

Normative Modeling with the Predictive Clinical Neuroscience Toolkit (PCNtoolkit)

Saige Rutherford and Andre F. Marquand

Abstract

In this chapter, we introduce normative modeling as a tool for mapping variation across large neuroimaging datasets. We provide practical guidance to illustrate how normative models can be used to map diverse patterns of individual differences found within the large datasets used to train the models. In other words, while normative modeling is a method often applied to big datasets containing thousands of subjects, it provides single subject inference and prediction. We use an open-source Python package, Predictive Clinical Neuroscience Toolkit (PCNtoolkit) and showcase several helpful tools (including an interface that does not require coding) to run a normative modeling analysis, evaluate the model fit, and visualize the results.

Key words Neuroimaging, normative modeling, PCNtoolkit, individual differences

1 Introduction

Normative modeling in neuroimaging refers to the statistical analysis of brain imaging data from a large sample of individuals in order to establish typical patterns of brain structure and function [2–7]. The overarching aim is to define a reference range for the given brain measurement (structure or function) in a certain sample and to create a reference standard against which to compare individuals, often with neurological or psychiatric conditions [8–17]. This is typically achieved by fitting flexible probabilistic regression models [See Glossary] to map centiles of variation in the population, akin to the use of growth charts in pediatric medicine (see example in Fig. 1). Owing to their ability to make predictions at the level of the individual participants, normative models can be used to detect subtle changes in brain structure or function that may indicate the early stages of a disease or to evaluate the effects of a certain treatment or intervention. These models can also be used

Robert Whelan and Hervé Lemaître (eds.), *Methods for Analyzing Large Neuroimaging Datasets*, Neuromethods, vol. 218, https://doi.org/10.1007/978-1-0716-4260-3_14, © The Author(s) 2025

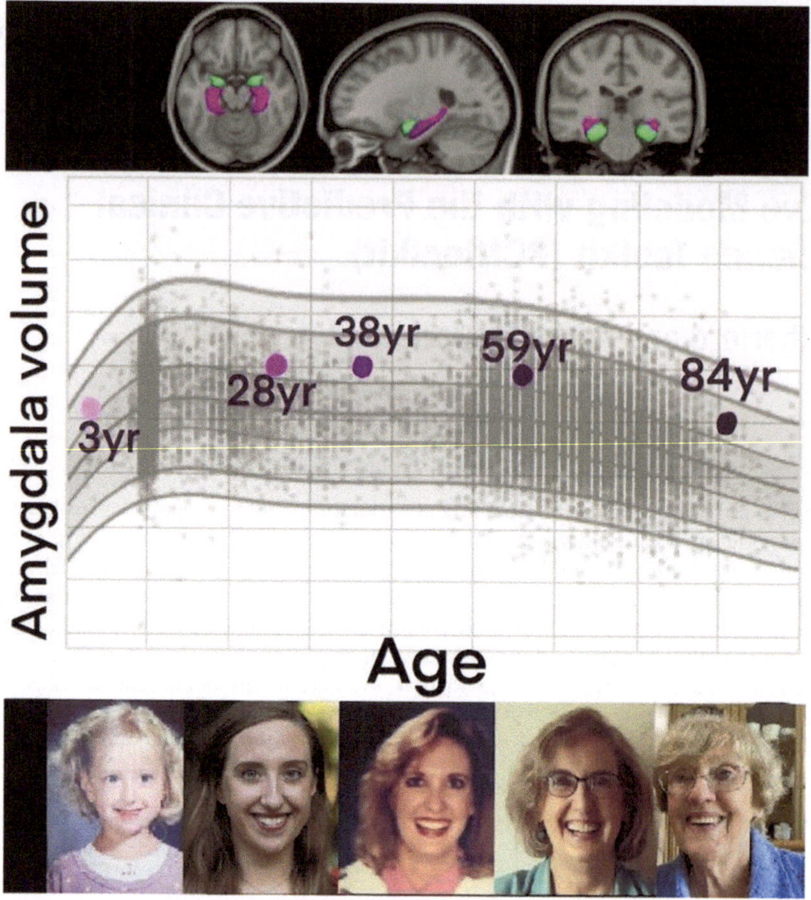

Fig. 1 Big data ($N = \sim 58{,}000$ subjects) normative model of amygdala volume across the human lifespan (ages two to 100) [1]

in clinical settings to evaluate brain function of individuals with suspected neurological or psychiatric disorders and to monitor the progression of the disorder over time.

2 Predictive Clinical Neuroscience Toolkit (PCNToolkit)

The Predictive Clinical Neuroscience (PCN) toolkit [18] is a Python package designed for multi-purpose tasks in clinical neuroimaging, including normative modelling, trend surface modelling in addition to providing implementations of a number of fundamental machine learning algorithms [See Glossary].

Normative modelling essentially aims to predict centiles of variance in a response variable (e.g., a region of interest or other neuroimaging-derived measure) on the basis of a set of covariates (e.g., age, clinical scores, diagnosis). In this paper, we take an applied perspective and provide guidance about how to perform

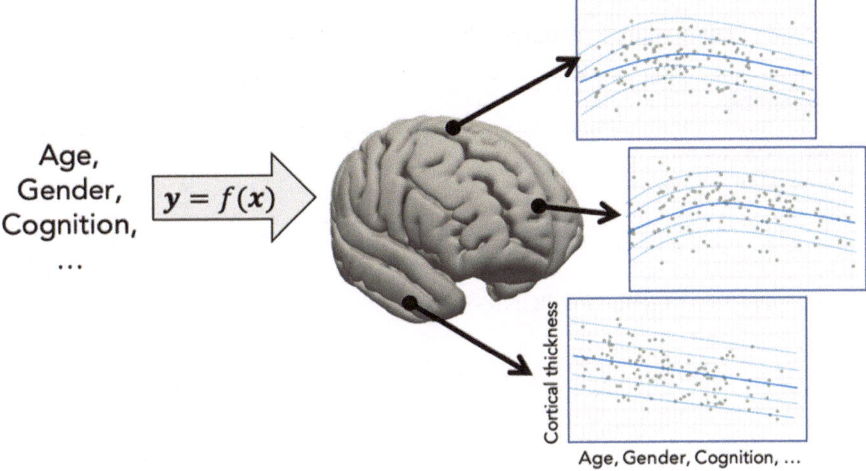

Fig. 2 Normative modeling example

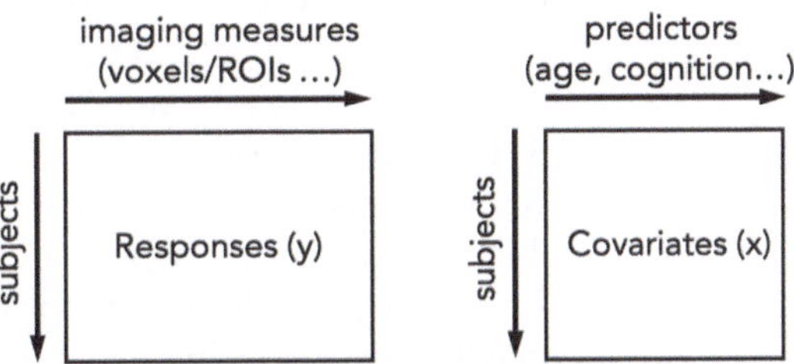

Fig. 3 Matrix representation of biological response variables and covariates

normative modelling in practice. We refer the reader to other review and protocol papers where in-depth conceptual and theoretical overviews of the approach can be found [19–21]. For example, the image below shows an example of a normative model that aims to predict vertex-wise cortical thickness data, essentially fitting a separate model for each vertex (Fig. 2).

In practice, normative modelling is done by regressing the biological response variables against a set of clinical or demographic covariates. In the instructions that follow, it is helpful to think of these as being stored in matrices as shown below (Fig. 3).

There are many options for this, but techniques that provide a distributional form for the centiles are appealing, since they help to estimate extreme centiles (where data are sparsest) more efficiently and can help to avoid centile crossings [22]. Bayesian methods are also beneficial in this regard because they also allow separation of modelling uncertainty from variation in the data. Many applications of normative modelling use Gaussian Process Regression [See Glossary], which is the default method in this toolkit. However, other

Table 1
Overview of available open-source resources for normative modeling

Resource	Description	Link
PCNportal	No code required interface for accessing pre-trained normative models	https://pcnportal.dccn.nl/
Gitter	Website for communication with PCNtoolkit developers	https://gitter.im/predictive-clinical-neuroscience/community
Read The Docs	Wiki resource page for the PCNtoolkit	https://pcntoolkit.readthedocs.io/en/latest/
GitHub	Code base for the PCNtoolkit. Contributions welcome!	https://github.com/amarquand/PCNtoolkit
Google Colab	Run python notebooks in a web browser without setup of python environment	Bayesian Linear Regression Hierarchical Bayesian regression

algorithms are available and scale better to estimating normative models on large datasets. These algorithms include Bayesian Linear Regression (BLR) and Hierarchical Bayesian Linear Regression (HBR). In the code tutorials included with this chapter, we implement normative models using BLR. Typically, each response variable (brain region) is estimated independently. In the sections that follow, we provide code tutorials for running a normative modeling analysis, with specific explanations of the modeling choices (these explanations are embedded in the relevant tutorials, which are available online via the links summarized in Table 1 and Fig. 4).

3 Code Tutorial 1: Transferring Pre-trained Big Data Normative Models

This code shows how to apply the coefficients from pre-estimated normative models to new data, such as regional cortical thickness from Freesurfer preprocessing. This can be done in two different ways: (1) using a new set of data derived from the same sites used to estimate the model and (2) on a completely different set of sites. In the latter case, we also need to estimate the site effect, which requires some calibration/adaptation data. As an illustrative example, we use a dataset derived from several OpenNeuro datasets and adapt the learned model to make predictions on these data [see Chapter 2]. This code can be run in your web browser using Google Colab here.

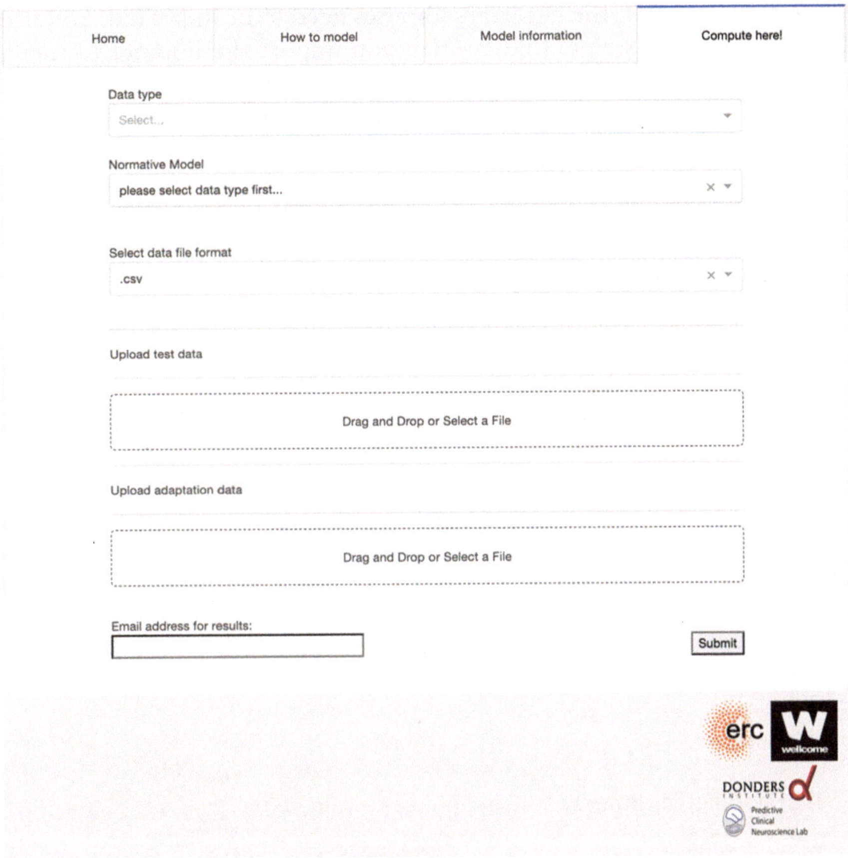

Home | How to model | Model information | Compute here!

Data type

Select...

Normative Model

please select data type first...

Select data file format

.csv

Upload test data

Drag and Drop or Select a File

Upload adaptation data

Drag and Drop or Select a File

Email address for results:

Submit

erc wellcome DONDERS INSTITUTE Predictive Clinical Neuroscience Lab

Fig. 4 Overview of the (no code required) PCNportal website for running normative modeling analysis

3.1 Using Lifespan Models to Make Predictions on New Data

3.1.1 The First Step Is to Install PCNtoolkit

```
#!/usr/bin/env python

get_ipython().system(' pip install pcntoolkit==0.28')

get_ipython().system(' git clone https://github.com/predictive-clinical-
neuroscience/braincharts.git')
```

3.1.2 Import Necessary Python Libraries

Next, the necessary libraries need to be imported. You need to be in the scripts folder when you import the libraries in the code block below, because there is a function specific to normative modeling—called *nm_utils*—that is in the scripts folder that we need to import.

```
import os

os.chdir('/content/braincharts/scripts/') # this path is setup for running on Google Colab.

Change it to match your local path if running locally


# Now we import the required libraries

import numpy as np

import pandas as pd

import pickle

from matplotlib import pyplot as plt

import seaborn as sns

from pcntoolkit.normative import estimate, predict, evaluate

from pcntoolkit.util.utils import compute_MSLL, create_design_matrix

from nm_utils import remove_bad_subjects, load_2d
```

3.1.3 Select and Unzip Model Folder

In this step, you will first unzip the models. You start by changing the directory. In this example, you will use the biggest sample as your training set (approx. $N = 58{,}000$ subjects from 82 sites). For more info on the other pre-trained models available in this repository, please refer to the accompanying paper [1].

```
# change the directory


os.chdir('/content/braincharts/models/')


#unzip the data

get_ipython().system(' unzip lifespan_57K_82sites.zip')
```

3.1.4 Set Paths and Directory Names

Next, you need to configure some basic variables, like where we want the analysis to be done (e.g., in a particular folder on your computer) and which lifespan model you want to use.

> **Note**
> We maintain a list of site IDs for each dataset, which describe the site names in the training and test data ('site_ids_tr' and 'site_ids_te'), plus also the adaptation data. The training site IDs are provided as a text file in the distribution and the test IDs are extracted automatically from the pandas dataframe (see below). If you use additional data from the sites (e.g., later waves from ABCD), it may be necessary to adjust the site names to match the names in the training set. See Rutherford et al. [1] for more details.

```
# Which model do we wish to use?

model_name = 'lifespan_57K_82sites'

site_names = 'site_ids_ct_82sites.txt'


# Where the analysis takes place

root_dir = '/content/braincharts'


# Where the data files live

data_dir = '/content/braincharts/docs'


# Where the models live

out_dir = os.path.join(root_dir, 'models', model_name)


# Load a set of site IDs from this model. This must match the training data

with open(os.path.join(root_dir,'docs', site_names)) as f:

    site_ids_tr = f.read().splitlines()
```

3.2 Loading Data

3.2.1 Test Data

First, you need to load the test data. For the purposes of this tutorial, you will make predictions for a multi-site transfer dataset, derived from OpenNeuro.

```
test_data = os.path.join(data_dir, 'OpenNeuroTransfer_ct_te.csv')

df_te = pd.read_csv(test_data)

# Extract a list of unique site ids from the test set
site_ids_te =  sorted(set(df_te['site'].to_list()))
```

3.2.2 Adaption Data

Next, you need to load the adaptation data. If the data you wish to make predictions for are not derived from the same scanning sites as those in the training set, it is necessary to learn the site effect so that it can be accounted for it in the predictions. In order to do this in an unbiased way, it is necessary to use a separate dataset, which is referred to as 'adaptation data. This must contain data for all the same sites as in the test dataset (if not, a warning is displayed) and we assume these are coded in the same way, based on a the 'site-num' column in the dataframe.

```
adaptation_data = os.path.join(data_dir, 'OpenNeuroTransfer_ct_ad.csv')

df_ad = pd.read_csv(adaptation_data)

# Extract a list of unique site ids from the test set
site_ids_ad =  sorted(set(df_ad['site'].to_list()))

if not all(elem in site_ids_ad for elem in site_ids_te):
    print('Warning: some of the testing sites are not in the adaptation data')
```

3.3 Configure the Models to Fit

3.3.1 Select Brain Phenotypes

Now, you configure which *imaging derived phenotypes* (IDPs) you would like to process. This is just a list of column names in the dataframe you have loaded above. You can either load the whole set (i.e., all phenotypes for which you have models for) or you can specify a subset.

```
# Load the list of idps for left and right hemispheres, plus subcortical regions

with open(os.path.join(data_dir, 'phenotypes_ct_lh.txt')) as f:

    idp_ids_lh = f.read().splitlines()

with open(os.path.join(data_dir, 'phenotypes_ct_rh.txt')) as f:

    idp_ids_rh = f.read().splitlines()

with open(os.path.join(data_dir, 'phenotypes_sc.txt')) as f:

    idp_ids_sc = f.read().splitlines()



# We choose here to process all idps

idp_ids = idp_ids_lh + idp_ids_rh + idp_ids_sc



# ... or alternatively, we could just specify a list

idp_ids = [ 'Left-Thalamus-Proper', 'Left-Lateral-Ventricle', 'rh_MeanThickness_thickness']
```

3.3.2 Configure Model Parameters

Now, you should configure some parameters to fit the model. First, choose which columns of the pandas dataframe contain the covariates (age and sex). The site parameters are configured automatically later on by the 'configure_design_matrix()' function when looping through the IDPs in the list. The supplied coefficients are derived from a 'warped' Bayesian linear regression model, which uses a nonlinear warping function to model non-Gaussianity ('sinarcsinh') plus a non-linear basis expansion (a cubic b-spline basis set with 5 knot points, which is the default value in the PCNtoolkit package). For further details about the likelihood warping approach, see Rutherford et al. (2022) [1] and Fraza et al. (2022) [23]. Since you are sticking with the default value, you do not need to specify any parameters for this, but you do need to specify the limits. Below, you choose to pad the input by a few years either side of the input range and set a couple of options that control the estimation of the model.

```
# Which data columns do we wish to use as covariates?

cols_cov = ['age','sex']


# Limits for cubic B-spline basis

xmin = -5

xmax = 110


# Absolute Z threshold above which a sample is considered to be an outlier (without fitting any model)

outlier_thresh = 7
```

3.4 Making Predictions

The next step is to make predictions. The code below will make predictions for each IDP separately. This is done by extracting a column from the dataframe (i.e., specifying the IDP as the response variable) and saving it as a numpy array. Next, configure the covariates, which is a numpy data array having the number of rows equal to the number of datapoints in the test set. The columns are specified as follows:

- The covariate columns (here age and sex, coded as 0 = female/ 1 = male).
- Dummy coded columns for the sites in the training set (one column per site).
- Columns for the basis expansion (seven columns for the default parameterization).

Once these are saved as numpy arrays in ascii format (as here) or (alternatively) in pickle format, these are passed as inputs to the 'predict()' method in the PCNtoolkit normative modelling framework. These are written in the same format to the location specified by 'idp_dir.' At the end of this step, you will have a set of predictions and Z-statistics for the test dataset that you can take forward to further analysis.

> **Note**
> When you need to make predictions on new data, the procedure is more involved, since we need to prepare, process and store covariates, response variables, and site ids for the adaptation data.

```
for idp_num, idp in enumerate(idp_ids):
    print('Running IDP', idp_num, idp, ':')
    idp_dir = os.path.join(out_dir, idp)
    os.chdir(idp_dir)

    # Extract and save the response variables for the test set
    y_te = df_te[idp].to_numpy()
```

```
# Save the variables
resp_file_te = os.path.join(idp_dir, 'resp_te.txt')
np.savetxt(resp_file_te, y_te)

# Configure and save the design matrix
cov_file_te = os.path.join(idp_dir, 'cov_bspline_te.txt')
X_te = create_design_matrix(df_te[cols_cov],
                            site_ids = df_te['site'],
                            all_sites = site_ids_tr,
                            basis = 'bspline',
                            xmin = xmin,
                            xmax = xmax)
np.savetxt(cov_file_te, X_te)

# Check whether all sites in the test set are represented in the training set
if all(elem in site_ids_tr for elem in site_ids_te):
    print('All sites are present in the training data')

    # Just make predictions
    yhat_te, s2_te, Z = predict(cov_file_te,
                                alg='blr',
                                respfile=resp_file_te,
```

```
                model_path=os.path.join(idp_dir,'Models'))
else:
    print('Some sites missing from the training data. Adapting model')

    # Save the covariates for the adaptation data
    X_ad = create_design_matrix(df_ad[cols_cov],
                                site_ids = df_ad['site'],
                                all_sites = site_ids_tr,
                                basis = 'bspline',
                                xmin = xmin,
                                xmax = xmax)
    cov_file_ad = os.path.join(idp_dir, 'cov_bspline_ad.txt')
    np.savetxt(cov_file_ad, X_ad)

    # Save the responses for the adaptation data
    resp_file_ad = os.path.join(idp_dir, 'resp_ad.txt')
    y_ad = df_ad[idp].to_numpy()
    np.savetxt(resp_file_ad, y_ad)

    # Save the site ids for the adaptation data
    sitenum_file_ad = os.path.join(idp_dir, 'sitenum_ad.txt')
    site_num_ad = df_ad['sitenum'].to_numpy(dtype=int)
    np.savetxt(sitenum_file_ad, site_num_ad)
```

```
# Save the site ids for the test data
sitenum_file_te = os.path.join(idp_dir, 'sitenum_te.txt')
site_num_te = df_te['sitenum'].to_numpy(dtype=int)
np.savetxt(sitenum_file_te, site_num_te)
```

```
yhat_te, s2_te, Z = predict(cov_file_te,
                    alg = 'blr',
                    respfile = resp_file_te,
                    model_path = os.path.join(idp_dir,'Models'),
                    adaptrespfile = resp_file_ad,
                    adaptcovfile = cov_file_ad,
                    adaptvargroupfile = sitenum_file_ad,
                    testvargroupfile = sitenum_file_te)
```

3.5 Plotting

3.5.1 Configure Age Range of Plots - Dummy Data

In this step, you plot the centiles of variation estimated by the normative model. You do this by making use of a set of dummy covariates that span the whole range of the input space (for age) for a fixed value of the other covariates (e.g., sex) so that we can make predictions for these dummy data points, then plot them. We configure these dummy predictions using the same procedure as we used for the real data. We can use the same dummy data for all the IDPs we wish to plot.

```
# Which sex do we want to plot?
sex = 1 # 1 = male 0 = female
if sex == 1:
    clr = 'blue';
else:
    clr = 'red'

# Create dummy data for visualization
print('configuring dummy data ...')
xx = np.arange(xmin, xmax, 0.5)
X0_dummy = np.zeros((len(xx), 2))
X0_dummy[:,0] = xx
X0_dummy[:,1] = sex

# Create the design matrix
X_dummy = create_design_matrix(X0_dummy, xmin=xmin, xmax=xmax, site_ids=None,
all_sites=site_ids_tr)

# Save the dummy covariates
cov_file_dummy = os.path.join(out_dir,'cov_bspline_dummy_mean.txt')
np.savetxt(cov_file_dummy, X_dummy)
```

3.5.2 Plot Real Data

Here, you will plot the normative models. First, we loop through the IDPs, plotting each one separately. The outputs of this step are a set of quantitative regression metrics for each IDP and a set of centile curves which we plot the test data against. This part of the code is relatively complex because we need to keep track of many quantities for the plotting. We also need to remember whether the data need to be warped or not. By default, in PCNtoolkit, predictions in the form of 'yhat,' 's2' are always in the warped (Gaussian) space. If we want predictions in the input (non-Gaussian) space, then we need to warp them with the inverse of the estimated warping function. This can be done using the function 'nm.blr.warp.warp_predictions(),'

Note

It is necessary to update the intercept for each of the sites. For purposes of visualization, here we do this by adjusting the median of the data to match the dummy predictions but note that all the quantitative metrics are estimated using the predictions that are adjusted properly using a learned offset (or adjusted using a hold-out adaptation set, as above).

Note For the calibration data we require at least two data points of the same sex in each site to be able to estimate the variance. Of course, in a real example, you would want many more than just two since we need to get a reliable estimate of the variance for each site.

```
sns.set(style='whitegrid')


for idp_num, idp in enumerate(idp_ids):
```

```
print('Running IDP', idp_num, idp, ':')

idp_dir = os.path.join(out_dir, idp)

os.chdir(idp_dir)


# Load the true data points

yhat_te = load_2d(os.path.join(idp_dir, 'yhat_predict.txt'))

s2_te = load_2d(os.path.join(idp_dir, 'ys2_predict.txt'))

y_te = load_2d(os.path.join(idp_dir, 'resp_te.txt'))


# Set up the covariates for the dummy data

print('Making predictions with dummy covariates (for visualisation)')

yhat, s2 = predict(cov_file_dummy,

                   alg = 'blr',

                   respfile = None,

                   model_path = os.path.join(idp_dir,'Models'),

                   outputsuffix = '_dummy')


# Load the normative model

with open(os.path.join(idp_dir,'Models', 'NM_0_0_estimate.pkl'), 'rb') as handle:

    nm = pickle.load(handle)


# Get the warp and warp parameters

W = nm.blr.warp
```

```
warp_param = nm.blr.hyp[1:nm.blr.warp.get_n_params()+1]

# First, we warp predictions for the true data and compute evaluation metrics
med_te = W.warp_predictions(np.squeeze(yhat_te), np.squeeze(s2_te), warp_param)[0]
med_te = med_te[:, np.newaxis]
print('metrics:', evaluate(y_te, med_te))

# Then, we warp dummy predictions to create the plots
med, pr_int = W.warp_predictions(np.squeeze(yhat), np.squeeze(s2), warp_param)

# Extract the different variance components to visualise
beta, junk1, junk2 = nm.blr._parse_hyps(nm.blr.hyp, X_dummy)
s2n = 1/beta # variation (aleatoric uncertainty)
s2s = s2-s2n # modelling uncertainty (epistemic uncertainty)

# Plot the data points
y_te_rescaled_all = np.zeros_like(y_te)
for sid, site in enumerate(site_ids_te):
    # Plot the true test data points
    if all(elem in site_ids_tr for elem in site_ids_te):
        # All data in the test set are present in the training set

        # First, we select the data points belonging to this particular site
```

```
        idx = np.where(np.bitwise_and(X_te[:,2] == sex, X_te[:,sid+len(cols_cov)+1] !=0))[0]
        if len(idx) == 0:
            print('No data for site', sid, site, 'skipping...')
            continue

        # Then directly adjust the data
        idx_dummy = np.bitwise_and(X_dummy[:,1] > X_te[idx,1].min(), X_dummy[:,1] <
X_te[idx,1].max())
        y_te_rescaled = y_te[idx] - np.median(y_te[idx]) + np.median(med[idx_dummy])
    else:
        # We need to adjust the data based on the adaptation dataset

        # First, select the data point belonging to this particular site
        idx = np.where(np.bitwise_and(X_te[:,2] == sex, (df_te['site'] == site).to_numpy()))[0]

        # Load the adaptation data
        y_ad = load_2d(os.path.join(idp_dir, 'resp_ad.txt'))
        X_ad = load_2d(os.path.join(idp_dir, 'cov_bspline_ad.txt'))
        idx_a = np.where(np.bitwise_and(X_ad[:,2] == sex, (df_ad['site'] ==
site).to_numpy()))[0]
        if len(idx) < 2 or len(idx_a) < 2:
            print('Insufficent data for site', sid, site, 'skipping...')
            continue
```

```
        # Adjust and rescale the data
        y_te_rescaled, s2_rescaled = nm.blr.predict_and_adjust(nm.blr.hyp,
                                                                X_ad[idx_a,:],
                                                                np.squeeze(y_ad[idx_a]),
                                                                Xs=None,
                                                                ys=np.squeeze(y_te[idx]))
    # Plot the (adjusted) data points
    plt.scatter(X_te[idx,1], y_te_rescaled, s=4, color=clr, alpha = 0.1)

   # Plot the median of the dummy data
   plt.plot(xx, med, clr)

   # Fill the gaps in between the centiles
   junk, pr_int25 = W.warp_predictions(np.squeeze(yhat), np.squeeze(s2), warp_param,
percentiles=[0.25,0.75])
   junk, pr_int95 = W.warp_predictions(np.squeeze(yhat), np.squeeze(s2), warp_param,
percentiles=[0.05,0.95])
   junk, pr_int99 = W.warp_predictions(np.squeeze(yhat), np.squeeze(s2), warp_param,
percentiles=[0.01,0.99])
   plt.fill_between(xx, pr_int25[:,0], pr_int25[:,1], alpha = 0.1,color=clr)
   plt.fill_between(xx, pr_int95[:,0], pr_int95[:,1], alpha = 0.1,color=clr)
   plt.fill_between(xx, pr_int99[:,0], pr_int99[:,1], alpha = 0.1,color=clr)
```

```
# Make the width of each centile proportional to the epistemic uncertainty
junk, pr_int25l = W.warp_predictions(np.squeeze(yhat), np.squeeze(s2-0.5*s2s),
warp_param, percentiles=[0.25,0.75])
junk, pr_int95l = W.warp_predictions(np.squeeze(yhat), np.squeeze(s2-0.5*s2s),
warp_param, percentiles=[0.05,0.95])
junk, pr_int99l = W.warp_predictions(np.squeeze(yhat), np.squeeze(s2-0.5*s2s),
warp_param, percentiles=[0.01,0.99])
junk, pr_int25u = W.warp_predictions(np.squeeze(yhat), np.squeeze(s2+0.5*s2s),
warp_param, percentiles=[0.25,0.75])
junk, pr_int95u = W.warp_predictions(np.squeeze(yhat), np.squeeze(s2+0.5*s2s),
warp_param, percentiles=[0.05,0.95])
junk, pr_int99u = W.warp_predictions(np.squeeze(yhat), np.squeeze(s2+0.5*s2s),
warp_param, percentiles=[0.01,0.99])
plt.fill_between(xx, pr_int25l[:,0], pr_int25u[:,0], alpha = 0.3,color=clr)
plt.fill_between(xx, pr_int95l[:,0], pr_int95u[:,0], alpha = 0.3,color=clr)
plt.fill_between(xx, pr_int99l[:,0], pr_int99u[:,0], alpha = 0.3,color=clr)
plt.fill_between(xx, pr_int25l[:,1], pr_int25u[:,1], alpha = 0.3,color=clr)
plt.fill_between(xx, pr_int95l[:,1], pr_int95u[:,1], alpha = 0.3,color=clr)
plt.fill_between(xx, pr_int99l[:,1], pr_int99u[:,1], alpha = 0.3,color=clr)

# Plot actual centile lines
plt.plot(xx, pr_int25[:,0],color=clr, linewidth=0.5)
```

```
    plt.plot(xx, pr_int25[:,1],color=clr, linewidth=0.5)
    plt.plot(xx, pr_int95[:,0],color=clr, linewidth=0.5)
    plt.plot(xx, pr_int95[:,1],color=clr, linewidth=0.5)
    plt.plot(xx, pr_int99[:,0],color=clr, linewidth=0.5)
    plt.plot(xx, pr_int99[:,1],color=clr, linewidth=0.5)

    plt.xlabel('Age')
    plt.ylabel(idp)
    plt.title(idp)
    plt.xlim((0,90))
    plt.savefig(os.path.join(idp_dir, 'centiles_' + str(sex)),  bbox_inches='tight')
    plt.show()

os.chdir(out_dir)
```

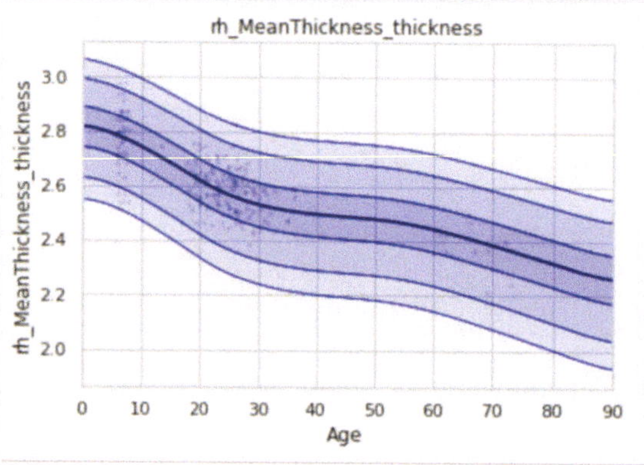

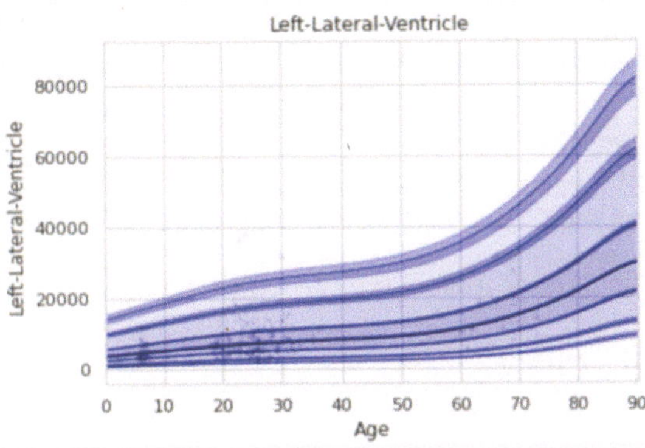

3.6 Organize the Output Files into a Single CSV File

Here, you will explore an example output folder of a single model (one ROI). It is useful to understand what each of these output files represents. Look at the variable names and comments in the code block above:

```
#folder contents

get_ipython().system(' ls rh_MeanThickness_thickness/')
```

You should check that the number of deviation scores matches the number of subjects in the test set. There should be one deviation score per subject (one line per subject), which you can verify by counting the line numbers in the Z_predict.txt file:

```
# lines count

get_ipython().system(' cat rh_MeanThickness_thickness/Z_predict.txt | wc')
```

The deviation scores are output as a text file in separate folders. You will want to summarize the deviation scores across all models estimates so you organize them into a single file and merge the deviation scores into the original data file.

```
get_ipython().system(' mkdir deviation_scores')

get_ipython().system(' for i in *; do if [[ -e ${i}/Z_predict.txt ]]; then cp ${i}/Z_predict.txt deviation_scores/${i}_Z_predict.txt; fi; done')
```

```
z_dir = '/content/braincharts/models/' + model_name + '/deviation_scores/'

filelist = [name for name in os.listdir(z_dir)]

os.chdir(z_dir)

Z_df = pd.concat([pd.read_csv(item, names=[item[:-4]]) for item in filelist], axis=1)

df_te.reset_index(inplace=True)

Z_df['sub_id'] = df_te['sub_id']

df_te_Z = pd.merge(df_te, Z_df, on='sub_id', how='inner')

df_te_Z.to_csv('OpenNeuroTransfer_deviation_scores.csv', index=False)
```

4 Code Tutorial 2: Visualizing the Results

This tutorial walks through several examples that visualize the outputs created by the normative modeling analysis that was run in tutorial 1. Again, this code can be run in your web browser using Google Colab here.

4.1 Brain Space Visualization of Extreme Deviations

First, we count the number of extreme (positive and negative) deviations at each brain region and visualize the count for each hemisphere. You can click around in 3D space on the visualizations (Scroll in/out, move the brain around, etc.)

```
#!/usr/bin/env python

get_ipython().system(' git clone https://github.com/predictive-clinical-neuroscience/PCNtoolkit-demo.git')

import os
import pandas as pd
import numpy as np
import matplotlib.pyplot as plt
import seaborn as sns
```

```
from nilearn import plotting

import nibabel as nib

from nilearn import datasets


os.chdir('/content/PCNtoolkit-demo')


Z_df = pd.read_csv('data/Z_long_format.csv')



# Change this threshold to view more or less extreme deviations.

# Discuss what you think is an appropriate threshold and adjust the below variables

accordingly.

Z_positive = Z_df.query('value > 2')

Z_negative = Z_df.query('value < -2')


positive_left_z = Z_positive.query('hemi == "left"')

positive_right_z = Z_positive.query('hemi == "right"')

positive_sc_z = Z_positive.query('hemi == "subcortical"')

negative_left_z = Z_negative.query('hemi == "left"')

negative_right_z = Z_negative.query('hemi == "right"')

negative_sc_z = Z_negative.query('hemi == "subcortical"')


positive_left_z2 =

positive_left_z['ROI_name'].value_counts().rename_axis('ROI').reset_index(name='counts')
```

```
positive_right_z2 =
positive_right_z['ROI_name'].value_counts().rename_axis('ROI').reset_index(name='counts')
positive_sc_z2 =
positive_sc_z['ROI_name'].value_counts().rename_axis('ROI').reset_index(name='counts')
negative_left_z2 =
negative_left_z['ROI_name'].value_counts().rename_axis('ROI').reset_index(name='counts')
negative_right_z2 =
negative_right_z['ROI_name'].value_counts().rename_axis('ROI').reset_index(name='counts')
negative_sc_z2 =
negative_sc_z['ROI_name'].value_counts().rename_axis('ROI').reset_index(name='counts')


destrieux_atlas = datasets.fetch_atlas_surf_destrieux()
fsaverage = datasets.fetch_surf_fsaverage()


# The parcellation is already loaded into memory
parcellation_l = destrieux_atlas['map_left']
parcellation_r = destrieux_atlas['map_right']


nl = pd.read_csv('data/nilearn_order.csv')


atlas_r = destrieux_atlas['map_right']
atlas_l = destrieux_atlas['map_left']
```

```
nl_ROI = nl['ROI'].to_list()

nl_positive_left = pd.merge(nl, positive_left_z2, on='ROI', how='left')
nl_positive_right = pd.merge(nl, positive_right_z2, on='ROI', how='left')

nl_positive_left['counts'] = nl_positive_right['counts'].fillna(0)
nl_positive_right['counts'] = nl_positive_right['counts'].fillna(0)

nl_positive_left = nl_positive_left['counts'].to_numpy()
nl_positive_right = nl_positive_right['counts'].to_numpy()

a_list = list(range(1, 76))
parcellation_positive_l = atlas_l
for i, j in enumerate(a_list):
    parcellation_positive_l = np.where(parcellation_positive_l == j, nl_positive_left[i],
parcellation_positive_l)

a_list = list(range(1, 76))
parcellation_positive_r = atlas_r
for i, j in enumerate(a_list):
    parcellation_positive_r = np.where(parcellation_positive_r == j, nl_positive_right[i],
parcellation_positive_r)
```

```
view_pos_r = plotting.view_surf(fsaverage.infl_right, parcellation_positive_r,
threshold=None, symmetric_cmap=False, cmap='plasma', bg_map=fsaverage.sulc_right)
```

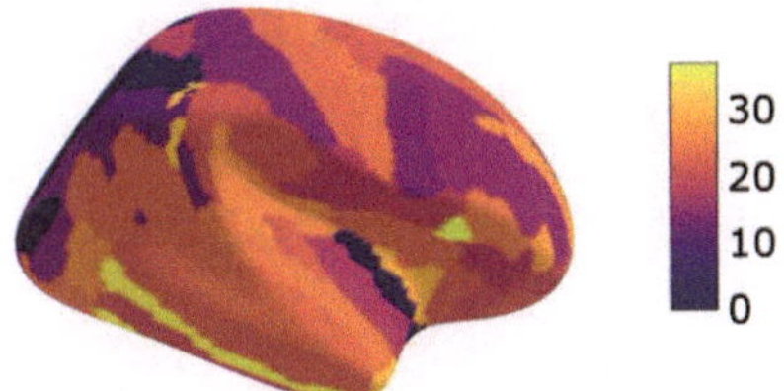


```
view_pos_l = plotting.view_surf(fsaverage.infl_left, parcellation_positive_l, threshold=None,
symmetric_cmap=False, cmap='plasma', bg_map=fsaverage.sulc_left)
```

```
nl_negative_left = pd.merge(nl, negative_left_z2, on='ROI', how='left')
nl_negative_right = pd.merge(nl, negative_right_z2, on='ROI', how='left')


nl_negative_left['counts'] = nl_negative_left['counts'].fillna(0)
nl_negative_right['counts'] = nl_negative_right['counts'].fillna(0)


nl_negative_left = nl_negative_left['counts'].to_numpy()
nl_negative_right = nl_negative_right['counts'].to_numpy()
```

```
a_list = list(range(1, 76))
parcellation_negative_l = atlas_l
for i, j in enumerate(a_list):
    parcellation_negative_l = np.where(parcellation_negative_l == j, nl_negative_left[i],
parcellation_negative_l)
```

```
a_list = list(range(1, 76))
parcellation_negative_r = atlas_r
for i, j in enumerate(a_list):
    parcellation_negative_r = np.where(parcellation_negative_r == j, nl_negative_right[i],
parcellation_negative_r)

view_neg_r = plotting.view_surf(fsaverage.infl_right, parcellation_negative_r,
threshold=None, symmetric_cmap=False, cmap='plasma', bg_map=fsaverage.sulc_right)
```

```
view_neg_l = plotting.view_surf(fsaverage.infl_left, parcellation_negative_l, threshold=None,
symmetric_cmap=False, cmap='plasma', bg_map=fsaverage.sulc_left)
```

4.2 Violin Plots of the Extreme Deviations

Here, you can count the number of 'extreme' deviations that each person has (both positive and negative) and summarize the distribution of extreme deviations for healthy controls and patients with schizophrenia.

```
Z_df = pd.read_csv('data/fcon1000_te_Z.csv')

deviation_counts = Z_df.loc[:, Z_df.columns.str.contains('Z_predict')]

deviation_counts['positive_count'] = deviation_counts[deviation_counts >= 2].count(axis=1)

deviation_counts['negative_count'] = deviation_counts[deviation_counts <= -2].count(axis=1)

deviation_counts['participant_id'] = Z_df['sub_id']
deviation_counts['group_ID'] = Z_df['group']
deviation_counts['site_ID'] = Z_df['site']
```

```
deviation_counts['all_counts'] = deviation_counts['positive_count'] +
deviation_counts['negative_count']

fig, ax = plt.subplots(figsize=(6,6))
sns.violinplot(data=deviation_counts, y="all_counts", x="group_ID", inner='box', ax=ax);
plt.legend=False
```

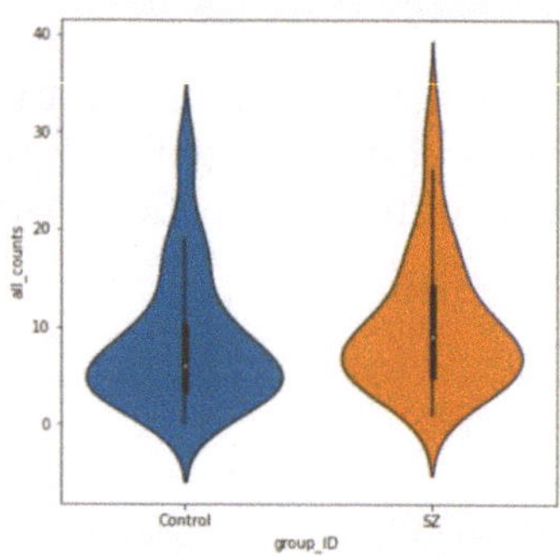

5 Conclusion

5.1 Evaluation

There are multiple results created from the normative model analysis. First, the evaluation metrics for each model (brain region) are saved to a CSV file in Subheading 3. The evaluation metrics can be visualized in numerous formats, such as histograms/density plots, scatter plots with fitted centiles, or brain-space visualizations. Several examples of these visualizations were shown in Subheading 4 on visualization. Quality checking the normative model evaluation metrics should be done to ensure proper model estimation. If a model fits well to the data, the evaluation metrics should follow a Gaussian distribution. Beyond the summary metrics, there are individual metrics that can be helpful for interpretation because they quantify the uncertainty of each individual's predicted value (for every brain region).

5.2 Limitations

As datasets grow in size, there is a need to use automated quality metrics. This means there could unintentionally be poor quality data included in the training set. Whenever possible, users should consider manually quality checking their own data. If the dataset is too large to check every subject, consider randomly checking a portion of the dataset and using the automated quality metrics to inform a threshold for which subjects should be manually visually inspected.

Another consideration, when training big data normative models, is that there are going to be differences in the available data modalities collected across studies and sites. The commonly available data needs to be considered when deciding which studies to include and which covariates to use in modeling. If the goal is to share the model, using uncommon covariates or brain measures will affect the utility and accessibility of the model.

It is also important to consider that normative modeling may not always be the best approach in all settings, as it is dependent on the chosen reference population (usually but not necessarily taken to be population of healthy controls), and it might not be appropriate in certain cases, for example, if there are substantial differences in the target cohort that are unrelated to the clinical condition of interest. In some situations, such as when studying rare diseases, there may not be enough data available to establish a normative model, or the population of healthy controls may not be representative of the population being studied. However, we acknowledge that we are not obliged to fit the normative model only using healthy data. It is equally valid to fit a normative model using patient and control data, just note that changing the reference cohort (training set) demographics changes the interpretation of the centiles. If you are modeling 'healthy' lifespan populations, the sample size will likely be large (on the order of thousands) because of the availability of publicly shared data from healthy controls. In contrast, if you want to model a specific clinical population, the sample size will be smaller due to availability of data. A smaller dataset that appropriately addresses the given research question is suitable. There is no 'one size fits all' approach to normative modeling.

5.3 Post-hoc Analysis

There are multiple possible downstream analyses that can be performed after a normative model has been fit, evaluated, and visualized. While there are too many diverging paths to cover them all within this chapter, we highlight recent work on post-hoc normative modeling possibilities. These include sub-typing using clustering algorithms [17] and stratification [8]. For example, such approaches applied to autism spectrum disorder (ASD) have shown particular promise for parsing the biological heterogeneity underlying this condition. For example Zabihi et al. [17] showed that a subset of individuals with autism have widespread patterns of increased cortical thickness relative to population norms, whereas others have widespread patterns of decreased cortical thickness. We refer to Rutherford et al. [24] for a detailed overview of the possibilities of downstream analyses that can be conducted using normative models and an in-depth comparison between normative modeling outputs (deviation scores) and raw data using different data modalities (structural and functional MRI) across several tasks (multivariate prediction (regression and classification) and case-control group different testing). Owing to this flexibility, and the ability to move beyond group level inferences to individual prediction, we consider that normative modeling is a promising method for understanding variation in large datasets.

References

1. Rutherford S, Fraza C, Dinga R, Kia SM, Wolfers T, Zabihi M, Berthet P, Worker A, Verdi S, Andrews D, Han LK, Bayer JM, Dazzan P, McGuire P, Mocking RT, Schene A, Sripada C, Tso IF, Duval ER, Chang S-E, Penninx BW, Heitzeg MM, Burt SA, Hyde LW, Amaral D, Wu Nordahl C, Andreasssen OA, Westlye LT, Zahn R, Ruhe HG, Beckmann C, Marquand AF (2022) Charting brain growth and aging at high spatial precision. eLife 11:e72904. https://doi.org/10.7554/eLife.72904
2. de Boer AAA, Kia SM, Rutherford S, Zabihi M, Fraza C, Barkema P, Westlye LT, Andreassen OA, Hinne M, Beckmann CF, Marquand A (2022) Non-gaussian normative modelling with hierarchical bayesian regression. bioRxiv. https://doi.org/10.1101/2022.10.05.510988
3. Dinga R, Fraza CJ, Bayer JMM, Kia SM, Beckmann CF, Marquand AF (2021) Normative modeling of neuroimaging data using generalized additive models of location scale and shape. bioRxiv. https://doi.org/10.1101/2021.06.14.448106
4. Fraza CJ, Dinga R, Beckmann CF, Marquand AF (2021) Warped Bayesian linear regression for normative modelling of big data. NeuroImage 245:118715. https://doi.org/10.1016/j.neuroimage.2021.118715
5. Kia SM, Huijsdens H, Dinga R, Wolfers T, Mennes M, Andreassen OA, Westlye LT, Beckmann CF, Marquand AF (2020) Hierarchical Bayesian regression for multi-site normative modeling of neuroimaging data. In: Martel AL, Abolmaesumi P, Stoyanov D, Mateus D, Zuluaga MA, Zhou SK, Racoceanu D, Joskowicz L (eds) Medical image computing and computer assisted intervention—MICCAI 2020. Springer International Publishing, Cham, pp 699–709
6. Kia SM, Huijsdens H, Rutherford S, de Boer A, Dinga R, Wolfers T, Berthet P, Mennes M, Andreassen OA, Westlye LT, Beckmann CF, Marquand AF (2022) Closing the life-cycle of normative modeling using federated hierarchical Bayesian regression. PLoS One 17:e0278776. https://doi.org/10.1371/journal.pone.0278776
7. Kia SM, Marquand A (2018) Normative modeling of neuroimaging data using scalable multi-task Gaussian processes. arXiv:1806.01047. https://doi.org/10.48550/arXiv.1806.01047
8. Floris DL, Wolfers T, Zabihi M, Holz NE, Zwiers MP, Charman T, Tillmann J, Ecker C, Dell'Acqua F, Banaschewski T, Moessnang C, Baron-Cohen S, Holt R, Durston S, Loth E, Murphy DGM, Marquand A, Buitelaar JK, Beckmann CF, Ahmad J, Ambrosino S, Auyeung B, Banaschewski T, Baron-Cohen S, Baumeister S, Beckmann CF, Bölte S, Bourgeron T, Bours C, Brammer M, Brandeis D, Brogna C, de Bruijn Y, Buitelaar JK, Chakrabarti B, Charman T, Cornelissen I, Crawley D, Dell'Acqua F, Dumas G, Durston S, Ecker C, Faulkner J, Frouin V, Garcés P, Goyard D, Ham L, Hayward H, Hipp J, Holt R, Johnson MH, Jones EJH, Kundu P, Lai M-C, Liogier d'Ardhuy X, Lombardo MV, Loth E, Lythgoe DJ, Mandl R, Marquand A, Mason L, Mennes M, Meyer-Lindenberg A, Moessnang C, Mueller N, Murphy DGM, Oakley B, O'Dwyer L, Oldehinkel M, Oranje B, Pandina G, Persico AM, Ruggeri B, Ruigrok A, Sabet J, Sacco R, San José Cáceres A, Simonoff E, Spooren W, Tillmann J, Toro R, Tost H, Waldman J, Williams SCR, Wooldridge C, Zwiers MP (2021) Atypical brain asymmetry in autism—a candidate for clinically meaningful stratification. Biol Psychiatry Cogn Neurosci Neuroimaging 6:802–812. https://doi.org/10.1016/j.bpsc.2020.08.008
9. Pinaya WHL, Scarpazza C, Garcia-Dias R, Vieira S, Baecker L, da Costa PF, Redolfi A, Frisoni GB, Pievani M, Calhoun VD, Sato JR, Mechelli A (2021) Using normative modelling to detect disease progression in mild cognitive impairment and Alzheimer's disease in a cross-

sectional multi-cohort study. Sci Rep 11: 15746. https://doi.org/10.1038/s41598-021-95098-0

10. Remiszewski N, Bryant JE, Rutherford SE, Marquand AF, Nelson E, Askar I, Lahti AC, Kraguljac NV (2022) Contrasting case-control and normative reference approaches to capture clinically relevant structural brain abnormalities in patients with first-episode psychosis who are antipsychotic naive. JAMA Psychiatry 79: 1133–1138. https://doi.org/10.1001/jamapsychiatry.2022.3010
11. Verdi S, Marquand AF, Schott JM, Cole JH (2021) Beyond the average patient: how neuroimaging models can address heterogeneity in dementia. Brain J Neurol 144:2946–2953. https://doi.org/10.1093/brain/awab165
12. Wolfers T, Arenas AL, Onnink AMH, Dammers J, Hoogman M, Zwiers MP, Buitelaar JK, Franke B, Marquand AF, Beckmann CF (2017) Refinement by integration: aggregated effects of multimodal imaging markers on adult ADHD. J Psychiatry Neurosci 42: 386–394. https://doi.org/10.1503/jpn.160240
13. Wolfers T, Doan NT, Kaufmann T, Alnæs D, Moberget T, Agartz I, Buitelaar JK, Ueland T, Melle I, Franke B, Andreassen OA, Beckmann CF, Westlye LT, Marquand AF (2018) Mapping the heterogeneous phenotype of schizophrenia and bipolar disorder using normative models. JAMA Psychiatry 75:1146–1155. https://doi.org/10.1001/jamapsychiatry.2018.2467
14. Wolfers T, Beckmann CF, Hoogman M, Buitelaar JK, Franke B, Marquand AF (2020) Individual differences v. the average patient: mapping the heterogeneity in ADHD using normative models. Psychol Med 50:314–323. https://doi.org/10.1017/S0033291719000084
15. Wolfers T, Rokicki J, Alnaes D, Berthet P, Agartz I, Kia SM, Kaufmann T, Zabihi M, Moberget T, Melle I, Beckmann CF, Andreassen OA, Marquand AF, Westlye LT (2021) Replicating extensive brain structural heterogeneity in individuals with schizophrenia and bipolar disorder. Hum Brain Mapp 42:2546–2555. https://doi.org/10.1002/hbm.25386
16. Zabihi M, Oldehinkel M, Wolfers T, Frouin V, Goyard D, Loth E, Charman T, Tillmann J, Banaschewski T, Dumas G, Holt R, Baron-Cohen S, Durston S, Bölte S, Murphy D, Ecker C, Buitelaar JK, Beckmann CF, Marquand AF (2019) Dissecting the heterogeneous cortical anatomy of autism Spectrum disorder using normative models. Biol Psychiatry Cogn Neurosci Neuroimaging 4:567–578. https://doi.org/10.1016/j.bpsc.2018.11.013
17. Zabihi M, Floris DL, Kia SM, Wolfers T, Tillmann J, Arenas AL, Moessnang C, Banaschewski T, Holt R, Baron-Cohen S, Loth E, Charman T, Bourgeron T, Murphy D, Ecker C, Buitelaar JK, Beckmann CF, Marquand A, EU-AIMS LEAP Group (2020) Fractionating autism based on neuroanatomical normative modeling. Transl Psychiatry 10:384. https://doi.org/10.1038/s41398-020-01057-0
18. Marquand A, Rutherford S, Kia SM, Wolfers T, Fraza C, Dinga R, Zabihi M (2021) PCNToolkit
19. Marquand AF, Rezek I, Buitelaar J, Beckmann CF (2016) Understanding heterogeneity in clinical cohorts using normative models: beyond case-control studies. Biol Psychiatry 80:552–561. https://doi.org/10.1016/j.biopsych.2015.12.023
20. Marquand AF, Kia SM, Zabihi M, Wolfers T, Buitelaar JK, Beckmann CF (2019) Conceptualizing mental disorders as deviations from normative functioning. Mol Psychiatry 24: 1415–1424. https://doi.org/10.1038/s41380-019-0441-1
21. Rutherford S, Kia SM, Wolfers T, Fraza C, Zabihi M, Dinga R, Berthet P, Worker A, Verdi S, Ruhe HG, Beckmann CF, Marquand AF (2022) The normative modeling framework for computational psychiatry. Nat Protoc 17:1711–1734. https://doi.org/10.1038/s41596-022-00696-5
22. Borghi E, de Onis M, Garza C, Van den Broeck J, Frongillo EA, Grummer-Strawn L, Van Buuren S, Pan H, Molinari L, Martorell R, Onyango AW, Martines JC, WHO Multicentre Growth Reference Study Group (2006) Construction of the World Health Organization child growth standards: selection of methods for attained growth curves. Stat Med 25:247–265. https://doi.org/10.1002/sim.2227
23. Fraza C, Zabihi M, Beckmann CF, Marquand AF (2022) The extremes of normative modelling. bioRxiv. https://doi.org/10.1101/2022.08.23.505049
24. Rutherford S, Barkema P, Tso IF, Sripada C, Beckmann CF, Ruhe HG, Marquand AF (2022) Evidence for embracing normative modeling. Elife. https://doi.org/10.7554/eLife.85082

Chapter 15

Studying the Connectome at a Large Scale

Rory Boyle and Yihe Weng

Abstract

This chapter outlines a flexible connectome-based predictive modeling method that is optimised for large neuroimaging datasets via the use of parallel computing and by adding the capability to account for possible site- and scanner-related heterogeneity in multi-site neuroimaging datasets. We present the decision points that need to be made when conducting a connectome-based predictive modeling analysis and we provide full code to conduct an analysis on public data. To date, connectome-based predictive modeling has been applied to predict different cognitive and behavioral phenotypes with many studies reporting accurate predictions that generalized to external datasets.

Key words Magnetic resonance imaging, Connectomic predictive modeling, Neuroimaging

1 Introduction

Connectome-based predictive modeling (CPM) is a data-driven approach that enables the prediction of behavioural and cognitive phenotypes from functional MRI (fMRI) connectivity data [1, 2]. CPM has been applied to successfully predict individual differences in various cognitive phenotypes including global cognition [3], attention [4, 5], executive function [6], fluid intelligence [1, 7, 8], processing speed [7], cognitive reserve [9], and creative ability [10]. CPM has also enabled accurate prediction of individual differences in behavioural phenotypes such as anxiety [11], depression [12], feelings of loneliness [13] and stress [14], childhood aggression I [15] and social impairments [16], and abstinence from use of substances including opioids [17] and cocaine [18]. For a comprehensive overview, see the original protocol paper which outlines the method and explains the benefits of the approach [2]. This chapter will outline a flexible CPM method which is optimised for large neuroimaging datasets via the use of parallel computing [*see* Glossary] and that enables researchers to account for possible site and scanner-related heterogeneity in multi-site neuroimaging datasets [19, 20] by controlling for site and/or

Robert Whelan and Hervé Lemaître (eds.), *Methods for Analyzing Large Neuroimaging Datasets*, Neuromethods, vol. 218,
https://doi.org/10.1007/978-1-0716-4260-3_15, © The Author(s) 2025

scanner type as a covariate or by using leave-site-out cross-validation [*see* Glossary]. To date, CPM has been used for the prediction of continuous variables but classification (i.e. prediction of binary variables) is also possible with modification of the code presented here.

In short, CPM identifies the most relevant functional connections ('edges') for a phenotype of interest across the whole brain (*see* Fig. 1). Within a cross-validation framework, edges are thresholded based on their correlation with the phenotype such that only edges where connectivity is significantly associated with the phenotype are retained. Connectivity can be either positively or negatively associated with the phenotype (positive and negative edges, respectively). Summary network strength values (single scalar values) are created by summing connectivity strength from all suprathreshold positive edges (positive network strength) and negative edges (negative network strength) separately. These values represent the connectivity strength of edges where stronger connectivity is associated with higher and lower values of the phenotype, respectively. Combined network strength, reflecting the overall connectivity strength across all edges associated with the phenotype, is calculated by subtracting negative network strength from positive network strength.

CPM is a promising strategy for objective measurement of behavioural and cognitive phenotypes for a number of reasons. First, CPM provides scalar values that summarise the strength of connectivity in phenotype-related networks and that, given sufficient accuracy, could be used as an objective measure to track changes in a phenotype over time. Second, CPM uses whole-brain functional connectivity in a data-driven manner and is therefore not constrained by a priori hypothesised regions or networks of interest. This may improve predictive accuracy of phenotypes, as individual differences in cognition are better predicted by whole-brain data versus data from specific regions of interest [21]. Third, CPM enables useful neurobiological insights through data visualisation and computational methods. Circle plots and/or glass brain plots can highlight specific brain regions critical for the phenotype. Computational lesions, where specific regions or networks are removed from the connectivity matrix prior to running CPM (*see* Subheading 7), can be applied to quantify the importance of specific brain regions or networks for a given phenotype [11, 22]. Finally, CPM has been widely shown to create measures of cognitive and behavioural phenotypes that generalise across datasets [5, 7, 18, 23] and from task-based fMRI to resting-state fMRI [5] and vice-versa [8]. That is, the CPM edges identified in one dataset can be used to make an accurate prediction in an independent dataset or on different types of data. This is a highly favourable feature of the approach as it may enable the measure to

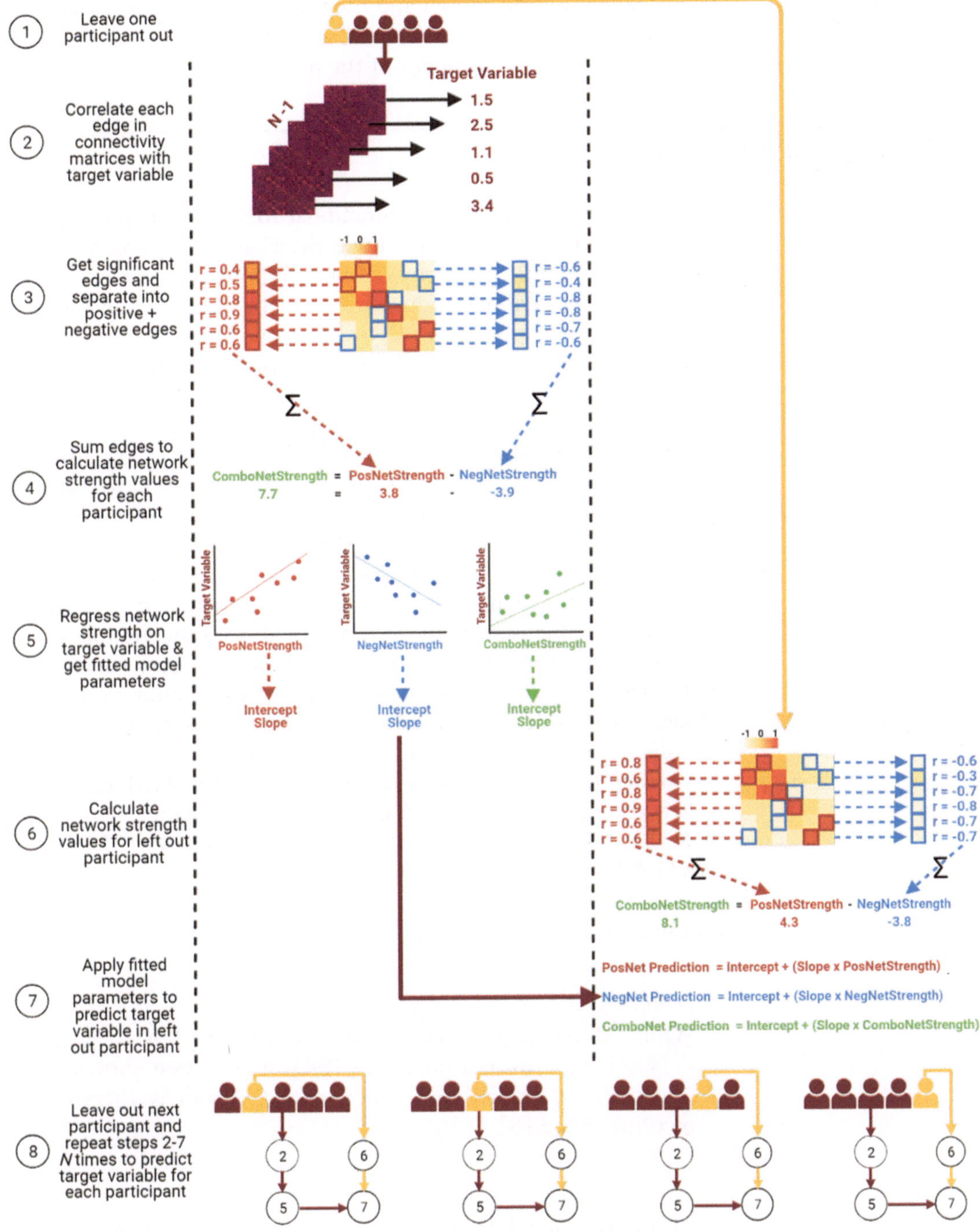

Fig. 1 Schematic of CPM with leave-one-out cross-validation. ComboNet combined network, PosNet positive network, NegNet negative network. (Image created with BioRender.com. BioRender license number: XZ23CG3Q4W)

be developed in one dataset and then shared with other researchers for use on their own data, even if their data is different from the data used in the development of the model.

2 Starting Point for Data

CPM requires pre-processed functional MRI data in the form of connectivity matrices which contain the Fisher z-transformed correlation between the time courses of each node pair. Task-based or resting-state functional connectivity can be used. While CPM is most commonly applied to fMRI data, connectivity matrices from other modalities can be used, as demonstrated in recent studies using EEG [24, 25].

As a rule of thumb, larger sample sizes (i.e., with several hundred observations) are desirable in models using internal cross-validation because small sample sizes can lead to variable and overoptimistic estimates of accuracy [26, 27]. However, CPM with training set samples sizes as small as n = 25 have developed accurate predictions in external test sets [23, 28]. It should be noted that both of these studies predicted a sustained attention variable, sensitivity (*d'*), obtained from the gradual-onset continuous performance task [29, 30] which has very strong reliability [5, 31]. When datasets are small, external validation of the model on an independent dataset may be necessary to provide a reasonable estimate of accuracy and generalisability.

Generally, researchers using the CPM with fMRI data have followed standard preprocessing pipelines. For instance, Rosenberg et al. (2016) [5] used a "36P" nuisance regression [32, 33] in MATLAB and SPM whereby 6 motion parameters, the mean WM, CSF and global signals, their derivatives, quadratic terms, and squares of derivatives were regressed from the data.

Note

While global signal regression [*see* Glossary] is a much debated topic in neuroimaging [34], it has been shown to improve the predictive accuracy of CPM and is therefore recommended [8].

Data are generally temporally smoothed with a zero-mean unit-variance Gaussian filter [4, 5, 12, 18, 23, 28]. However, data have also been band-pass filtered (0.01–0.1 Hz, 0.008–0.09 Hz) and smoothed with 4–8 mm FWHM Gaussian kernels [3, 4, 6, 11, 13, 35]. Other studies have applied a high-pass filter and 8 mm FWHM Gaussian kernel to the data [10] or a low-pass filter [8].

The 268-node Shen atlas [36] is the standard parcellation scheme used for CPM and this atlas enables easy visualisation of the resulting connectomes using the BioImageSuite Web Connectivity Viewer. However, parcellation schemes have not been shown to affect the predictive accuracy of CPM as Greene et al. [8] found that accuracy did not substantially change when using 600- or 250-node parcellations.

Standard CPM does not have a method for handling missing data. However, techniques have been described to handle cases with missing behavioural data (i.e. missing data for the target variable) using imputation [37]. The code outlined in this chapter can handle two scenarios where there is missing data: (1) missing nodes in time series across all participants and (2) participants with missing connectivity matrices, target variable, or covariates. In the former scenario, restricted (or lesioned) connectivity matrices can be used (*see* Subheading 7) in place of the full connectivity matrices. In the second scenario, the code will remove any participants with missing data in any form (i.e. connectivity matrix, target variable, or covariates).

3 Data Storage and Computing

The code provided here can be run on a single computer using MATLAB. The code was written using MATLAB version R2020a. While earlier versions of MATLAB may also execute the code successfully, it has not been tested on earlier versions and minor modifications may be required where newer functions are not available in older MATLAB versions. The Parallel Computing Toolbox is required for parallelised functions. For data visualisation, a web browser is required. For 100 participants using leave-one-out cross-validation (LOOCV) and 1000 iterations of a random permutation test, 181.55 MB is required (*see* Table 1).

4 Software and Coding

A beginner level of coding knowledge is needed to run this code. A generalized analysis script is provided. The user is only required to modify the file paths so that the input data can be loaded and to provide an output directory and the number of participants with connectivity matrices. Specific options/decision points can be modified by changing the strings/flags of the relevant variables and each decision point is accompanied by clearly commented code to aid the user in making such modifications. No further user input or coding knowledge is needed to execute the remainder of the code. Modification of the CPM method requires intermediate or advanced coding knowledge (e.g., using a robust regression instead of a linear regression).

Table 1
File size for run_flexible_CPM for 100 participants

File	File type	Provided/Created	Size
Connectivity matrices	.csv	Provided	127 MB
Target variable .csv	.csv	Provided	1 KB
Covariates .csv	.csv	Provided	2 KB
Input data .mat	.mat	Created	52.2 MB
Target variable .mat	.mat	Created	1 KB
Output data .mat	.mat	Created	168 KB
Positive mask	.txt	Created	1.09 MB
Negative mask	.txt	Created	1.09 MB

Note: CPM with 100 participants using LOOCV and 1000 iterations of a random permutation test. If k-fold cross-validation with multiple iterations is used, additional data will be saved in the output data .mat file but this also will require relatively little additional data storage

5 Computational Expense

On a Dell OptiPlex 5060 with an Intel Core i7 processor and 32 GB RAM, it takes 3 h 7.5 min to execute the code for CPM with 100 participants using leave-one-out cross-validation with adjustment for 2 covariates at the edge selection step, a *p*-value feature selection threshold of 0.01, and 1000 iterations of a random permutation test (*see* Table 2).

For repeated ten-fold cross-validation, CPM runtime scales proportionally with an approximate 1:0.8 ratio (increase in participants: increase in runtime) as an average increase of 830.5% in the number of participants corresponds to an average increase in runtime of 676.25% (*see* Table 3). Using the more computationally expensive LOOCV [38], CPM runtime scales less efficiently with an approximate 1:9 ratio as an average of 830.5% increase in the number of participants corresponds to a 7245.22% increase in runtime (*see* Table 3).

6 Method

Brief Overview

As shown in Fig. 1, there are eight steps in a CPM analysis. Following completion of the CPM, there are up to three further steps required depending on a researcher's aims (i.e., visualisation, permutation testing, and external validation). There are eight different decision points to be made when applying CPM (*see* Fig. 2). Taking

Table 2
Time taken per step of CPM_Neuromethods_analysis_script.m for 100 participants

Code section	Name of section	Total seconds
1	Load data and prepare variables	13.6626
2	Specify inputs	0.0018
3	Preallocation of arrays	0.0004
4	Run and evaluate CPM, and extract parameters	24.6629
5	Extract selected edges	0.2561
6	Store and save parameters, predictions, selected edges, and model results	0.4676
7	Create masks for visualisation	0.3945
8	Run permutation test	11,206

Note: CPM with 100 participants using leave-one-out cross-validation with adjustment for 2 covariates at the edge selection step, a p-value feature selection threshold of 0.01 and 1000 iterations of a random permutation test

Table 3
Increase in runtime of CPM in proportion to increase in participants using LOOCV vs. repeated ten-fold cross-validation (with 100 iterations)

	LOOCV		Repeated ten-fold CV	
N	Total secs	Increase in runtime (%)	Total secs	Increase in runtime (%)
10	146	–	195	–
100	11,245	7613	1241	537
861	784,593	6877	9854	694

Note: Both LOOCV and repeated 10-fold cross-validation analyses use CPM with adjustment for 2 covariates at the edge selection step, a p-value feature selection threshold of 0.01 and 1000 iterations of a random permutation test. CPM with LOOCV is not parallelized whereas CPM with repeated k-fold cross-validation is parallelized (requires MATLAB's Parallel Computing Toolbox)

the example of CPM with LOOCV, with adjustment for two covariates at edge selection, a *p*-value feature selection threshold of 0.01 and 1000 iterations of a random permutation test, a brief overview of the method is as follows:

6.1 *Cross-Validation* (*see* Fig. 1, **step 1**). The dataset is split up into a 'train' and 'test' set. In the case of LOOCV, a single participant is used as the test set whereas in ten-fold cross-validation, 10 equally sized subsets of the data are created and a single subset is used as the test set.

6.2 *Edge Selection: Relate Edges to Behavior* (*see* Fig. 1, **step 2**). Each edge in the connectivity matrix is correlated with the target variable. Pearson's correlation is typically used. If covariates are being adjusted for at edge selection, then a partial

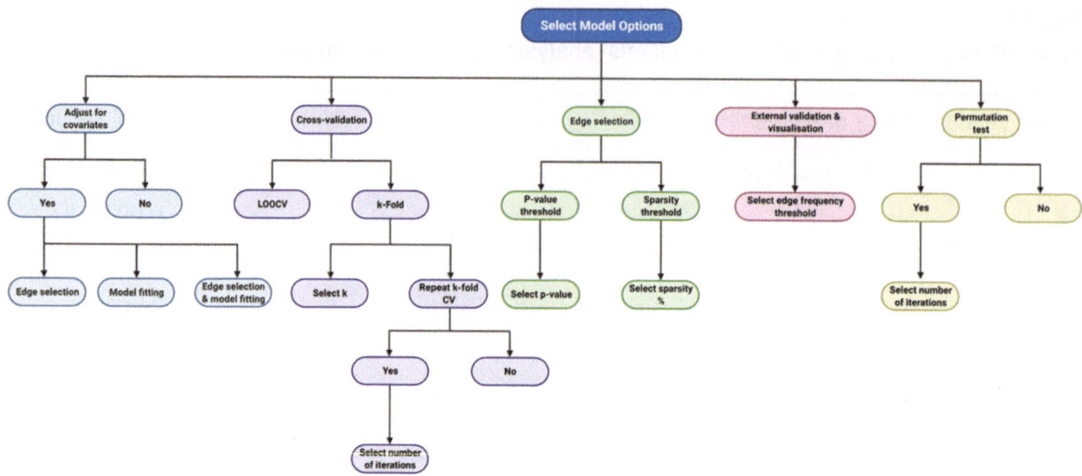

Fig. 2 Decision tree for CPM analysis. (Image created with BioRender.com. BioRender license number: LW23CG3FI3)

Pearson's correlation is used, controlling for the covariates in the relationship between connectivity in each edge and the target variable. Spearman's correlation may also be used at this point if the target variable values are not normally distributed [2, 39]. Robust regression could also be used to attenuate outlier effects [2].

6.3 *Edge Selection: Threshold Edges* (*see* Fig. 1, **step 3**). Edges are separated into a positive network (edges where connectivity is positively associated with the target variable) and a negative network (edges where connectivity is negatively associated with the target variable). Edges are then thresholded so that only the most strongly correlated edges are retained for the model. A typical threshold is to retain edges with a p-value <0.01 for the correlation between connectivity and the target variable [5]. Another option is to use a *sparsity threshold*, where the x% most strongly correlated edges are retained in each network [8]. The indices of the selected edges are retained.

6.4 *Network Strength Calculation* (*see* Fig. 1, **step 4**). In both networks, the connectivity strength of all edges is summed to calculate positive and negative network strength. Combined network strength is calculated by subtracting negative network strength from positive network strength.

6.5 *Fit Linear Model* (*see* Fig. 1, **step 5**). In three linear regressions (a separate regression for each network strength value), the network strength values are regressed on the target variable. Other regression methods have also been used, including partial least squares regression which can account for multicollinearity [23] and support vector regression with a radial basis function which can account for non-linear relationships [6]. If covariates are being adjusted for model fitting, then the covariates are included in each regression. The model parameters (i.e. model intercept, network strength slope, and slopes for covariates if included) are extracted and retained for **step** 7.

6.6 *Network Strength Calculation in Left Out Participant* (*see* Fig. 1, **step 6**). The thresholded edges from **step 3** are extracted from the left-out participant's connectivity matrix. The positive, negative, and combined network strength values are then calculated using these edges as described in **step 4**.

6.7 *Apply Fitted Model to Left Out Participant* (*see* Fig. 1, **step** 7). For each of the three networks, the network strength values, along with the model parameters fitted in **step 5**, are then used in the linear regression equation:

$$Y = a + bX$$

where Y = network strength predicted value, a = fitted intercept, b = fitted slope, X = network strength value.

If covariates are being adjusted for at model fitting, then the covariates are also included in the regression equation along with the fitted model parameters:

$$Y = a + bX + c^1Z^1 + c^2Z^2 + c^3Z^3$$

where Y = network strength predicted value, a = fitted intercept, b = fitted slope for network strength, X = network strength value, c^1 = fitted slope for covariate 1, Z^1 = covariate 1, c^2 = fitted slope for covariate 2, Z^2 = covariate 2, c^3 = fitted slope for covariate 3, Z^3 = covariate 3.

The predicted values are retained for each of the three networks.

6.8 *Repeat for Each Participant* (*see* Fig. 1, **step 8**). The next participant is left out and **steps 2–7** are repeated until each participant is left out once.

6.9 *Store Edges, Parameters, and Predicted Values.* The edges selected in each iteration are stored as a binary mask where 1 = selected, 0 = not selected in an $m*m*N$ array (where m = number of nodes, e.g. 268, N = number of participants). The fitted parameters are averaged such that the mean fitted intercept and mean slopes for network strength values are obtained and then stored. If covariates are adjusted for at model fitting, then the mean slopes for covariates are also calculated and stored. The predicted values for each network are stored.

6.10 *Data Visualization*. Edges selected in a particular % of folds are extracted from the m^*m^*N array. For LOOCV, the standard is to use 100% of folds (i.e. only retain edges that were selected in every fold [2, 5, 35, 39]. However, for other cross-validation schemes (e.g. ten-fold cross-validation), less conservative thresholds could be used. Two .txt files are created containing a positive network mask and negative network mask, each containing 1 in the location where an edge was selected and 0 where it was not. These .txt files can then be loaded into BioImageSuite Web Connectivity Viewer to display the underlying connectomes.

Note

To visualise all connections in both networks, in your web browser, go to https://bioimagesuiteweb.github.io/webapp/connviewer.html and apply the following steps:

(i) *Parcellations > Use the Shen 268 Atlas (assuming this parcellation was used)*

(ii) *Parcellations > Use Yale Network definitions*

(iii) *File > Load Positive Matrix > Select positive network mask .txt file.*

(iv) *File > Load Negative Matrix > Select negative network mask .txt file.*

In Connectivity Control on right hand side, ensure All is selected for Mode and Both is selected for Lines to Draw > Mode > All

6.11 *Permutation Testing*. An empirical null distribution for the correlation between the predicted values and the target variable values is created by randomly shuffling the target variable and repeating CPM p times. The standard is to use $p = 1000$, such that 1000 iterations of random permutation testing are conducted [5, 8, 35], although 5000 iterations have also been used [39]. Permuted p-values are obtained for each network strength prediction as the proportion of permuted correlations that are greater than or equal to the true correlation for that network strength model.

6.12 *External Validation*. The trained model can then be applied to independent data to firmly test generalizability, using the edges and parameters stored in **step 9**. For a model that used LOOCV, only the edges that were selected in all folds are used. Using these edges, network strength values are calculated in the test set (i.e., repeat **step 6**). Then, using the stored model parameters, the fitted model is applied to network strength values in the test set (i.e., repeat **step 7**).

7 Application of Computational Lesions

After running CPM and generating predictions of a phenotype, researchers might be further interested in estimating the relative importance of specific region(s) of interest or functional networks within the positive and negative connectomes for the phenotype of interest. BioImageSuite Web Connectivity Viewer (*see above:* 6.10. *Data visualization*) provides some neurobiological insight via the degree of each node or network. Simulating computational lesions is another method for estimating the relative importance of regions/networks to a given phenotype [11, 22]. This method involves removing nodes within a specific region of interest or functional network from the connectivity matrix (i.e., simulating a computational lesion) before running CPM with the 'lesioned' connectivity matrix. The predictive accuracy for the "lesioned" connectivity matrix can then be compared to the full connectivity matrix. Code is available here for applying computational lesions to functional networks and here for applying computational lesions to specific region(s) of interest.

Steiger's Z test [40] can be used to statistically compare the predictive power of the full connectivity matrix versus the 'lesioned' matrix to establish the statistical significance of the lesioned region/network's contribution to the prediction of the phenotype [11, 22]. Furthermore, for a set of theoretically relevant regions or networks, researchers could repeat CPM for each region/network in the set, test their statistical significance, and rank the regions/networks in terms of the loss in predictive accuracy. This may help to identify the most theoretically important region/network for a given phenotype.

Note: Missing Data
Computational lesions could also be applied to nodes within specific region(s) which may have missing data across participants. For example consider the case where there is poor scanner coverage of the cerebellum for a large number of participants. In this scenario, a researcher may decide to remove nodes from the cerebellum. This can be done *after* preprocessing by applying a computational lesion to nodes from the cerebellum and then using the resulting connectivity matrices for the main CPM analysis.

8 Extension of CPM to Model Non-linear Brain-Phenotype Associations

A limitation to the method outlined thus far is that it will detect only linear relationships between functional connectivity and the phenotype of interest. Spearman's correlation can be used in place of Pearson's correlation at the edge selection step, to account for nonlinear but monotonic relationships between connectivity and the target variable [2]. For instance, in a modified CPM, a partial Spearman's correlation was used in the edge selection step to relate functional connectivity to an ordinal variable (a thirty-two item Frailty Index) [41]. Henneghan et al. (2020) [6] have previously adapted CPM replacing linear regression with a support vector regression with a radial basis function kernel, which account for non-linear relationships [42]. The adapted model outperformed a standard linear CPM for the prediction of self-reported memory and executive function ability, although performance was only slightly numerically better for executive function [6]. As such, while the standard CPM described here may be limited to linear relationships, modified versions of CPM can be used to model nonlinear relationships between connectivity and the target variable.

9 Imaging-Specific Issues (Head Motion)

Head motion is a major issue in functional connectivity [43] and must be carefully considered when using CPM. Preprocessing pipelines should include nuisance regression of motion parameters and global signal regression (as described in Starting point for data). Careful quality control should be implemented whereby, ideally, images are visually inspected for motion-related artefacts and excluded if artefacts are detected.

Participants with excessive head motion should be excluded from CPM analyses. There are different metrics with different thresholds used. Average head motion can be calculated with mean framewise displacement (FWD) [*see* Glossary] and conservative thresholds of mean FWD < 0.1 mm [8], mean FWD < 0.15 mm [23], to mean FWD < 0.2 mm [39], have been used. In certain populations such as clinical [44, 45], child [46], or older adult populations [4, 47], more liberal thresholds may be necessary to prevent excessive data loss due to greater head motion [48]. Excessive head motion in a single run can be calculated using *maximum* FWD where a conservative threshold of max FWD < 0.15 mm [8, 23] has been used. Excessive head motion can also be measured by large movements in translation and rotation axes, were participants with a single run with >2 mm translation or >3° translation can be excluded [5].

Following these motion controls, researchers should assess the correlation between FWD and the target variable. If this correlation is statistically significant, exclusion of further high-motion participants may be necessary [2] using more conservative thresholds. Additional checks include verifying that mean FWD is not associated with the functional connectivity matrices [49] nor the connectomes and predicted values obtained from CPM [8, 39]. If significant associations are identified, options include excluding edges that are correlated with mean FWD across participants [23, 49]; including mean FWD as a covariate in CPM (i.e. in **step 2**: edge selection and/or **step 5**: fit linear model); including mean FWD as a covariate when evaluating model performance [49]; and regressing out mean FWD from both the target variable and the connectivity matrices within the CPM with cross-validation.

10 Decision Points

In the step-by-step example below, there are eight different decision points (*see* Fig. 2).

Here, we briefly discuss additional decision points that arise during preprocessing and if alternative or more advanced modelling techniques are used (as discussed in Potential pitfalls/problems). Indeed, a potential challenge with CPM—common to many neuroimaging methods—is the considerable amount of analytical flexibility involved. In addition to the step-by-step example given here, there are further possible decision points that are not included in this code. These decision points occur at different stages of the analysis. In preprocessing, there are many different pipelines, confound removal, and motion correction strategies that could be used. There are also different options for filtering, smoothing, and functional parcellation. Although the standard measure of functional connectivity, the Fisher-z transformed Pearson's correlation between time courses in each pair of nodes, is used for CPM, different measures have also been used, including accordance, discordance, and information flow [23, 28]. At edge selection, Spearman's correlation or robust regression could be used instead of Pearson's correlation to relate connectivity strength in each edge to the target variable [2, 39]. At model fitting, alternative methods such as partial least squares regression [23] and support vector regression with a radial basis function kernel could also be used instead of linear regression [6]. Finally, when evaluating model performance, Spearman's correlation could be used instead of Pearson's correlation to assess the association between predicted and observed values [8].

This number of decisions inherent in CPM provides researchers with several additional "researcher degrees of freedom" [50] whereby many different models and parameters (e.g. *p*-value thresholds) could be tested but only significant models are reported. On the other hand, this degree of analytical flexibility may enable CPM to cater for different use cases (e.g. different data types, regression models, and populations of interest) and therefore may ensure wide usability of the model. Thorough and honest reporting of the analytical decisions and analyses will enable other researchers to account for the analytical flexibility associated with a given model when considering results from that model. Pre-registration of the analysis plan is another solution to this problem [26]. External validation can also ensure that analytical flexibility is not enabling 'false positive' models which can accurately predict some outcomes despite there being no true association in the underlying data [26].

In addition to numerous decision points, there are some arbitrary decisions involved in specifying the model. For instance, if a model using multiple iterations of k-fold cross-validation and permutation testing is specified, then the researcher must select the number of iterations of k-fold cross-validation, the *p*-value or sparsity threshold [*see* Glossary] for edge selection, and the number of iterations of permutation testing. While a reasonable option is to use previously used parameters for these decisions, these parameters may not be optimal for a given dataset and it is unclear if the selection of these parameters was empirically driven or based on heuristics/rules of thumb. For some decisions, data-driven methods can be used to select the parameter. For instance, researchers have attempted to obtain optimal thresholds by testing CPM with different *p*-values and selecting the model with the higher *r* value [8, 35]. However, data-driven methods to optimise the selection threshold in LOOCV may result in a lower *p*-value that optimises the model for the training set by selecting only the most strongly correlated edges. However, very few edges might be retained and as such generalisability to external datasets may be impaired. To balance the accuracy and generalisability, an additional validation dataset may be required. This would enable the generalisability of the model to be assessed without using the test set and thereby 'double dipping'.

11 Interpreting and Reporting Results

For each network, the code in the step-by-step example provides three performance metrics and a *p*-value for assessing statistical significance:

- *r*: Pearson's correlation between the predicted values and the actual values.

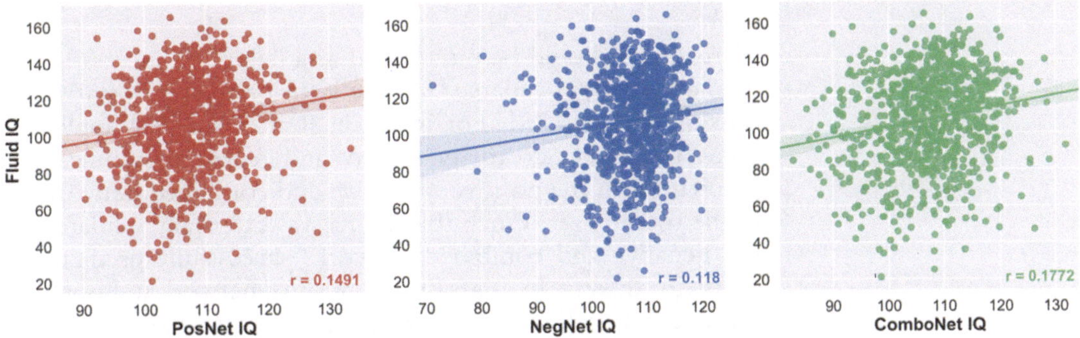

Fig. 3 Fluid IQ values versus positive-, negative-, and combined-network strength predicted IQ values

- R^2: Proportion of variance explained in the actual values by the predicted values. This is the coefficient of determination, calculated using an ordinary least squares regression, as recommended by Poldrack et al. (2020) [26].
- *MAE*: Mean absolute error between the predicted values and the actual values.
- *p*: Permuted *p*-value obtained via random permutation testing.

Researchers should, at a minimum, report the *r* and *p*-value and display a scatterplot of predicted values (x-axis) vs. actual values of the target variable (y-axis) for each network (*see* Fig. 3). Where possible, researchers should externally validate their model by applying the trained CPM to an independent dataset and report the *r* and *p*-value for the correlation between the predicted and actual values in that dataset. This provides a true and rigorous assessment of model generalizability and this is especially important given that external validation on held-out data results in lower estimates of predictive accuracy [51].

Most, if not all, CPM papers have reported external validation performance to-date. Ideally, researchers should also report R^2 as it provides an easily interpretable metric of performance with respect to the actual outcome (i.e., the amount of variance explained in the target variable). Reporting *MAE* is also useful for comparing different machine learning models.

Note
Occasionally, models can result in negative correlations between the network strength predicted values and the target variable. However, this is not a meaningful association as all network strength predicted values should be positively associated with the target variable. Therefore, negative

(continued)

correlations should be interpreted as a failure to successfully predict the target variable [8] even if the prediction is statistically significant [35]. For instance, Ren et al. [35] applied CPM to predict creativity anxiety and reported significant predictions in the positive, negative, and combined networks in the training set. When this model was externally validated, the negative and combined networks successfully predicted creativity anxiety. In contrast, the positive network did not as positive network strength predicted creativity anxiety scores were negatively correlated with observed creativity anxiety scores.

12 Step-by-Step Example

Download Code, Data, and Analysis Scripts

1. The user should first download the flexible_CPM code from the GitHub repository using this link.
2. The user should add this code to their MATLAB path.
3. The user can download the AOMIC data from this link [see Chapter 2].
4. When the AOMIC data is downloaded, the user can prepare the data by downloading and running the Neuromethods_prepare_csv_CPM.m script from this link.
5. The user should then download and open Neuromethods_main_analysis_CPM.m script using this link.

Note: Two additional scripts are provided for this book chapter in the GitHub repository

- Leave-site-out cross-validation script.
  - This script is explained in the final section of the chapter.
- Computational lesion analysis script.
  - This is a fully commented script which users can download and execute if they would like to apply CPM with computational lesions to obtain further neurobiological insights.
  - This is typically only useful if CPM provides accurate results with the full whole-brain connectivity data.

12.1 Load Data and Prepare Variables (Subheading 1)

```
%% 1) Load data and prepare variables
% Specify file paths
target_path = 'W:\AOMIC\doc\behav_data.csv';
covars_path = 'W:\AOMIC\doc\covariates.csv';
conn_mx_dir = 'W:\AOMIC\connectivity_matrices';
output_path = 'W:\AOMIC\output\kfold_10';

% Specify target variable (ensure string is same as listed in .csv file)
target_var_name = 'IQ';

% Specify name of model for saving output
model_name = 'IQ_CPM';

% Prepare input data
nsubs = 861;  % list number of participants
[cpm_predictors, file_order] = prep_predictors_CPM(conn_mx_dir, nsubs,...
        output_path);

% Load predictor variables (i.e. connectivity matrices)
all_mats = cpm_predictors;

% Prepare target variable
[cpm_target, final_ppts, ix_ppts_to_keep] = prep_target_CPM(target_path,...
        file_order, output_path);

% Load target variable and covariates to be included
data = readtable(covars_path);

% Get ppts with connectivity matrices, covariates, and target variable
subids=table2array(data(:,1));
log_ix = ismember(subids, final_ppts);
data = data(log_ix,:);  % drop ppts w/o conn mx/covariates/target variable

% Specify target variable
all_behav = cpm_target;


% Prep covariates - age, sex, fwd: covar_names = {'age', 'sex', 'mean_FWD'};
% if no covars to be included, use: covar_names = {}
% covariates must be named here as listed in .csv file
covar_names = {'age', 'sex'};
all_covars = table2array(data(:, covar_names));

% convert covar_names to string for saving
covar_str = "";
for i = 1:length(covar_names)
        if i == 1
        covar_str = append(covar_str, covar_names(i));
        else
        covar_str = append(covar_str, ' ', covar_names(i));
        end
end
```

1. To execute the code in Subheading 1, the user is required to have the following:
   - A .csv file of size (1 + N) * 2, where N = number of participants. The first column must contain participant IDs (participant IDs must be the same as named in connectivity matrix file) and the second column must contain target variable, e.g. fluid intelligence scores. The first row (header row) should contain variable names (e.g. subid and fluidIQ). This is specified by **target_path**.
   - A .csv file of size (1 + N) * m, where m = number of covariates. The first column must contain participant IDs (participant IDs must be the same as named in connectivity matrix file) and m columns should contain covariates. The first row (header row) should contain variable names (e.g. subid, age, sex). This is specified by **covars_path**.
   - A directory containing *only* .csv files with each participants 268*268 connectivity matrix. The .csv files should be named ParticipantID.csv (e.g. for participant ID = 112, . csv file should be named 112.csv). This is specified by **conn_mx_dir.**
   - An empty directory for storing output of the model. Create this directory prior to executing the code. This is specified by **output_path.**
2. The user is required to provide the following input:
   - **target_var_name**: string specifying the name of a target variable.
   - **model_name**: string specifying a name for saving output files.
   - **nsubs**: double containing the number of participants with connectivity matrices.
   - **covar_names**: cell array of strings containing names of covariates (covariate names must be the same as in the header row of the .csv file specified by covars_path. If no covariates are to be used, then provide an empty cell array).
3. This section will prepare the input data and target variable for CPM using the following functions:
   - **prep_target_CPM( ):**
   - **prep_predictors_CPM( ):**

12.2 Specify Inputs (Subheading 2)

```
%% 2) Specify model inputs
% Specify method for dealing with confounds. adjust_stage = 'relate'
% adjusts for confounds via partial correlation during feature selection.
% adjust_stage = 'fit' adjusts for confounds via multiple regression during
% model fitting. adjust_stage = 'both' adjusts for confounds at both above
% steps. adjust_stage = '' does not adjust for confounds.
adjust_stage = 'relate';

% Specify cross-validation scheme - LOOCV: k=n, 10-fold CV: k=10, etc.
% for LOOCV: k = length(all_behav);
k = 10;

% Specify iterations (may want to run multiple iterations of k-fold CV)
% for LOOCV: iterations = 1
iterations=100;

% Specify feature selection type. 'p-value' will threshold edges based on
% p-value of correlation between edges and target variable. 'sparsity' will
% threshold edges by selecting the X % most strongly correlated edges to
% target variable.
thresh_type = 'p-value';

% Specify feature selection threshold. If 'p-value', enter p-value for
% correlations between edges and target variable. If 'sparsity', enter % of
% most highly correlated edges to be retained.
% Note: % should be in decimal (i.e. between 0 and 1, 5% = 0.05)
thresh = 0.01;

% Specify edge frequency threshold (i.e. how many folds edge must be
% significantly correlated (i.e. below thresh) with target variable in
% order to be selected for application to the test set)
% Note: % should be in decimal (i.e. between 0 and 1, 100% = 1)
freq_thresh = 1;

% Specify if permutation test to be conducted (will greatly increase
% runtime of code); run_permutation = 'yes' or 'no'
run_permutation = 'yes';

% if yes, specify number of iterations of random permutation
% (e.g. perm_iterations = 1000;). if no, save as []
perm_iterations = 1000;

% Save info describing model inputs
model_info = struct('target_path',target_path,'covars_path',covars_path, ...
        'target_variable', target_var_name, 'N', length(all_behav),'covars',...
        covar_str, 'adjust_stage', {adjust_stage}', 'k', k, 'iterations',...
        iterations, 'fs_thresh_type',thresh_type,'fs_thresh',thresh,'freq_thresh',...
      freq_thresh,'permutation_test', run_permutation, 'perm_iterations',...
      perm_iterations, 'output_path', output_path, 'model_name', model_name);

clearvars -except all_behav all_mats all_covars k thresh thresh_type freq_thresh...
      adjust_stage iterations model_info perm_iterations run_permutation
```

- Subheading 2 requires user to specify the following options (as outlined in Fig. 2):
  - **adjust_stage**: string specifying a method for dealing with confounds.
  - **k**: double specifying number of cross-validation folds.
  - **iterations**: double specifying number of iterations of cross-validation (1 when LOOCV is used).
  - **thresh_type**: string specifying a method for feature selection thresholding.
  - **thresh**: double specifying feature selection threshold.
  - **freq_thresh**: double specifying edge frequency threshold for visualisation and external validation.
  - **run_permutation**: string specifying option to run permutation test for statistical significance.
  - **perm_iterations**: double specifying number of random permutation iterations to run.
- This section will prepare the model inputs and save them in a structure array.

12.3 Preallocation of Arrays (Subheading 3)

```
%% 3) Preallocate arrays for storing results and parameters
% Preallocate arrays for storing CPM predicted values and results
[behav_pred_pos_all, behav_pred_neg_all, behav_pred_combined_all, ...
        R_pos, P_pos, R_neg, P_neg, R_combined, P_combined,...
        rsq_pos, rsq_neg, rsq_combined, mae_pos, mae_neg, mae_combined] = ...
        prep_results_arrays_CPM(all_mats, iterations);

% Preallocate arrays for storing CPM parameters
if strcmp(adjust_stage, 'relate')
        [~, ~, int_pos_ntwrk, int_neg_ntwrk,...
        int_combined_ntwrk, slope_pos_ntwrk, slope_neg_ntwrk, ...
        slope_combined_ntwrk, ~, ~, ~] = prep_parameters_arrays_CPM(all_mats,...
        all_covars, k, iterations);
else
        [~, ~, int_pos_ntwrk, int_neg_ntwrk,...
        int_combined_ntwrk, slope_pos_ntwrk, slope_neg_ntwrk, ...
        slope_combined_ntwrk, slope_pos_covars, slope_neg_covars,...
        slope_combined_covars] = prep_parameters_arrays_CPM(all_mats,...
        all_covars, k, iterations);
end
```

- This section preallocates arrays for computational efficiency with the functions:
  - **prep_results_arrays_CPM( ):** preallocates arrays for storing CPM predicted values and results
  - **prep_parameters_arrays_CPM( ):** preallocates arrays for storing model parameters

12.4 Run CPM (Subheading 4)

```
%% 4) Run CPM, evaluate model performance, extract selected edges + model parameters
parfor i = 1:iterations
        fprintf('\n Running iteration # %6.3f\n',i);

        % Run CPM
        [behav_pred_pos_all(:,i), behav_pred_neg_all(:,i), ...
        behav_pred_combined_all(:,i), parameters_pos, parameters_neg,...
        parameters_combined, pos_mask, neg_mask, no_node, no_covars] = ...
        run_flexible_CPM(all_behav, all_mats, all_covars, k, ...
        thresh_type, thresh, adjust_stage); %#ok<ASGLU>

        % Evaluate model performance
        [R_pos(i), P_pos(i), R_neg(i), P_neg(i), R_combined(i), P_combined(i),...
        rsq_pos(i), rsq_neg(i), rsq_combined(i), mae_pos(i), mae_neg(i),...
        mae_combined(i)] = evaluate_CPM(all_behav, behav_pred_pos_all(:,i),...
        behav_pred_neg_all(:,i), behav_pred_combined_all(:,i));

        % Extract model parameters
        if strcmp(adjust_stage, 'relate')
        [int_pos_ntwrk(i), int_neg_ntwrk(i), int_combined_ntwrk(i),...
        slope_pos_ntwrk(i), slope_neg_ntwrk(i), slope_combined_ntwrk(i),...
        ~, ~, ~] = extract_parameters_CPM(parameters_pos, parameters_neg,...
        parameters_combined, adjust_stage, no_covars);
        else
        [int_pos_ntwrk(i), int_neg_ntwrk(i), int_combined_ntwrk(i),...
        slope_pos_ntwrk(i), slope_neg_ntwrk(i), slope_combined_ntwrk(i),...
        slope_pos_covars(i,:), slope_neg_covars(i,:), ...
        slope_combined_covars(i,:)] = ...
        extract_parameters_CPM(parameters_pos, parameters_neg,...
        parameters_combined,adjust_stage, no_covars);
        end

% Store positive and negative edges masks - accounting for parallel loop
        c_pos_mask_all{i,1} = pos_mask;
        c_neg_mask_all{i,1} = neg_mask;
end

% Add positive and negative edge masks
pos_mask_all = [];
neg_mask_all = [];
for j = 1:iterations
        pos_mask_all = cat(3, c_pos_mask_all{j,1}, pos_mask_all);
        neg_mask_all = cat(3, c_neg_mask_all{j,1}, neg_mask_all);
end
```

- Subheading 4 performs the CPM using the following functions:
  - **run_flexible_CPM( ):** executes the CPM
  - **evaluate_CPM( ):** evaluates model performance
  - **extract_parameters_CPM( ):** extracts model parameters
- If multiple iterations are specified (e.g. if running 100 iterations of k-fold cross-validation), then this section is looped over and the CPM is conducted separately within each iteration.

12.5 Extract Selected Edges (Subheading 5)

```
%% 5) Extract selected edges
% set number of folds across all iterations of model
k_all = k*iterations;

% no_node not available after parfor loop so call again here
no_node = size(all_mats,1);

% get indices of edges in each network that are correlated with the
% target variable in >= number of total folds specified by freq_thresh
[pos_edges_all, neg_edges_all, pos_edges_thresh, neg_edges_thresh] = ...
        extract_edges_CPM(pos_mask_all, neg_mask_all, no_node, k_all, freq_thresh);
```

- This section extracts the selected edges using the function:
  - **extract_edges_CPM( ):** extracts indices of edges within the positive and negative edge masks.

12.6 Store and Save Parameters, Predictions, Selected Edges, and Model Results (Subheading 6)

```
%% 6) Store parameters, predicted values, edges, and results in structs and save data
if strcmp(adjust_stage, 'relate')
        parameters = struct('int_pos_ntwrk', int_pos_ntwrk, 'int_neg_ntwrk', ...
        int_neg_ntwrk, 'int_combined_ntwrk', int_combined_ntwrk,...
        'slope_pos_ntwrk', slope_pos_ntwrk, 'slope_neg_ntwrk', slope_neg_ntwrk,...
        'slope_combined_ntwrk', slope_combined_ntwrk,...
        'pos_mask_all', pos_mask_all, 'neg_mask_all', neg_mask_all);
else
        parameters = struct('int_pos_ntwrk', int_pos_ntwrk, 'int_neg_ntwrk', ...
        int_neg_ntwrk, 'int_combined_ntwrk', int_combined_ntwrk,...
        'slope_pos_ntwrk', slope_pos_ntwrk, 'slope_neg_ntwrk', slope_neg_ntwrk,...
        'slope_combined_ntwrk', slope_combined_ntwrk, 'slope_pos_covars',...
        slope_pos_covars, 'slope_neg_covars', slope_neg_covars,...
        'slope_combined_covars', slope_combined_covars,...
        'pos_mask_all', pos_mask_all, 'neg_mask_all', neg_mask_all);
end

predictions = struct('pos_preds_all_folds', behav_pred_pos_all, ...
        'neg_preds_all_folds', behav_pred_neg_all, 'combined_preds_all_folds',...
        behav_pred_combined_all);
```

```
edges = struct('pos_edges_all', pos_edges_all, 'neg_edges_all', neg_edges_all,...
        'pos_edges_thresh', pos_edges_thresh, 'neg_edges_thresh', neg_edges_thresh);

results = struct('R_pos', R_pos, 'R_neg', R_neg, 'R_combined', R_combined,...
        'rsq_pos', rsq_pos, 'rsq_neg', rsq_neg, 'rsq_combined', rsq_combined,...
        'mae_pos', mae_pos, 'mae_neg', mae_neg, 'mae_combined', mae_combined);
% Get mean parameters, predicted values, and results across iterations
if iterations > 1
        mean_parameters = structfun(@mean, parameters, 'UniformOutput', false);
        mean_predictions = structfun(@(x) mean(x, 2), predictions, 'UniformOutput', false);
        mean_results = structfun(@mean, results, 'UniformOutput', false);
end

% save data
variable_list = {'model_info', 'parameters', 'predictions', 'results', 'edges', ....
        'mean_parameters', 'mean_predictions', 'mean_results'};

save_file = [model_info.output_path filesep model_info.model_name '.mat'];
for variable_ix = 1:length(variable_list)
        if exist(variable_list{variable_ix})
        if exist(save_file)
        save(save_file, variable_list{variable_ix}, '-append');
        else
        save(save_file, variable_list{variable_ix});
        end
        end
end
```

- Subheading 6 creates four structures which are then saved in the pre-specified output folder:
  - *Parameters*: stores model parameters for external validation.
  - *Predictions*: stores predicted values of each network for visualisation and further analysis (i.e. association between predicted values and other variables of interest).
  - *Edges*: stores selected edges for external validation and visualisation.
  - *Results*: stores results of model performance.
  - Where multiple iterations of cross-validation were performed (e.g. 100 iterations of ten-fold cross-validation), this section calculates the mean of each value across all iterations for the structs parameters, predictions, results and saves them as:

    mean_parameters

    mean_predictions

    mean_results

12.7 Create Masks for Visualisation (Subheading 7)

```
%% 7) Create masks for visualisation
% Check if restricted or computationally lesioned connectivity matrix used
% and get indices of selected edges in full (i.e. 268 * 268 connectivity

% matrix).
% NOTE: THIS CODE ASSUMES SHEN ATLAS PARCELLATION USED.
if no_node < 268

        % Read in csv file mapping node indices in restricted/lesioned
        % connectivity matrix to node indices in original connectivity matrix
        % 1st column is original index, 2nd column is new index
        node_indices =
csvread('Y:\cogReserve\CPM_info\restricted_timeseries_node_indices.csv',1, 1);

        % map edge indices back to indices in full connectivity matrix
        [pos_edges_orig, neg_edges_orig] = get_original_edge_indices_CPM(...
        pos_edges_thresh, neg_edges_thresh, node_indices);

        % create binary edge masks (in 268*268 connectivity matrix)
        [pos_edge_mask, neg_edge_mask] = ...
        create_masks_CPM(pos_edges_orig, neg_edges_orig, 268);

else % if using a full timeseries
        pos_edges_orig = pos_edges_thresh(:, [1 3 4]);
        neg_edges_orig = neg_edges_thresh(:, [1 3 4]);
        [pos_edge_mask, neg_edge_mask] = ...
        create_masks_CPM(pos_edges_orig, neg_edges_orig, 268);
end

% save binary edge masks for visualisation in bioimagesuite
pos_mask_file = [model_info.output_path filesep model_info.model_name
'_pos_mask.txt'];
neg_mask_file = [model_info.output_path filesep model_info.model_name
'_neg_mask.txt'];
save(pos_mask_file, 'pos_edge_mask', '-ascii');
save(neg_mask_file, 'neg_edge_mask', '-ascii');

% save info on edges wrt to their position in the original 268*268
% connectivity matrix
orig_edge_ix = [model_info.output_path filesep model_info.model_name
'_orig_edge_ix.mat'];
save(orig_edge_ix, 'pos_edges_orig', 'pos_edge_mask', 'neg_edges_orig',
'neg_edge_mask');
```

- Subheading 7 creates binary masks for the positive and negative networks for visualisation in BioImageSuiteWeb. This section assumes the Shen atlas parcellation is used. If fewer than 268 nodes are used for each participant (e.g. only non-cerebellar and brainstem nodes were used due to poor coverage of the cerebellum and brainstem or a computational lesion was applied to the connectivity matrix), then this section will create masks and then insert them back into a full Shen atlas mask (i.e. including empty cerebellar & brainstem nodes or including the lesioned nodes).

12.8 Run Permutation Test (Subheading 8)

```
%% 8) Run permutation test and save
if strcmp(run_permutation, 'yes')
        if iterations > 1
        [perm_p_pos, perm_p_neg, perm_p_combined] = ...
        CPM_permutation_test_parallelised(all_behav, all_mats, ...
        all_covars, k, thresh_type,thresh, adjust_stage,...
        mean_results.R_pos, mean_results.R_neg, mean_results.R_combined,...
        perm_iterations);
        else
        [perm_p_pos, perm_p_neg, perm_p_combined] = ...
        CPM_permutation_test_parallelised(all_behav, all_mats, ...
        all_covars, k, thresh_type, thresh, adjust_stage, ...
        R_pos, R_neg, R_combined, perm_iterations);
        end

        % save permutation test results
        perm_results = struct('perm_p_pos', perm_p_pos, 'perm_p_neg', ...
        perm_p_neg, 'perm_p_combined', perm_p_combined);

        save(save_file, 'perm_results', '-append');
end
```

- This section runs a permutation test (if specified) and saves the output to the pre-specified output directory, using the function:
  - **CPM_permutation_test_parallelised( ):**

13 Reporting Results

The CPM outlined here generated statistically significant predictions of fluid intelligence on unseen participants from the same dataset (*see* Fig. 3 and Table 4). The combined connectome explained the most variance (3.2%) in fluid intelligence. The positive connectome generated relatively more accurate predictions of fluid intelligence versus the negative connectome. In line with this, the positive connectome consisted of a much denser set of connections that were significantly related to fluid IQ than the negative network (*see* Fig. 4). The positive connectome consisted of a relatively large number of connections within the fronto-parietal network and between the fronto-parietal and medial-frontal networks (*see* Fig. 5).

Table 4
CPM performance for prediction of fluid IQ

Positive network strength			Negative network strength			Combined network strength		
r	R^2	*MAE*	*r*	R^2	*MAE*	*r*	R^2	*MAE*
.149**	.022	20.571	.118*	.014	20.981	.177**	.032	20.583

Note: * < 0.01, ** < 0.001 (permutated *p*-value)

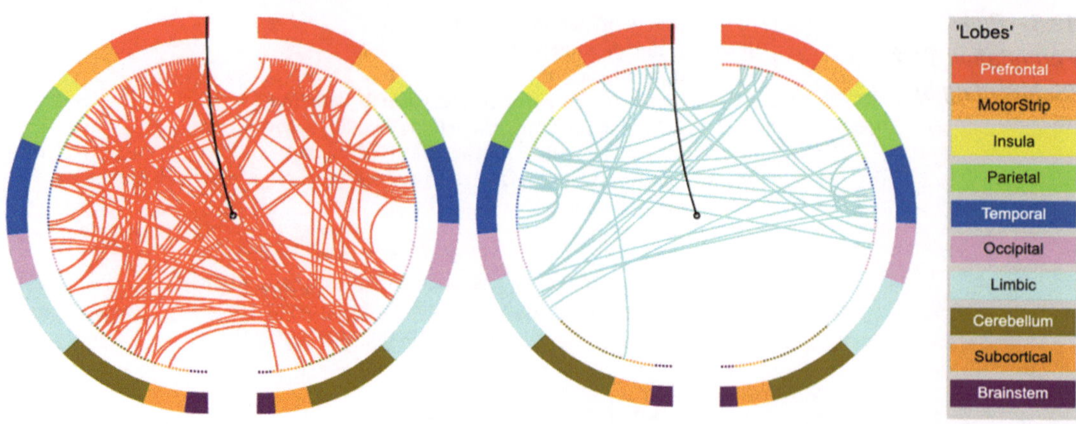

Fig. 4 Circle plots illustrating the positive (left; red) and negative (right; cyan) fluid IQ connectomes. *These circle plots are inverted such that the right side of each plot corresponds to the left hemisphere and the left side to the right hemisphere*

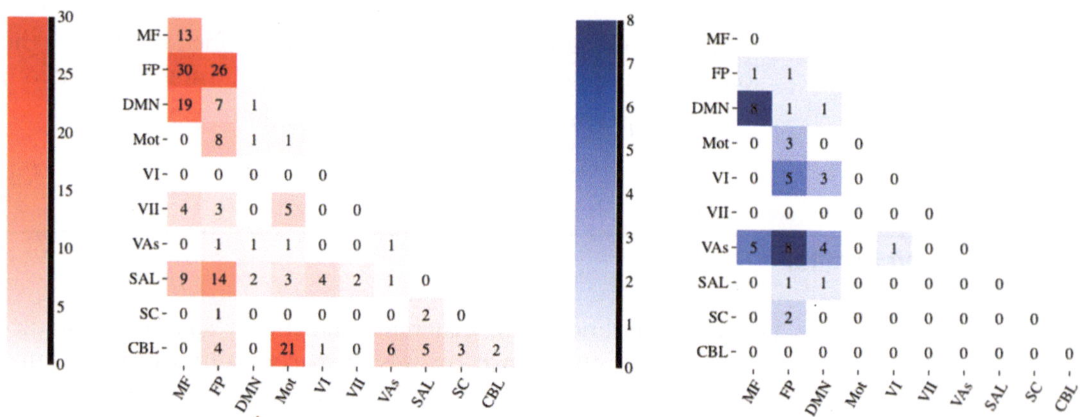

Fig. 5 Connectivity matrices summarising patterns of functional connectivity within and between different canonical networks in the positive (left; red) and negative (right; blue) connectomes. Note: MF Medial Frontal Network, FP Frontoparietal Network, DMN Default Mode Network, Mot Motor Network, Vis I Visual I Network, Vis II Visual II Network, VAs Visual Association Network, SAL Salience Network, SC Subcortical Network, CBL Cerebellar Network

14 Leave-Site-Out Cross-Validation Example

This example follows the same steps as outlined above but implements CPM with leave-site-out cross-validation for multi-site data. The steps are identical except that some additional inputs that are required (i.e. site information) and some CPM functions are replaced with functions adapted for leave-site-out cross-validation. The relevant differences are highlighted below so the user can implement leave-site-out cross-validation.

1. As the AOMIC data is from a single site, the user will need to assign different sites to the AOMIC data to *simulate* a multisite dataset. This can be done by creating a .csv file as outlined here:
   - Create a .csv file of size (1 + N) * 2, where N = number of participants. The first column must contain participant IDs (participant IDs must be the same as named in connectivity matrix file) and the second column must contain the site numbers, e.g. 1–5. The first row (header row) should contain variable names (e.g. subid and sites). Participant IDs must be the same as named in the connectivity matrices.
2. The user should then download and open the Neuromethods_leaveSiteOut_CPM.m script using this link.
3. To execute the code, the user is required to provide the .csv file created in the previous step (in addition to the files outlined in the 'Specify inputs' step in the main example):
   - The .csv file is specified by **sites_path**.
4. This script will then run, as outlined in the above example, but with two new functions:
   - **run_flexible_CPM_leaveSiteOut( ):** replaces **run_flexible_CPM( )**
   - **CPM_permutation_test_leaveSiteOut_parallelised( )** replaces **CPM_permutation_test_parallelised( )**

15 Computational Lesion Example

Users can also download and execute the Neuromethods_computational_lesion_analysis.m script from this link if they would like to apply computation lesions to CPM to obtain further neurobiological insights. This script is fully commented with instructions.

References

1. Finn ES, Shen X, Scheinost D, Rosenberg MD, Huang J, Chun MM, Papademetris X, Constable RT (2015) Functional connectome fingerprinting: identifying individuals based on patterns of brain connectivity. Nat Neurosci 18:1664–1671. https://doi.org/10.1038/nn.4135
2. Shen X, Finn ES, Scheinost D, Rosenberg MD, Chun MM, Papademetris X, Constable RT (2017) Using connectome-based predictive modeling to predict individual behavior from brain connectivity. Nat Protoc 12:506–518. https://doi.org/10.1038/nprot.2016.178
3. Lin Q, Rosenberg MD, Yoo K, Hsu TW, O'Connell TP, Chun MM (2018) Resting-state functional connectivity predicts cognitive impairment related to Alzheimer's disease. Front Aging Neurosci 10:94. https://doi.org/10.3389/fnagi.2018.00094
4. Fountain-Zaragoza S, Samimy S, Rosenberg MD, Prakash RS (2019) Connectome-based models predict attentional control in aging adults. NeuroImage 186:1–13. https://doi.org/10.1016/j.neuroimage.2018.10.074
5. Rosenberg MD, Finn ES, Scheinost D, Papademetris X, Shen X, Constable RT, Chun

MM (2016) A neuromarker of sustained attention from whole-brain functional connectivity. Nat Neurosci 19:165–171. https://doi.org/10.1038/nn.4179

6. Henneghan AM, Gibbons C, Harrison RA, Edwards ML, Rao V, Blayney DW, Palesh O, Kesler SR (2020) Predicting patient reported outcomes of cognitive function using connectome-based predictive modeling in breast cancer. Brain Topogr 33:135–142. https://doi.org/10.1007/s10548-019-00746-4
7. Gao S, Greene AS, Constable RT, Scheinost D (2019) Combining multiple connectomes improves predictive modeling of phenotypic measures. NeuroImage 201:116038. https://doi.org/10.1016/j.neuroimage.2019.116038
8. Greene AS, Gao S, Scheinost D, Constable RT (2018) Task-induced brain state manipulation improves prediction of individual traits. Nat Commun 9:2807. https://doi.org/10.1038/s41467-018-04920-3
9. Boyle R, Connaughton M, McGlinchey E, Knight SP, De Looze C, Carey D, Stern Y, Robertson IH, Kenny RA, Whelan R (2023) Connectome-based predictive modelling of cognitive reserve using task-based functional connectivity. Eur J Neurosci 57:490–510. https://doi.org/10.1111/ejn.15896
10. Beaty RE, Kenett YN, Christensen AP, Rosenberg MD, Benedek M, Chen Q, Fink A, Qiu J, Kwapil TR, Kane MJ, Silvia PJ (2018) Robust prediction of individual creative ability from brain functional connectivity. Proc Natl Acad Sci USA 115:1087–1092. https://doi.org/10.1073/pnas.1713532115
11. Wang Z, Goerlich KS, Ai H, Aleman A, Luo Y-J, Xu P (2021) Connectome-based predictive modeling of individual anxiety. Cereb Cortex 31:3006–3020. https://doi.org/10.1093/cercor/bhaa407
12. Ju Y, Horien C, Chen W, Guo W, Lu X, Sun J, Dong Q, Liu B, Liu J, Yan D, Wang M, Zhang L, Guo H, Zhao F, Zhang Y, Shen X, Constable RT, Li L (2020) Connectome-based models can predict early symptom improvement in major depressive disorder. J Affect Disord 273:442–452. https://doi.org/10.1016/j.jad.2020.04.028
13. Feng C, Wang L, Li T, Xu P (2019) Connectome-based individualized prediction of loneliness. Soc Cogn Affect Neurosci 14:353–365. https://doi.org/10.1093/scan/nsz020
14. Goldfarb EV, Rosenberg MD, Seo D, Constable RT, Sinha R (2020) Hippocampal seed connectome-based modeling predicts the feeling of stress. Nat Commun 11:2650. https://doi.org/10.1038/s41467-020-16492-2
15. Ibrahim K, Noble S, He G, Lacadie C, Crowley MJ, McCarthy G, Scheinost D, Sukhodolsky DG (2022) Large-scale functional brain networks of maladaptive childhood aggression identified by connectome-based predictive modeling. Mol Psychiatry 27:985–999. https://doi.org/10.1038/s41380-021-01317-5
16. Dufford AJ, Kimble V, Tejavibulya L, Dadashkarimi J, Ibrahim K, Sukhodolsky DG, Scheinost D (2022) Predicting transdiagnostic social impairments in childhood using connectome-based predictive modeling. medRxiv. https://doi.org/10.1101/2022.04.07.22273518
17. Lichenstein SD, Scheinost D, Potenza MN, Carroll KM, Yip SW (2021) Dissociable neural substrates of opioid and cocaine use identified via connectome-based modelling. Mol Psychiatry 26:4383–4393. https://doi.org/10.1038/s41380-019-0586-y
18. Yip SW, Scheinost D, Potenza MN, Carroll KM (2019) Connectome-based prediction of cocaine abstinence. Am J Psychiatry 176:156–164. https://doi.org/10.1176/appi.ajp.2018.17101147
19. Chen AA, Beer JC, Tustison NJ, Cook PA, Shinohara RT, Shou H, Initiative the ADN (2020) Removal of scanner effects in covariance improves multivariate pattern analysis in neuroimaging data. https://doi.org/10.1101/858415
20. Yamashita A, Yahata N, Itahashi T, Lisi G, Yamada T, Ichikawa N, Takamura M, Yoshihara Y, Kunimatsu A, Okada N, Yamagata H, Matsuo K, Hashimoto R, Okada G, Sakai Y, Morimoto J, Narumoto J, Shimada Y, Kasai K, Kato N, Takahashi H, Okamoto Y, Tanaka SC, Kawato M, Yamashita O, Imamizu H (2019) Harmonisation of resting-state functional MRI data across multiple imaging sites via the separation of site differences into sampling bias and measurement bias. PLoS Biol 17:e3000042. https://doi.org/10.1371/journal.pbio.3000042
21. Zhao W, Palmer CE, Thompson WK, Chaarani B, Garavan HP, Casey BJ, Jernigan TL, Dale AM, Fan CC (2021) Individual differences in cognitive performance are better predicted by global rather than localized BOLD activity patterns across the cortex. Cereb Cortex 31:1478–1488. https://doi.org/10.1093/cercor/bhaa290
22. Feng C, Yuan J, Geng H, Gu R, Zhou H, Wu X, Luo Y (2018) Individualized prediction

of trait narcissism from whole-brain resting-state functional connectivity. Hum Brain Mapp 39:3701–3712. https://doi.org/10.1002/hbm.24205

23. Yoo K, Rosenberg MD, Hsu W-T, Zhang S, Li C-SR, Scheinost D, Constable RT, Chun MM (2018) Connectome-based predictive modeling of attention: comparing different functional connectivity features and prediction methods across datasets. NeuroImage 167:11–22. https://doi.org/10.1016/j.neuroimage.2017.11.010
24. Hakim N, Awh E, Vogel EK, Rosenberg MD (2020) Predicting cognitive abilities across individuals using sparse EEG connectivity. https://doi.org/10.1101/2020.07.22.216705
25. Kabbara A, Robert G, Khalil M, Verin M, Benquet P, Hassan M (2022) An electroencephalography connectome predictive model of major depressive disorder severity. Sci Rep 12:6816. https://doi.org/10.1038/s41598-022-10949-8
26. Poldrack RA, Huckins G, Varoquaux G (2020) Establishment of best practices for evidence for prediction: a review. JAMA Psychiatry 77:534–540. https://doi.org/10.1001/jamapsychiatry.2019.3671
27. Varoquaux G (2018) Cross-validation failure: small sample sizes lead to large error bars. NeuroImage 180:68–77. https://doi.org/10.1016/j.neuroimage.2017.06.061
28. Kumar S, Yoo K, Rosenberg MD, Scheinost D, Constable RT, Zhang S, Li C-SR, Chun MM (2019) An information network flow approach for measuring functional connectivity and predicting behavior. Brain Behav 9. https://doi.org/10.1002/brb3.1346
29. Esterman M, Noonan SK, Rosenberg M, Degutis J (2013) In the zone or zoning out? Tracking behavioral and neural fluctuations during sustained attention. Cereb Cortex 23:2712–2723. https://doi.org/10.1093/cercor/bhs261
30. Rosenberg M, Noonan S, DeGutis J, Esterman M (2013) Sustaining visual attention in the face of distraction: a novel gradual-onset continuous performance task. Atten Percept Psychophys 75:426–439. https://doi.org/10.3758/s13414-012-0413-x
31. Fortenbaugh FC, Rothlein D, McGlinchey R, DeGutis J, Esterman M (2018) Tracking behavioral and neural fluctuations during sustained attention: a robust replication and extension. NeuroImage 171:148–164. https://doi.org/10.1016/j.neuroimage.2018.01.002
32. Ciric R, Wolf DH, Power JD, Roalf DR, Baum GL, Ruparel K, Shinohara RT, Elliott MA, Eickhoff SB, Davatzikos C, Gur RC, Gur RE, Bassett DS, Satterthwaite TD (2017) Benchmarking of participant-level confound regression strategies for the control of motion artifact in studies of functional connectivity. NeuroImage 154:174–187. https://doi.org/10.1016/j.neuroimage.2017.03.020
33. Satterthwaite TD, Elliott MA, Gerraty RT, Ruparel K, Loughead J, Calkins ME, Eickhoff SB, Hakonarson H, Gur RC, Gur RE, Wolf DH (2013) An improved framework for confound regression and filtering for control of motion artifact in the preprocessing of resting-state functional connectivity data. NeuroImage 64:240–256. https://doi.org/10.1016/j.neuroimage.2012.08.052
34. Murphy K, Fox MD (2017) Towards a consensus regarding global signal regression for resting state functional connectivity MRI. NeuroImage 154:169–173. https://doi.org/10.1016/j.neuroimage.2016.11.052
35. Ren Z, Daker RJ, Shi L, Sun J, Beaty RE, Wu X, Chen Q, Yang W, Lyons IM, Green AE, Qiu J (2021) Connectome-based predictive modeling of creativity anxiety. NeuroImage 225:117469. https://doi.org/10.1016/j.neuroimage.2020.117469
36. Shen X, Tokoglu F, Papademetris X, Constable RT (2013) Groupwise whole-brain parcellation from resting-state fMRI data for network node identification. NeuroImage 82:403–415. https://doi.org/10.1016/j.neuroimage.2013.05.081
37. Liang Q, Jiang R, Adkinson BD, Rosenblatt M, Mehta S, Foster ML, Dong S, You C, Negahban S, Zhou HH, Chang J, Scheinost D (2024) Rescuing missing data in connectome-based predictive modeling. Imaging Neuroscience 2:1–16. https://doi.org/10.1162/imag_a_00071
38. Rad KR, Maleki A (2020) A scalable estimate of the out-of-sample prediction error via approximate leave-one-out cross-validation. J R Stat Soc Ser B Stat Methodol 82:965–996. https://doi.org/10.1111/rssb.12374
39. Gao M, Wong CHY, Huang H, Shao R, Huang R, Chan CCH, Lee TMC (2020) Connectome-based models can predict processing speed in older adults. NeuroImage 223:117290. https://doi.org/10.1016/j.neuroimage.2020.117290
40. Steiger JH (1980) Tests for comparing elements of a correlation matrix. Psychol Bull 87:245–251. https://doi.org/10.1037/0033-2909.87.2.245

41. Zúñiga RG, Davis JRC, Boyle R, De Looze C, Meaney JF, Whelan R, Kenny RA, Knight SP, Ortuño RR (2023) Brain connectivity in frailty: insights from the Irish longitudinal study on ageing (TILDA). Neurobiol Aging 124:1–10. https://doi.org/10.1016/j.neurobiolaging.2023.01.001
42. Ramedani Z, Omid M, Keyhani A, Shamshirband S, Khoshnevisan B (2014) Potential of radial basis function based support vector regression for global solar radiation prediction. Renew Sust Energ Rev 39:1005–1011. https://doi.org/10.1016/j.rser.2014.07.108
43. Power JD, Barnes KA, Snyder AZ, Schlaggar BL, Petersen SE (2012) Spurious but systematic correlations in functional connectivity MRI networks arise from subject motion. NeuroImage 59:2142–2154. https://doi.org/10.1016/j.neuroimage.2011.10.018
44. Goto M, Abe O, Miyati T, Yamasue H, Gomi T, Takeda T (2016) Head motion and correction methods in resting-state functional MRI. Magn Reson Med Sci 15:178–186. https://doi.org/10.2463/mrms.rev.2015-0060
45. Makowski C, Lepage M, Evans AC (2019) Head motion: the dirty little secret of neuroimaging in psychiatry. J Psychiatry Neurosci 44:62–68. https://doi.org/10.1503/jpn.180022
46. Horien C, Fontenelle S, Joseph K, Powell N, Nutor C, Fortes D, Butler M, Powell K, Macris D, Lee K, Greene AS, McPartland JC, Volkmar FR, Scheinost D, Chawarska K, Constable RT (2020) Low-motion fMRI data can be obtained in pediatric participants undergoing a 60-minute scan protocol. Sci Rep 10:21855. https://doi.org/10.1038/s41598-020-78885-z
47. Geerligs L, Tsvetanov KA, Null C-C, Henson RN (2017) Challenges in measuring individual differences in functional connectivity using fMRI: the case of healthy aging. Hum Brain Mapp 38:4125–4156. https://doi.org/10.1002/hbm.23653
48. Power JD, Mitra A, Laumann TO, Snyder AZ, Schlaggar BL, Petersen SE (2014) Methods to detect, characterize, and remove motion artifact in resting state fMRI. NeuroImage 84:320–341. https://doi.org/10.1016/j.neuroimage.2013.08.048
49. Rosenberg MD, Hsu W-T, Scheinost D, Todd Constable R, Chun MM (2018) Connectome-based models predict separable components of attention in novel individuals. J Cogn Neurosci 30:160–173. https://doi.org/10.1162/jocn_a_01197
50. Wicherts JM, Veldkamp CLS, Augusteijn HEM, Bakker M, van Aert RCM, van Assen MALM (2016) Degrees of freedom in planning, running, analyzing, and reporting psychological studies: a checklist to avoid p-hacking. Front Psychol 7:1832. https://doi.org/10.3389/fpsyg.2016.01832
51. Cwiek A, Rajtmajer SM, Wyble B, Honavar V, Grossner E, Hillary FG (2022) Feeding the machine: challenges to reproducible predictive modeling in resting-state connectomics. Netw Neurosci 6:29–48. https://doi.org/10.1162/netn_a_00212

Chapter 16

Deep Learning Classification Based on Raw MRI Images

Sebastian Moguilner and Agustin Ibañez

Abstract

In this chapter, we describe a step-by-step implementation of an automated anatomical MRI feature extractor based on artificial intelligence machine learning for classification. We applied the *DenseNet*—a state-of-the-art convolutional neural network producing more robust results than previous deep learning network architectures—to data from male ($n = 400$) and female ($n = 400$), age-, and education- matched healthy adult subjects. Moreover, we illustrate how an occlusion sensitivity analysis provides meaningful insights about the relevant information that the neural network used to make accurate classifications. This addresses the "black-box" limitations inherent in many deep learning implementations. The use of this approach with a specific dataset demonstrates how future implementations can use raw MRI scans to study a range of outcome measures, including neurological and psychiatric disorders.

Key words Deep Learning, MRI, Artificial intelligence, Convolutional neural network, Occlusion sensitivity analysis

1 Introduction

Medical imaging computer-aided diagnosis has been a field of intense research over the last decades. Its primary motivation is to reduce diagnostic errors, automatize procedures with large datasets, and provide additional insights to better interpret the images. However, research translation from the lab to the real world (i.e., from "bench to bedside") has often been limited by a failure to generalize to novel datasets. In particular, rule-based algorithms lack the flexibility needed to handle heterogeneous samples. Current developments in computer vision, stemming from the field of artificial intelligence [see Glossary], provide an innovative solution to create flexible decision-making pipelines automatically. Moreover, thanks to recent improvements in dedicated computing hardware, it is now possible to handle large neuroimaging datasets.

Deep learning [see Glossary] computer vision methods, based on convolutional neural networks (CNN) [see Glossary], are characterized by their flexibility in evaluating images without prior orientation, metric, or shape conventions that are usually set prior

Robert Whelan and Hervé Lemaître (eds.), *Methods for Analyzing Large Neuroimaging Datasets*, Neuromethods, vol. 218, https://doi.org/10.1007/978-1-0716-4260-3_16, © The Author(s) 2025

image processing [1]. In this way, possible biases in the preprocessing steps such as image filtering, segmentation, rotation, and smoothing are eliminated because the input consists of unprocessed (raw) data, from which the most relevant image features are automatically extracted [2]. This approach has already proven successful in image recognition in general [3], in medical radiology [4], and neuroradiological domains such as in dementia characterization [5, 6]. However, open procedures with step-by-step detailed examples on how to implement a Deep Learning neural network on open access data with interpretable results are scarce.

This chapter will describe the implementation of a state-of-the-art deep learning algorithm called the *DenseNet* [7]. We show a step-by-step example to use the DenseNet on brain anatomical MRI images to classify male ($n = 400$) and female (n-400), age- and education- matched healthy adult subjects. Sex differences in the human brain are of great interest for studies of neuropsychological traits [8] and the differential prevalence of psychiatric [9] and neurological disorders such as Alzheimer's disease [10]. Despite its high relevance in epidemiological research, biological sex differences have been defined as an under-explored and often controversial subject in the neuroscientific literature [11, 12]. Crucially, previous studies looking at sex-related brain differences suffered from low statistical power due to small sample sizes, having a mean sample size of 130 participants [13]. This context calls for new and unbiased approaches able to handle more extensive databases, in which computer-vision algorithms with automatic feature extraction [see Glossary] on big-data contexts may help to gain further insights on brain differences that are undetectable by the human eye and/or by traditional machine learning [see Glossary] approaches.

2 Methods

2.1 The DenseNet Deep Learning Algorithm

The DenseNet is a state-of-the-art deep learning CNN employed in computer vision tasks aided by artificial intelligence. In this network architecture, each layer is connected to every other layer in a feed-forward [see Glossary] fashion, producing networks that are substantially deeper, more accurate, and more efficient to train (see Fig. 1c for the DenseNet diagram).

Consider $a^{[0]}$ as the input volume passed through a CNN, with L as the number of layers in the network, and g the non-linear transformation of lth layer. Traditional feed-forward convolutional networks connect the output of the lth layer as input to the $(l + 1)$th layer, giving the following transition layer: $a^{[l]} = g(a^{[l-1]})$. Other CNN architectures such as the ResNet [3] bypasses the non-linear transformations with an identity function:

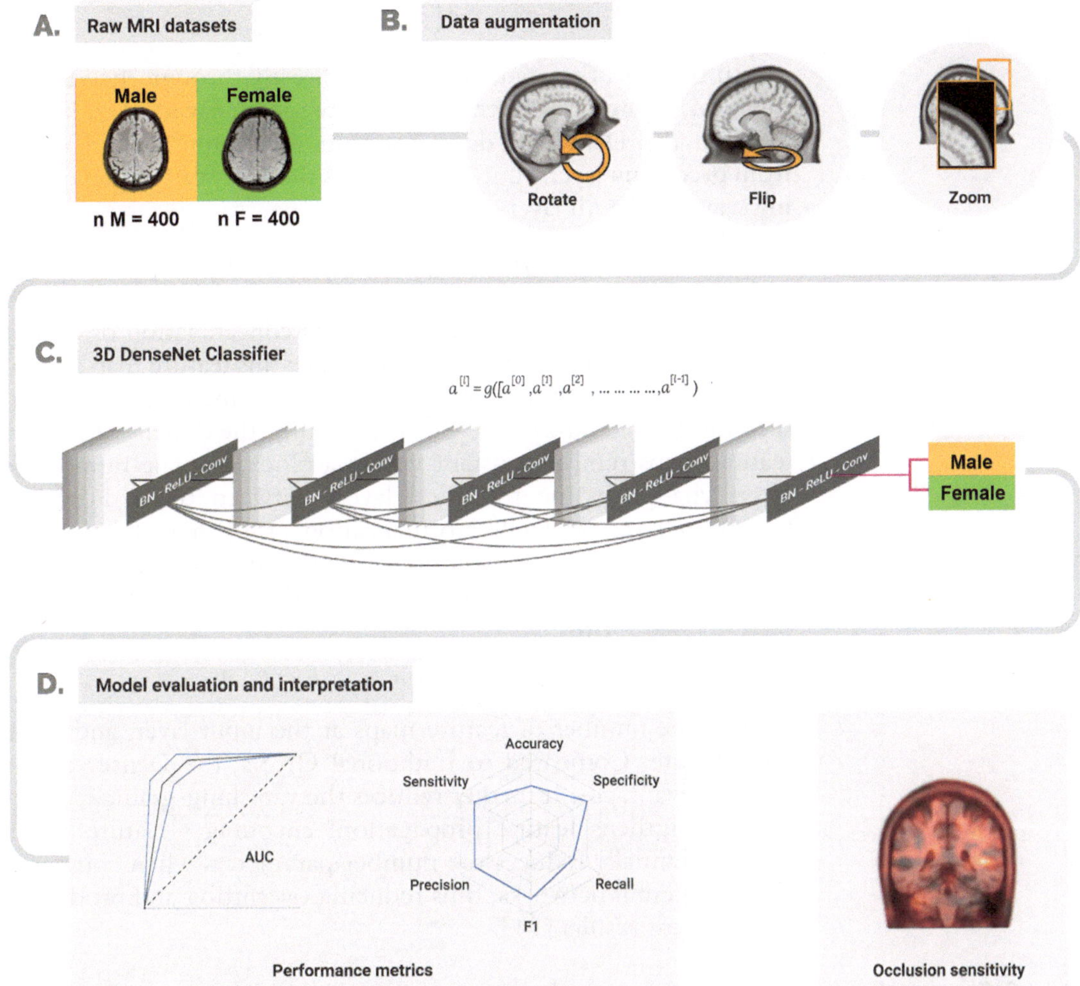

Fig. 1 Step-by-step pipeline. (**a**) Number of raw MRI datasets employed in the analysis (before data augmentation) for male and female groups and related demographic matching. (**b**) Data preparation and augmentation pipeline consisting of random volume rotations, flipping, zooms and an image size scaling. The random augmentation process increased the sample size by a factor of 10. (**c**) 3D DenseNet network architecture consisting of a sequence of Dense Blocks and transition layers consisting of a Batch Normalization (BN), a rectified linear unit (ReLU), and a convolution transformation, ending in a prediction layer to produce the output. (**d**) Model evaluation interpretation, with the performance metrics consisting of the ROC [see Glossary] curve and an AUC [see Glossary] report, a radar plot showing the accuracy, sensitivity, specificity, precision, recall, and F1 metrics. An *occlusion sensitivity analysis* [see Glossary] was developed to obtain the most relevant image information used for the classification

$$a^{[l]} = g\left(a^{[l-1]}\right) + a^{[l-1]} \quad (1)$$

In this way, the gradient can flow directly through the identity function, from early layers to the subsequent layers. Unlike the ResNet, the DenseNet does not sum the output feature maps from preceding layers, but concatenates them instead in the following way for the lth layer:

$$a^{[l]} = g\left(\left[a^{[0]}, a^{[1]}, a^{[2]}, \ldots, a^{[l-1]}\right]\right) \quad (2)$$

where ($[a^{[0]}, a^{[1]}, a^{[2]}, \ldots, a^{[l-1]}]$) is the concatenation of output feature maps of the preceding layers. Since this feature map grouping requires the feature maps to have equal dimensions, the DenseNet is divided into Dense Blocks, in which the dimensions of the feature maps remain constant within a block, but the number of filters change in the transition layers between them. For each transition layer, a Batch Normalization (BN), a rectified linear unit (ReLU), and a convolution followed by average pooling is applied. Another important feature of DenseNet is the growth rate, defined for the

$$l\text{th feature map as } k_0 + k \times (l - 1),$$

with k_0 the number of feature maps at the input layer, and k the growth rate. Compared to traditional CNNs, the DenseNet has several advantages: It further reduces the vanishing-gradient problem, strengthens feature propagation, encourages feature reuse, and substantially reduces the number parameters when compared to other neural networks, thus reducing overfitting and producing more robust results [14].

2.2 Software and Coding

To run the code of this chapter, several library dependencies should be installed. However, all of them are contained in the MONAI PyTorch-based API (https://github.com/Project-MONAI). Project Medical Open Network for AI (MONAI) is an initiative to share and develop best practices for AI in healthcare imaging across academia and enterprise researchers. This collaboration has expanded to include academic and industry leaders throughout the medical imaging field. Project MONAI has released multiple open-source PyTorch-based frameworks for annotating, building, training, deploying, and optimizing AI workflows in healthcare. These standardization frameworks provide high-quality, user-friendly software that facilitates reproducibility and easy integration. The suite of libraries, tools, and SDKs within MONAI provides a robust and common foundation that covers the end-to-end medical AI life cycle, from annotation through deployment.

2.3 Requirements and Setup

The current example requires the installation of the matplotlib library and the Jupyter Notebook GUI. These can be installed with:

python -m pip install -U pip

python -m pip install -U matplotlib

python -m pip install -U notebook

Note

To run the notebook from Google Colab, a GPU setup is needed. To use GPU resources through Colab, please remember to change the runtime type to GPU:

1. *From the Runtime menu select Change runtime type*
2. *Choose GPU from the drop-down menu*

Click SAVE This will reset the notebook and may ask you if you are a robot (these instructions assume you are not).

Running:

nvidia-smi

in a cell, it will verify this has worked and show you what kind of hardware you have access to.

To install the current MONAI milestone release:

pip install monai

To install the weekly preview release:

pip install monai-weekly

The packages installed using pip install could be removed by:

pip uninstall -y monai

pip uninstall -y monai-weekly

From conda-forge, to install the current milestone release:

conda install -c conda-forge monai

You can verify the installation by:

python -c 'import monai; monai.config.print_config()'

If the installation is successful, this command will print out the MONAI version information.

2.4 Dataset

The open dataset included in this example comprises of healthy subjects' (Males $n = 400$, Females $n = 400$) T1-weighted data, compliant with the Brain Imaging Data Structure (BIDS), was

Table 1
Demographic statistical results for the database

	Female $n = 400$	Male $n = 400$	Statistics
Age (years)	22.86 (1.72)	22.85 (1.67)	$F = 0.08$ $p = 0.92^{a}$, $d = 0.006$
Education level (low/medium/high)	40:184:176	42:176:182	$\chi^2 = 0.32$, $p = 0.84^{b}$

Results are presented as mean (SD). Age data was assessed through independent two-sample t test. Level of education analyzed via Pearson's chi-squared (χ^2) test. Effects sizes were calculated through Cohen's d (d)
[a]*p*-values calculated via independent two-sample t-test
[b]*p*-values calculated via chi-squared test (χ2)

downloaded from the Amsterdam Open MRI Collection (AOMIC) (https://openneuro.org/datasets/ds003097/versions/1.2.1) [see Chapter 2]. The samples were matched on age and education level (Table 1). This dataset was collected between 2010 and 2012. The faculty's ethical committee of the University of Amsterdam approved this study before data collection started (EC number: 2010-BC-1345). Prior to the experiment, subjects were informed about the goal and scope of the research, the MRI procedure, safety measures, general experimental procedures, privacy and data sharing concerns, and voluntary nature of the project. Before the start of the experiment, subjects signed an informed consent form and were screened for MRI safety.

All MRI structural data was obtained on the same Philips 3 T Intera scanner. At the start of each scan session, a low-resolution survey scan was made, which was used to determine the location of the field-of-view. For all structural (T1-weighted) the slice stack was not angulated, and the following parameters were used: FOV (RL/AP/FH; mm) = 160 × 256 × 256, Voxel size (mm) = 1 × 1 × 1, TR/TE (ms) = 8.1/3.7, Flip angle (deg.) = 8, Acquisition direction = Sagittal, Duration = 5 min 58 s.

3 Step-by-Step Code Script

3.1 Load Libraries and Dependencies

In this section, we begin installing MONAI via the pip package management system and importing the necessary libraries and dependencies. Some of them may not be installed on your system so you should proceed with installing them via the *pip* command.

Script 1. Installation of packages and importing libraries

```
!python -c "import monai" || pip install -q "monai-weekly[nibabel, tqdm]"

import logging
import os
import sys
import tempfile
import shutil

import matplotlib.pyplot as plt
import torch
from torch.utils.tensorboard import SummaryWriter

import numpy as np
import glob
import pandas as pd

import monai
from monai.apps import download_and_extract
from monai.config import print_config
from monai.data import CacheDataset, DataLoader, ImageDataset
from monai.data.meta_obj import MetaObj
from monai.data.utils import decollate_batch
from monai.transforms import EnsureChannelFirst, Compose, RandRotate, Resize,
ScaleIntensity, RandFlip, EnsureType, RandZoom, Activations, AsDiscrete, ToNumpy
from monai.metrics import ROCAUCMetric
from sklearn.metrics import classification_report, confusion_matrix

import matplotlib.tri as tri
from scipy import ndimage

import sklearn.metrics as metrics
from matplotlib.pyplot import figure
import matplotlib.pyplot as plt
import random
from random import sample

logging.basicConfig(stream=sys.stdout, level=logging.INFO)
print_config()
```

3.2 Load the T1-Weighted Images from the Downloaded Database and Extract Labels

Now we begin building a list of the nii.gz anatomical files using the glob library. The path should point to the folder where you downloaded the files. Then we load the demographic information table using Pandas. Finally, we extract only the labels (Male/Female) of the files that have matched demographic variables using a *for* loop.

Script 2. Data loading

```
### Load MRI data and demographic variables

images = glob.glob('/content/drive/MyDrive/DL/**/*1_T1w.nii.gz', recursive=True)
images.sort()

df = pd.read_csv('behavioral_data_matched.csv')

shortlist = []
labels = []

for i in range(len(images)):
  if images[i][29:37] in df['ID'].values:
    shortlist.append(images[i])

labels = df['sex'].values
```

3.3 Split the Training, Validation, and Testing Files

Having the matched image and labels dataset, we proceed to split the dataset into separate training, validation, and testing subsets. This process ensures that we are able to test the generalization of the results by using an independent test dataset. Then we define the data transformations to get a bigger training set. For this purpose, we apply random brain volume rotations (minimum degree = 0°, maximum = 180°), brain volume flipping (inverse mirror-like image), and image zoom (1× to 1.5×) to focus on different brain areas each time.

Script 3. Data split and augmentation settings

```
### Split the dataset into training, validation, and testing sets
imtrain = images[:300]
labtrain = labels[:300]
imval = images[300:400]
labval = labels[300:400]
imtest = images[400:800]
labtest = labels[400:800]

# Define transforms
train_transforms = Compose([ScaleIntensity(), EnsureChannelFirst(), Resize((50, 100, 100))])
val_transforms = Compose([ScaleIntensity(), EnsureChannelFirst(), Resize((50, 100, 100))])
train_ds = ImageDataset(image_files=imtrain, labels=labtrain, transform=train_transforms)
```

```
############ DATA AUGMENTATION
aug_transforms = Compose([ScaleIntensity(), EnsureChannelFirst(), Resize((50, 100,
100)), RandRotate(), RandFlip(), RandZoom()])
aug_ds = ImageDataset(image_files=imtrain, labels=labtrain, transform=aug_transforms)
aug_ds2 = ImageDataset(image_files=imtrain, labels=labtrain, transform=aug_transforms)
aug_ds3 = ImageDataset(image_files=imtrain, labels=labtrain, transform=aug_transforms)
aug_ds4 = ImageDataset(image_files=imtrain, labels=labtrain, transform=aug_transforms)
aug_ds5 = ImageDataset(image_files=imtrain, labels=labtrain, transform=aug_transforms)
aug_ds6 = ImageDataset(image_files=imtrain, labels=labtrain, transform=aug_transforms)
aug_ds7 = ImageDataset(image_files=imtrain, labels=labtrain, transform=aug_transforms)
aug_ds8 = ImageDataset(image_files=imtrain, labels=labtrain, transform=aug_transforms)
aug_ds9 = ImageDataset(image_files=imtrain, labels=labtrain, transform=aug_transforms)
aug_ds10 = ImageDataset(
   image_files=imtrain, labels=labtrain, transform=aug_transforms)
train_ds = torch.utils.data.ConcatDataset([train_ds, aug_ds, aug_ds2, aug_ds3, aug_ds4,
aug_ds5, aug_ds6, aug_ds7, aug_ds8, aug_ds9, aug_ds10])
############
# create a training data loader
train_loader = DataLoader(train_ds, batch_size=5, shuffle=True, num_workers=2,
pin_memory=torch.cuda.is_available())

# create a validation data loader
val_ds = ImageDataset(image_files=imval, labels=labval, transform=val_transforms)
val_loader = DataLoader(val_ds, batch_size=2, num_workers=2,
pin_memory=torch.cuda.is_available())

# create a test data loader
test_ds = ImageDataset(image_files=imtest, labels=labtest, transform=val_transforms)
test_loader = DataLoader(test_ds, batch_size=1, num_workers=2,
pin_memory=torch.cuda.is_available())
```

3.4 Create the DenseNet Model

In this setup, we will setup the DenseNet model 121 (*see* Table 2 for network architecture details). We set the initial learning to $1e^{-5}$, the batch size parameter was set to 50 samples, and the maximum number of epochs to 10. The Adam optimizer [see Glossary] will be used to minimize the Cross Entropy loss of the 3D-DenseNet during training process [15].

Table 2
DenseNet 121 architecture details

Layers	Output size	Transformation
Convolution	112 × 112 × 112	7 × 7 × 7 convolution, stride 2
Pooling	56 × 56 × 56	3 × 3 × 3 max pooling, stride 2
Dense block 1	56 × 56 × 56	Input concatenation
Transition layer 1	56 × 56 × 56 28 × 28 × 28	1 × 1 × 1 convolution 2 × 2 × 2 average pooling, stride 2
Dense block 2	28 × 28 × 28	Input concatenation
Transition layer 2	28 × 28 × 28 14 × 14 × 14	2 × 2 × 2 average pooling, stride 2 1 × 1 × 1 convolution
Dense block 3	14 × 14 × 14	Input concatenation
Transition layer 3	14 × 14 × 14 7 × 7 × 7	2 × 2 × 2 average pooling, stride 2 1 × 1 × 1 convolution
Dense block 4	7 × 7 × 7	Input concatenation
Classification layer	1 × 1 × 1	7 × 7 × 7 global average pooling Fully connected softmax

Script 4. DenseNet model definition

```
# Create DenseNet121, CrossEntropyLoss and Adam optimizer
device = torch.device("cuda" if torch.cuda.is_available() else "cpu")

model = monai.networks.nets.DenseNet121(
    spatial_dims=3, in_channels=1, out_channels=2).to(device)
loss_function = torch.nn.CrossEntropyLoss()
optimizer = torch.optim.Adam(model.parameters(), 1e-5)

# start training
val_interval = 2
best_metric = -1
best_metric_epoch = -1
epoch_loss_values = []
metric_values = []
writer = SummaryWriter()
max_epochs = 10
for epoch in range(max_epochs):
    print("-" * 10)
    print(f"epoch {epoch + 1}/{max_epochs}")
    model.train()
    epoch_loss = 0
    step = 0
```

```
        for batch_data in train_loader:
            step += 1
            inputs, labels = batch_data[0].to(device), batch_data[1].to(device)
            optimizer.zero_grad()
            outputs = model(inputs)
            loss = loss_function(outputs, labels)
            loss.backward()
            optimizer.step()
            epoch_loss += loss.item()
            epoch_len = len(train_ds) // train_loader.batch_size
            print(f"{step}/{epoch_len}, train_loss: {loss.item():.4f}")
            writer.add_scalar("train_loss", loss.item(), epoch_len * epoch + step)
        epoch_loss /= step
        epoch_loss_values.append(epoch_loss)
        print(f"epoch {epoch + 1} average loss: {epoch_loss:.4f}")
        if (epoch + 1) % val_interval == 0:
            model.eval()
            with torch.no_grad():
                num_correct = 0.0
                metric_count = 0
                for val_data in val_loader:
                    # val_images, val_labels = val_data["img"].to(
                    #     device), val_data["label"].to(device)
                    val_images, val_labels = val_data[0].to(device), val_data[1].to(device)
                    val_outputs = model(val_images)
                    value = torch.eq(val_outputs.argmax(dim=1), val_labels)
                    metric_count += len(value)
                    num_correct += value.sum().item()
                metric = num_correct / metric_count
                metric_values.append(metric)
                if metric > best_metric:
                    best_metric = metric
                    best_metric_epoch = epoch + 1
                    torch.save(model.state_dict(),
                               "best_metric_model_classification3d_array.pth")
                    print("saved new best metric model")
                print(
                    "current epoch: {} current accuracy: {:.4f} "
                    "best accuracy: {:.4f} at epoch {}".format(
                        epoch + 1, metric, best_metric, best_metric_epoch
                    )
                )
                writer.add_scalar("val_accuracy", metric, epoch + 1)
    print(
        f"train completed, best_metric: {best_metric:.4f} "
        f"at epoch: {best_metric_epoch}")
    writer.close()
```

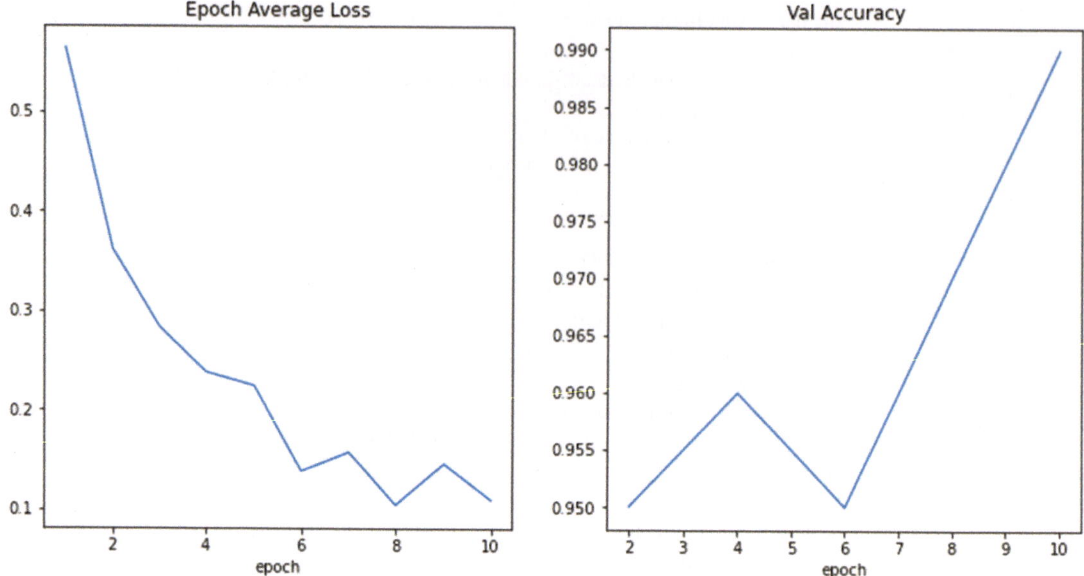

Fig. 2 Epoch average loss, indicating the error across validation runs and validation accuracy plot to check if the network is learning from the input data

3.5 Plot the Average Loss and Validation Accuracy During Training Epochs

Now that training has ended, the next step consists in studying how the DenseNet network is learning to classify the input files correctly. To this end, we will plot the average loss, reflecting classification error during validation across epochs and the validation accuracy (Fig. 2).

Script 5. Plotting average loss and validation accuracy

```
### Plot loss and accuracy curves during training

plt.figure("train", (12, 6))
plt.subplot(1, 2, 1)
plt.title("Epoch Average Loss")
x = [i + 1 for i in range(len(epoch_loss_values))]
y = epoch_loss_values
plt.xlabel("epoch")
plt.plot(x, y)
plt.subplot(1, 2, 2)
plt.title("Val Accuracy")
x = [val_interval * (i + 1) for i in range(len(metric_values))]
y = metric_values
plt.xlabel("epoch")
plt.plot(x, y)
plt.show()
```

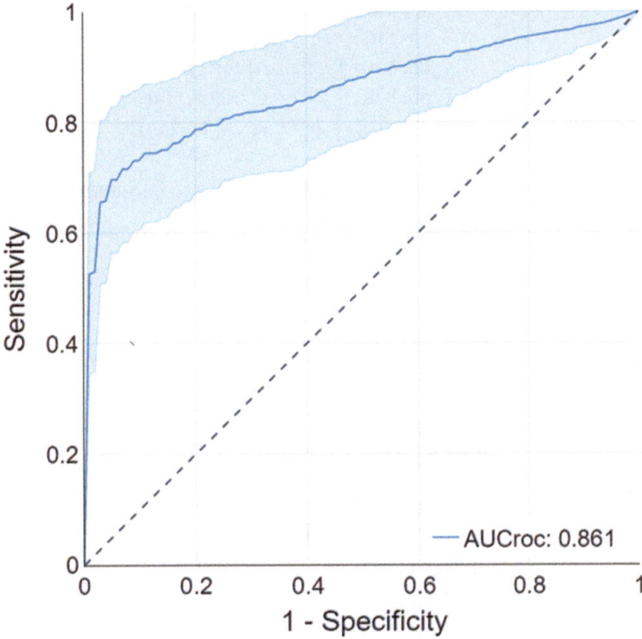

Fig. 3 ROC curve indicating the classification performance of male vs. female discrimination in the testing datasets at various thresholds of sensitivity (true positive rate) and 1-specificity (false positive rate)

3.6 Plot the Test-Set ROC Curve and Get the Area Under the Curve

To have a broader insight regarding the network generalization on a testing dataset, we will plot the receiver operating characteristic (ROC) curve (Fig. 3), representing the diagnostic ability when a discrimination threshold is varied.

Script 6. ROC curve plot

```
### Plot ROC

auc_metric = ROCAUCMetric()
y_pred_trans = Compose([EnsureType(), Activations(softmax=True)])
num_class = 2
y_trans = Compose([EnsureType(), AsDiscrete(to_onehot=num_class)])


for i in range(2):
  test_ds = ImageDataset(image_files = imtest, labels = labtest, transform=val_transforms)
  test_loader = DataLoader(test_ds, batch_size=1, num_workers=2,
pin_memory=torch.cuda.is_available())

  with torch.no_grad():
      y_pred = torch.tensor([], dtype=torch.float32, device=device)
      y = torch.tensor([], dtype=torch.long, device=device)
      for test_data in test_loader:
        test_images, test_labels = (
          test_data[0].to(device),
          test_data[0].to(device),
        )
```

```
        y_pred = torch.cat([y_pred, model(test_images)], dim=0)
        y = torch.cat([y, test_labels], dim=0)
    y_onehot = [y_trans(i) for i in decollate_batch(y)]
    y_pred_act = [y_pred_trans(i) for i in decollate_batch(y_pred)]
    auc_metric(y_pred_act, y_onehot)

numps = Compose([EnsureType(),ToNumpy()])
preds = numps(y_pred_act)
yver = numps(y_true)

a = np.array(preds)
pred = a[:,1]

figure(figsize=(3, 2), dpi=300)
fpr, tpr, threshold = metrics.roc_curve(y_true, pred)
roc_auc = metrics.auc(fpr, tpr)

plt.title('ROC')
plt.plot(fpr, tpr, 'b', label = 'AUC = %0.2f' % roc_auc)
plt.legend(loc = 'lower right')
plt.plot([0, 1], [0, 1],'r--')
plt.xlim([0, 1])
plt.ylim([0, 1])
plt.ylabel('True Positive Rate')
plt.xlabel('False Positive Rate')
plt.show()
```

3.7 Plot the Classification Report in a Radar Chart

Now we will get detailed classification report with multiple metrics (Fig. 4): accuracy, sensitivity, specificity, precision, recall, and F1. The last three are particularly important to deal with imbalanced datasets between classes as they provide unbiased performance results.

Script 7. Classification report

```
### Get performance report

y_true = []
y_pred = []

with torch.no_grad():
    for test_data in test_loader:
        test_images, test_labels = (
            test_data[0].to(device),
            test_data[1].to(device),
        )
        pred = model(test_images).argmax(dim=1)
        for i in range(len(pred)):
            y_true.append(test_labels[i].item())
            y_pred.append(pred[i].item())

class_names = ['Female', 'Male']

rep = classification_report(
    y_true, y_pred, target_names=class_names, digits=4, output_dict=True)

### Get radar chart

tp, fn, fp, tn = confusion_matrix(y_true, y_pred).ravel()
acc = (tp+tn)/(tp+tn+fp+fn)
sen = (tp)/(tp+fn)
sp = (tn)/(tn+fp)
```

```
markers = [0, 0.2, 0.4, 0.6, 0.8, 1.0]
proportions = [acc, sp, rep['weighted avg'].get('recall'), rep['weighted avg'].get('f1-score'),
rep['weighted avg'].get('precision'), sen]
labels = ['Accuracy', 'Specificity', 'Recall', 'F1', 'Precision', 'Sensitivity']

def make_radar_chart(name, stats, attribute_labels=labels,
               plot_markers=markers):

  labels = np.array(attribute_labels)

  angles = np.linspace(0, 2*np.pi, len(labels), endpoint=False)
  stats = np.concatenate((stats,[stats[0]]))
  angles = np.concatenate((angles,[angles[0]]))
  fig = plt.figure()
  ax = fig.add_subplot(111, polar=True)
  ax.plot(angles, stats, 'o-', linewidth=2)
  ax.fill(angles, stats, alpha=0.25)
  ax.set_thetagrids(angles * 180/np.pi, labels, fontsize = 18)
  plt.yticks(markers, fontsize = 18)
  ax.set_title(name, fontsize = 18)
  ax.grid(True)
  fig.set_size_inches(10,10, forward = False)

  return plt.show()

make_radar_chart("Model Performance", proportions)
```

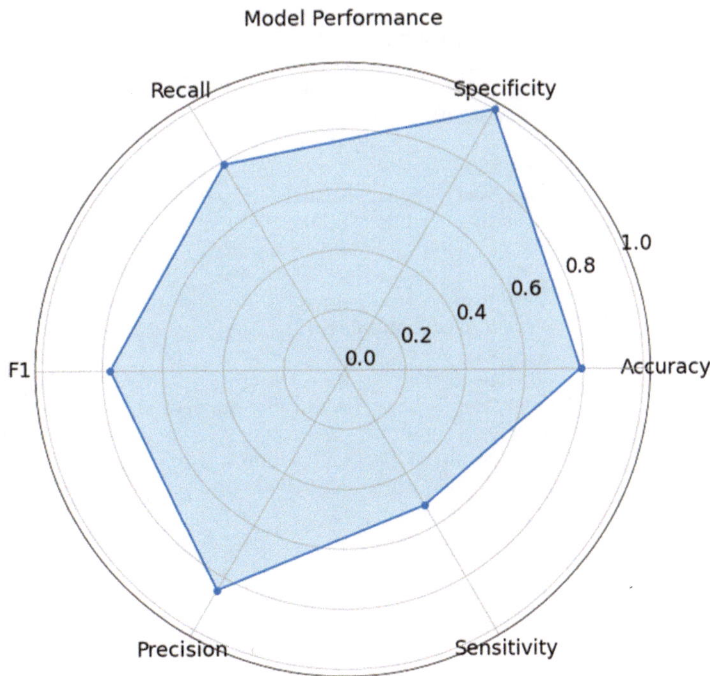

Fig. 4 The radar chart depicts the classification reports, providing a detailed profile of classification performance including accuracy, sensitivity, specificity, precision, recall and F1

3.8 Obtain the Occlusion Sensitivity Map

Finally, we are able to gain insights on what features the DenseNet has been utilizing when classifying subjects. This step is particularly important because we need to be sure that the classification is based on brain anatomy, and not other possible confounding factors such as skull bone structure. Furthermore, information about specific anatomical differences would provide insights regarding biological underpinnings of sex. To this end, we ran *occlusion sensitivity* analyses [16]. The occlusion sensitivity analysis is a technique in convolutional neural networks that is employed to understand what parts of an image are more relevant for deciding a classification output. During this process, small image areas are perturbed by superimposing an occluding mask composed of a gray square. This mask is moved across the image, and the change in probability score for a given class is measured as a function of mask position. When an important part of the image is occluded, the probability score for the predicted class will fall sharply, producing a map of classification relevance.

Script 8. Occlusion sensitivity output interpretation

```
### Run occlusion sensitivity analysis in the sagittal plane (other planes are available,
commented below)

msk = 3
stride = 3

for i in range(10):
  # Get a random image and its corresponding label
  img, label = get_next_im()
  print(label)

  # Get the occlusion sensitivity map
  occ_sens = monai.visualize.OcclusionSensitivity(nn_module=model, mask_size=msk,
n_batch=10, stride=stride)
  # Only get a single slice to save time.
  # For the other dimensions (channel, width, height), use
  # -1 to use 0 and img.shape[x]-1 for min and max, respectively
  depth_slice = img.shape[2] // 2

  # Sagital
  occ_sens_b_box = [depth_slice-1, depth_slice, -1, -1, -1, -1]

  occ_result, _ = occ_sens(x=img, b_box=occ_sens_b_box)
  occ_result = occ_result[0, label.argmax().item()][None]

  fig = plt.figure(figsize=(15,15))
  plt.xlim([0, 100])
  plt.ylim([0, 100])
  ax = fig.add_subplot(111)
  im = occ_result

  img2 = plt.imshow(ndimage.rotate(img[0, 0, depth_slice, ...].detach().cpu(),-90),
interpolation='nearest', cmap='gray', origin='lower')
  img3 = plt.imshow(ndimage.rotate(np.squeeze(im[0][0].detach().cpu()),-90),
interpolation='nearest', cmap='gist_heat', origin='lower', alpha = 0.4)
  fig.colorbar(img3)
  plt.show()
```

4 Discussion on Model Example Interpretation

The example displayed in Fig. 5 shows an occlusion sensitivity cluster located in the orbitofrontal cortex. This area seems to be relevant for sex differences. A recent study in more than 2000 MRI scans [17] showed that females, on average, had relatively greater volume in the orbitofrontal cortex when compared to male subjects. This result has been also replicated using state-of-the-art techniques applied to GMV in a large sample ($n = 2838$) and in two independent cohorts, with a large age range, with all data acquired from the same MRI scanner [18]. Importantly, the orbitofrontal cortex is relevant to sex differences in geriatric depression [19], tau deposition in heterozygotes [20], leftward functional connectivity asymmetry [21], and even microsatellite polymorphisms of steroid hormone receptors [22]. Future studies may investigate if these brain anatomical differences are influenced by different factors, including biological heterogeneity or by social conventions generating plastic brain changes through the lifespan.

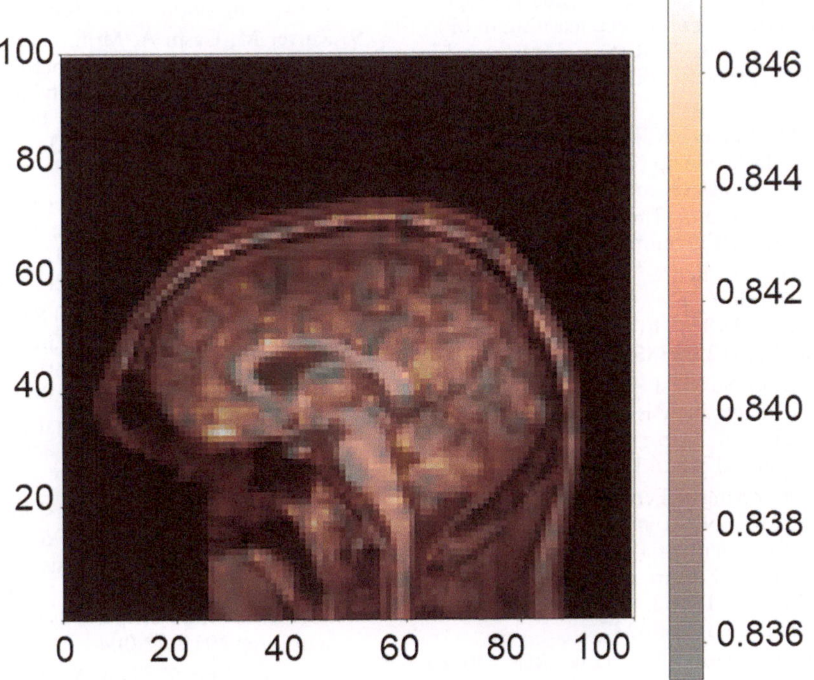

Fig. 5 Sagittal plane T1-weighted MRI image overlaid with the output of the occlusion sensitivity map

5 Conclusions

In this chapter, we described a step-by-step implementation example of a deep learning pipeline based on raw data. After increasing the sample using augmentation, we trained the DenseNet to accurately classify male and female subjects in the test set. In addition, via the occlusion sensitivity analysis, anatomical insights were gained, plus evidence that the classifier was using brain tissue sources and not possible confounding factors. We hope this framework will guide future implementations of different protocols that use neuroimaging data for classification.

References

1. Ching T, Himmelstein DS, Beaulieu-Jones BK, Kalinin AA, Do BT, Way GP, Ferrero E, Agapow P-M, Zietz M, Hoffman MM, Xie W, Rosen GL, Lengerich BJ, Israeli J, Lanchantin J, Woloszynek S, Carpenter AE, Shrikumar A, Xu J, Cofer EM, Lavender CA, Turaga SC, Alexandari AM, Lu Z, Harris DJ, DeCaprio D, Qi Y, Kundaje A, Peng Y, Wiley LK, Segler MHS, Boca SM, Swamidass SJ, Huang A, Gitter A, Greene CS (2018) Opportunities and obstacles for deep learning in biology and medicine. J R Soc Interface 15: 20170387. https://doi.org/10.1098/rsif.2017.0387
2. Pedersen M, Verspoor K, Jenkinson M, Law M, Abbott DF, Jackson GD (2020) Artificial intelligence for clinical decision support in neurology. Brain Commun 2:fcaa096. https://doi.org/10.1093/braincomms/fcaa096
3. He K, Zhang X, Ren S, Sun J (2015) Deep residual learning for image recognition. arXiv:1512.03385. https://doi.org/10.48550/arXiv.1512.03385
4. Shen D, Wu G, Suk H-I (2017) Deep learning in medical image analysis. Annu Rev Biomed Eng 19:221–248. https://doi.org/10.1146/annurev-bioeng-071516-044442
5. Ahmed MR, Zhang Y, Feng Z, Lo B, Inan OT, Liao H (2019) Neuroimaging and machine learning for dementia diagnosis: recent advancements and future prospects. IEEE Rev Biomed Eng 12:19–33. https://doi.org/10.1109/RBME.2018.2886237
6. Huys QJM, Maia TV, Frank MJ (2016) Computational psychiatry as a bridge from neuroscience to clinical applications. Nat Neurosci 19:404–413. https://doi.org/10.1038/nn.4238
7. Huang G, Liu Z, Van Der Maaten L, Weinberger KQ (2017) Densely connected convolutional networks. In: 2017 IEEE conference on computer vision and pattern recognition (CVPR), pp 2261–2269
8. Baez S, Flichtentrei D, Prats M, Mastandueno R, García AM, Cetkovich M, Ibáñez A (2017) Men, women. . .who cares? A population-based study on sex differences and gender roles in empathy and moral cognition. PLoS One 12:e0179336. https://doi.org/10.1371/journal.pone.0179336
9. Rutter M, Caspi A, Moffitt TE (2003) Using sex differences in psychopathology to study causal mechanisms: unifying issues and research strategies. J Child Psychol Psychiatry 44:1092–1115. https://doi.org/10.1111/1469-7610.00194
10. Mazure CM, Swendsen J (2016) Sex differences in Alzheimer's disease and other dementias. Lancet Neurol 15:451–452. https://doi.org/10.1016/S1474-4422(16)00067-3
11. Beery AK, Zucker I (2011) Sex bias in neuroscience and biomedical research. Neurosci Biobehav Rev 35:565–572. https://doi.org/10.1016/j.neubiorev.2010.07.002
12. Cahill L (2006) Why sex matters for neuroscience. Nat Rev Neurosci 7:477–484. https://doi.org/10.1038/nrn1909
13. Ruigrok ANV, Salimi-Khorshidi G, Lai M-C, Baron-Cohen S, Lombardo MV, Tait RJ, Suckling J (2014) A meta-analysis of sex differences in human brain structure. Neurosci Biobehav Rev 39:34–50. https://doi.org/10.1016/j.neubiorev.2013.12.004
14. Xiao B, Yang Z, Qiu X, Xiao J, Wang G, Zeng W, Li W, Nian Y, Chen W (2022) PAM-DenseNet: a deep convolutional neural network for computer-aided COVID-19 diagnosis. IEEE Trans Cybern 52:12163–12174. https://doi.org/10.1109/TCYB.2020.3042837

15. Kandel I, Castelli M, Popovič A (2020) Comparative study of first order optimizers for image classification using convolutional neural networks on histopathology images. J Imaging 6:92. https://doi.org/10.3390/jimaging6090092
16. Dyrba M, Hanzig M, Altenstein S, Bader S, Ballarini T, Brosseron F, Buerger K, Cantré D, Dechent P, Dobisch L, Düzel E, Ewers M, Fliessbach K, Glanz W, Haynes J-D, Heneka MT, Janowitz D, Keles DB, Kilimann I, Laske C, Maier F, Metzger CD, Munk MH, Perneczky R, Peters O, Preis L, Priller J, Rauchmann B, Roy N, Scheffler K, Schneider A, Schott BH, Spottke A, Spruth EJ, Weber M-A, Ertl-Wagner B, Wagner M, Wiltfang J, Jessen F, Teipel SJ, ADNI, AIBL, DELCODE study groups (2021) Improving 3D convolutional neural network comprehensibility via interactive visualization of relevance maps: evaluation in Alzheimer's disease. Alzheimers Res Ther 13:191. https://doi.org/10.1186/s13195-021-00924-2
17. Liu S, Seidlitz J, Blumenthal JD, Clasen LS, Raznahan A (2020) Integrative structural, functional, and transcriptomic analyses of sex-biased brain organization in humans. Proc Natl Acad Sci USA 117:18788–18798. https://doi.org/10.1073/pnas.1919091117
18. Lotze M, Domin M, Gerlach FH, Gaser C, Lueders E, Schmidt CO, Neumann N (2019) Novel findings from 2,838 adult brains on sex differences in gray matter brain volume. Sci Rep 9:1671. https://doi.org/10.1038/s41598-018-38239-2
19. Lavretsky H, Kurbanyan K, Ballmaier M, Mintz J, Toga A, Kumar A (2004) Sex differences in brain structure in geriatric depression. Am J Geriatr Psychiatry 12:653–657. https://doi.org/10.1176/appi.ajgp.12.6.653
20. Yan S, Zheng C, Paranjpe MD, Li Y, Li W, Wang X, Benzinger TLS, Lu J, Zhou Y (2021) Sex modifies APOE ε4 dose effect on brain tau deposition in cognitively impaired individuals. Brain 144:3201–3211. https://doi.org/10.1093/brain/awab160
21. Liang X, Zhao C, Jin X, Jiang Y, Yang L, Chen Y, Gong G (2021) Sex-related human brain asymmetry in hemispheric functional gradients. NeuroImage 229:117761. https://doi.org/10.1016/j.neuroimage.2021.117761
22. Tan GC-Y, Chu C, Lee YT, Tan CCK, Ashburner J, Wood NW, Frackowiak RS (2020) The influence of microsatellite polymorphisms in sex steroid receptor genes ESR1, ESR2 and AR on sex differences in brain structure. NeuroImage 221:117087. https://doi.org/10.1016/j.neuroimage.2020.117087

[illegible]

List of Resources

- Amazon Web Services (AWS): https://aws.amazon.com
- BIDS: https://bids.neuroimaging.io/
- BIDS apps: An index of BIDS applications. https://bids-apps.neuroimaging.io/apps/
- BIDS converters: https://bids.neuroimaging.io/benefits.html#mri-and-pet-converters
- BIDS starter kit: https://bids-standard.github.io/bids-starter-kit/index.html
- BIDS validators: https://bids-standard.github.io/bids-validator/
- BioImageSuite Web Connectivity Viewer: https://bioimagesuiteweb.github.io/webapp/connviewer.html
- Boto: https://boto3.amazonaws.com/v1/documentation/api/latest/index.html
- Carbon tracker toolboxes: https://ohbm-environment.org/carbon-tracker-toolboxes/
- CAT12 (Computational Anatomy Toolbox, https://neuro-jena.github.io/cat); here applied without license costs as a standalone version (https://neuro-jena.github.io/enigma-cat12/#standalone) or as a Singularity container (https://github.com/inm7-sysmed/ENIGMA-cat12-container).
- Code Carbon: https://mlco2.github.io/codecarbon/index.html
- Collaborative Informatics and Neuroimaging Suite(COINS): https://coins.trendscenter.org
- Coinstac: https://coinstac.org
- C-PAC https://fcp-indi.github.io/
- Cyberduck: https://cyberduck.io/
- Datalad: https://www.datalad.org/

Robert Whelan and Hervé Lemaître (eds.), *Methods for Analyzing Large Neuroimaging Datasets*, Neuromethods, vol. 218, https://doi.org/10.1007/978-1-0716-4260-3,

- Docker: https://docs.docker.com/get-started/
- Docker hub: https://hub.docker.com/
- EEGLAB: https://sccn.ucsd.edu/eeglab/index.php
- Experiment Impact Tracker (EIT): https://github.com/Breakend/experiment-impact-tracker
- ExploreDTI: https://www.exploredti.com/
- FAIR: https://www.go-fair.org/fair-principles/
- fMRIPrep: https://fmriprep.org/en/stable/
- Git: https://git-scm.com/
- Github desktop: https://desktop.github.com/
- Github flow: https://githubflow.github.io/
- GNU make: https://www.gnu.org/software/make/
- HED resources: https://www.hed-resources.org/en/latest/index.html
- Longitudinal Online Research and Imaging System (LORIS): http://www.loris.ca
- Matplotlib: tools for data visualization: https://matplotlib.org
- Matlab Compiler Runtime: https://www.mathworks.com/products/compiler/matlab-runtime.html
- MONAI: https://monai.io/about.html
- MRIQC: https://mriqc.readthedocs.io/en/stable/
- Nextflow: https://www.nextflow.io/
- NeuroElectroMagnetic Archive and compute Resource (NEMAR): https://nemar.org/
- Neuroscience Gateway (NSG): https://www.nsgportal.org/
- NeuroStars Forum: https://neurostars.org/
- Neurovault: https://neurovault.org
- Nilearn: package supporting neuroimaging analyses of structural and functional volumetric data. Includes tools for voxelwise statistical analyses, multi-voxel pattern analysis (MVPA), GLMs, clustering and parcellation, etc. https://nilearn.github.io/stable/index.html
- NiPreps: https://www.nipreps.org/
- Nipype: https://nipype.readthedocs.io/en/latest/
- Nltools: neuroimaging package for fMRI data analyses (e.g., resting-state, task-based, movie-watching) that incorporates code from nilearn and scikit-learn. https://nltools.org
- Numpy (numerical package for scientific and arithmetic computing): https://numpy.org/
- OpenNeuro: OpenNeuro.org

- Open Science Framework (OSF): http://osf.io
- Panda: supports data manipulation and analysis, particularly useful for reading in data in csv/xls format. https://pandas.pydata.org/
- Pingouin: simple statistical functions and graphics.in Python. https://pingouin-stats.org
- Pybids: https://bids-standard.github.io/pybids/analysis/index.html
- Repronim: https://www.repronim.org/
- Seaborn: tools for data visualization. https://seaborn.pydata.org
- Statsmodels: package in python to build statistical models and perform statistical tests. https://www.statsmodels.org/stable/index.html
- Scikit-learn: tools for machine learning, including classification, model selection, and dimensionality reduction. https://scikit-learn.org/stable/
- Scitran: https://scitran.github.io/
- Snakemake: https://snakemake.readthedocs.io/en/stable/
- Sourcetree: https://www.sourcetreeapp.com/
- Stackoverflow: https://stackoverflow.com
- Statsmodels: package in Python to build statistical models and perform statistical tests. https://www.statsmodels.org/stable/index.html
- Statistical Parametric Mapping (SPM): https://www.fil.ion.ucl.ac.uk/spm/
- Trunk based development: https://trunkbaseddevelopment.com/
- XNAT: https://central.xnat.org/

Glossary

Artificial Intelligence (AI) A branch of computer science that deals with the creation of intelligent machines that can simulate human intelligence, including the ability to think, reason, and make decisions.

Artifact Subspace Reconstruction (ASR) A technique used in signal processing and image processing to remove unwanted artifacts or noise from data by reconstructing a subspace that represents the signal of interest while suppressing the artifacts or noise.

Area under the Receiver Operating Characteristic (AROC; often abbreviated as AUC) A metric that evaluates the overall performance of a binary classification model by comparing the True Positive Rate against the False Positive Rate across various thresholds. A value of 1 indicates perfect discrimination, while 0.5 suggests no better performance than random guessing.

BIDS (Brain Imaging Data Structure) A standardized format for organizing and sharing neuroimaging data, including data from structural and functional MRI, EEG, and MEG studies.

BIDSApp (Brain Imaging Data Structure App) A software tool that automates the analysis of neuroimaging data in the BIDS format. BIDSApp is built on the BIDS specification and provides a standardized way to run a range of neuroimaging analysis pipelines in a consistent and reproducible manner.

Bind mounts A feature that allows a directory or file on a host system to be mounted into a different location within the same

Robert Whelan and Hervé Lemaître (eds.), *Methods for Analyzing Large Neuroimaging Datasets*, Neuromethods, vol. 218, https://doi.org/10.1007/978-1-0716-4260-3, © The Editor(s) (if applicable) and The Author(s) 2025

filesystem or a container. This provides direct access to the mounted directory or file, ensuring changes in either location are immediately reflected in the other.

Block design (cf. event-related design) A type of experimental design used in functional magnetic resonance imaging studies to investigate brain activity during a specific cognitive condition. In block design experiments, the task is repeated in a series of discrete time intervals or "blocks," each separated by a brief rest period.

BOLD (Blood Oxygen Level Dependent) The change in the magnetic properties of hemoglobin in the blood as it delivers oxygen to active neurons in the brain.

Bonferroni correction A statistical adjustment applied to control for the increased risk of Type I errors in multiple comparisons by dividing the desired significance level by the number of comparisons.

Circle plots A method often used in network analysis of visualizing data where nodes are arranged in a circle.

Coarse-grained parallelism See Granularity.

Confusion matrix A table used to assess the performance of a classification model by comparing actual vs. predicted labels. It includes True Positives (correctly predicted positives), True Negatives (correctly predicted negatives), False Positives (incorrectly predicted positives), and False Negatives (incorrectly predicted negatives).

Connectome A map of functional or structural connections in the brain.

Container A standardized unit of software that encapsulates code, runtime, system tools, and libraries, enabling consistent and portable deployment across different computing environments.

Convolutional Neural Networks (CNNs) A type of deep learning model commonly used for image and video analysis. CNNs are structured to automatically and adaptively learn spatial hierarchies of features through multiple layers of convolutional filters.

Connectome-based Predictive Modeling (CPM) A method that involves constructing predictive models that relate patterns of connectivity within the brain to specific variables of interest (e.g., cognitive abilities, personality traits, or patient status).

Central Processing Unit (CPU) The main processing component in a computer that performs the majority of the tasks required to run the computer's software and hardware.

Cross-validation (CV) A technique used in machine learning and statistical modeling to evaluate the performance of a predictive model by testing it on separate data. The goal of cross-validation is to assess how well a model can generalize to new data rather than just fit to the training data.

Curse of dimensionality A set of issues encountered when dealing with data that contains too many variables, factors, or features. For example, the presence of more variables than data points may lead to a poorly fitting model with a bad predictive value outside the sample dataset.

Data leakage This occurs when information from the test dataset is inadvertently used to create a model, leading to overly optimistic performance metrics.

Deep learning A subfield of machine learning that involves building and training artificial neural networks with many layers. These neural networks are designed to learn and recognize patterns in large datasets and are inspired by the structure and function of the human brain.

Derivatives Derived data or processed versions of original data, often involving transformations, pre-processing, or feature extraction, aimed at enhancing or facilitating analysis while maintaining traceability to the source data.

Digital Imaging and Communications in Medicine (DICOM) A standard for handling, storing, printing, and transmitting medical images and associated information.

Directed acyclic graph (DAG) A finite graph that consists of vertices connected by edges, where each edge has a direction, and there are no cycles,

	meaning you cannot start at any vertex and return to it by following a directed path.
Differential programming	A group of techniques that leverage automatic differentiation to efficiently and accurately compute derivatives, facilitating optimization tasks.
Diffusion-weighted Magnetic Resonance Imaging (dMRI)	A medical imaging technique that provides information about the movement of water molecules in tissues.
Diffusion tensor imaging (DTI)	A particular magnetic resonance imaging analysis technique that provides information about the direction, orientation, and magnitude of water diffusion in tissues, particularly in white matter tracts of the brain.
Dummy coding (cf. one-hot encoding)	Representing categorical variables with numerical values (0 or 1) to facilitate their inclusion in regression models. Each category of the categorical variable is assigned a binary variable (dummy variable), where 0 indicates the absence of the category and 1 indicates its presence.
Event-related design (cf. block design)	A type of experimental design used in functional magnetic resonance imaging to investigate brain activity in response to discrete events, such as a visual stimulus, a sound, or a motor response. In event-related designs, the cognitive task is not presented in blocks, but instead, individual events are presented in a random or pseudorandom order, with a variable interstimulus interval (ISI) or inter-trial interval (ITI) between each event.
External validation	Testing a predictive model on a new set of data to quantify its predictive value.
FAIR Principles	Prescribe characteristics of data and digital objects to maximize their reuse by the scientific community – that is, to maximize data sharing, exploration, reuse, and deposition by parties other than the original researcher.
False Discovery Rate	A statistical method used to control the proportion of falsely rejected null hypotheses among all rejected

hypotheses in multiple hypothesis testing.

False Positive Rate The probability of incorrectly rejecting the null hypothesis.

Feature Selection The process of reducing the number of predictive variables that enable you to create the most accurate machine learning model, to create the most parsimonious model possible.

Field-Programmable Gate Arrays (FPGAs) Integrated circuits made up of configurable blocks that can be programmed and re-programmed to the specific desires of the user post-manufacture.

File Paths Specifies the location of a file in a computer/website folder directory.

Flat cross-validation A method where the data is not grouped or stratified in any specific way before being split into training and validation sets. This approach treats the dataset as a single flat array, randomly dividing it into different subsets for training and testing without considering potential imbalances or clustering in the data.

Framewise displacement Metric quantifying the average head motion across frames or volumes in a time series, providing insights into the potential impact of subject motion on the data.

Functional connectivity Statistical correlation between the activity of different brain regions, indicating their synchronized functioning and potential communication pathways.

Gaussian Process Regression A non-parametric Bayesian approach to regression analysis.

Generalizability The ability of a machine learning model to accurately predict outcomes for previously unseen data.

Gini Scores Quantifies the impurity of a set of data by measuring the probability of misclassifying a randomly chosen element.

Glass brain plots A translucent image of the brain in standardized brain space that is useful in visualizing source localization.

Glyphs Typically, a graphical representation used in diffusion tensor imaging to visualize and analyze the diffusion properties of water molecules in biological tissues.

General Linear Model (GLM) A statistical framework used to model the relationship between a dependent variable and one or more independent variables.

Global Signal Regression The removal or suppression of the overall signal intensity across all brain regions to address potential confounds and enhance the specificity of localized neural activity measurements.

Graphics Processing Unit (GPU) A specialized processor designed to perform tasks related to computer graphics and image processing. Unlike Central Processing Units, which are optimized for general-purpose computing tasks, GPUs excel in handling large volumes of parallel data simultaneously because they contain thousands of cores that can execute multiple tasks concurrently.

Graphical User Interface (GUI) Unlike a text-based software program, a GUI is a software program that allows the user to interact with it by way of manipulating visual components like buttons, icons, menus, etc.

Graph Theory A branch of mathematics that deals with the study of graphs, which are mathematical structures used to model pairwise relationships between objects. A graph consists of nodes (e.g., brain regions) and edges (connections such as correlation between nodes) that connect pairs of vertices.

Granularity The extent to which a system or entity (e.g., a neuroimaging processing job) is broken down into smaller components (e.g., sub-tasks). Fine granularity means that components are small and detailed, while coarse granularity means they are larger and less detailed. Granularity impacts performance, scalability, and complexity in system design and management.

Grid Search A method for finding the optimal hyperparameters of a machine learning model that involves evaluating the model's performance for each combination of hyperparameters specified in

	the grid. Grid search helps determine the best set of hyperparameters that maximize the model's performance on a given dataset.
Hemodynamic Response Function (HRF)	The physiological changes in blood flow, volume, and oxygenation that occur in response to neural activity in the brain. HRF is often used as a model to estimate the underlying neural activity from fMRI (functional magnetic resonance imaging) signals.
Hyperparameters	Configurable settings or parameters external to the model that influence its learning process and performance, such as learning rate or number of hidden layers in a neural network.
Interoperability	The ability of two different software systems, applications, data stores, etc., to communicate effectively with one another.
Interpretability	How easily understood the predictions made by a machine learning model are.
JSON file	A file that stores simple data structures and objects in JavaScript Object Notation (JSON) format, which is a standard data interchange format. It is primarily used for transmitting data between a web application and a server. JSON files are lightweight, text-based, human-readable, and can be edited using a text editor.
K-fold Cross Validation	a type of cross-validation where the data is split into *k* number of subsets on which the predictive value of the machine learning model is subsequently evaluated.
Leave-one-out Cross Validation	A (computationally expensive) variation of k-fold cross-validation where k is equal to the number of data points in the dataset. As is suggested in the name, each fold removes a single, different data point.
Machine-actionable	Data that are consistently structured in a way that allows computers/machines to be programmed against said structure.
Machine-interpretable	Data that a computer/machine can contextualize and understand.
Machine learning	A subfield of artificial intelligence that involves developing algorithms and

statistical models that enable computer systems to automatically improve their performance on a specific task by learning from data.

Machine-readable Structured data that a machine/computer can automatically process. Examples of machine-readable data are JSON, XML, or CSV files.

Markdown A lightweight markup language with plain-text formatting syntax that is designed to be converted to Hypertext Markup Language (HTML) and is often used for formatting readme files, writing messages in online discussion forums, and creating rich text using a plain text editor.

Mass-univariate testing This involves conducting statistical tests, typically at each individual voxel or region separately. Typically, the large number of statistical tests is subsequently corrected for multiple comparisons (e.g., using the False Discovery Rate).

Mega-analysis The analysis of a large quantity of raw data which have been aggregated across a number of different studies/sources.

Meta-analysis The analysis of summary data from a number of different studies/sources. It involves synthesizing data from individual studies to obtain an overall effect size or estimate of the magnitude and direction of an effect.

Metadata Data that offers context about data without revealing its content (e.g., a column title in an excel spreadsheet will give you information about what the data in said column pertains to but won't reveal the exact data itself).

Massive Open Online Course (MOOCs) An open access online course created and shared with the intention of high levels of participation and engagement. Examples of companies that offer MOOCs are Coursera, edX, and Khan Academy.

Multiverse analysis Implementation of several different processing and analysis pipelines and thereafter combining results (either as a range of possibilities or by statistical aggregation).

Naming convention

	A framework for the descriptive naming of files that allows individuals to easily understand what information the files contain.
Network-Based Statistic (NBS)	A statistical method to identify significant subnetworks or clusters of connected brain regions that exhibit consistent differences between experimental conditions.
Nested cross-validation	A form of cross-validation often for the purpose of obtaining optimal model hyperparameters. It involves employing an outer and an inner cross-validation loop. The outer loop divides the dataset into training and test sets multiple times. Each iteration of the outer loop creates a different partition of the data. For each partition created by the outer loop, there is an inner loop in which further cross-validation is performed (e.g., to tune the model's hyperparameters).
Notebook	An interactive web-based computational platform.
Parallelization, parallel computing	A type of computing where a process is subdivided into several subprocesses to speed it up.
Permutation testing	A non-parametric method used to quantify the significance of a hypothesis by randomly shuffling observed data points between groups or conditions. Then, the observed test statistic is compared with a distribution of permuted test statistics generated under the null hypothesis.
Object-based storage	A data storage architecture that manages data as discrete units called objects, each containing the data itself, metadata, and a unique identifier, which allows for easy retrieval and scalability.
Orientation	When viewing medical images, neurological orientation presents the left hemisphere of the brain on the left of the image. In contrast, radiological orientation is flipped (i.e., the left hemisphere of the brain will appear on the right of the image)
Overfitting	When a machine learning or statistical model is designed with too much

	specificity in regards to the original dataset that its usefulness on subsequent datasets is diminished. This is because the model captures the underlying patterns in the training data and also noise and random fluctuations in those data.
Parallelization	The division of a particular computational task into independent sub-tasks that are executed concurrently.
Pre-registration	Submitting a detailed plan of study methods and analyses before the research begins, which is reviewed and registered by a third party.
Random search	A method for finding the optimal hyperparameters for your predictive model, where a grid of hyperparameters will be randomly sampled from a specified number of times and the best combination selected.
Regression model	See General linear model.
Remote directory	A folder stored on a server or another computer that can be accessed over a network or the internet.
Repository	In the context of data science, a library or archive where data is stored, often for open access.
Reproducibility	The ability to obtain the same results as a prior study using procedures that are closely matched with those used in the original research.
SHapley Additive exPlanations (SHAP)	Assigns each feature in a model an importance value by quantifying the impact of each feature's presence on the machine learning model using game theory.
Shell	A user interface for accessing an operating system's services, typically through a command-line interface or a graphical user interface.
Sparsity threshold	A predefined level that determines the extent to which coefficients or features in a model or representation are set to zero, promoting a sparse or reduced set of non-zero elements.
Sudo rights	superuser do. A command on Unix and Linux systems that gives extra privileges (root/administrator access rights) to regular users.
.tsv format; tab-separated file	A text-based file for the storage of data in tabular form.

Tag hierarchy	Ordering of Hypertext Markup Language (HTML) tags that defines the structure and flow of an HTML document.
Telemetry	The automated collection and transmission of data measurements from remote or inaccessible sources to an IT system in a different location for monitoring and analysis.
Tractography	A technique used in diffusion magnetic resonance imaging to reconstruct and visualize the three-dimensional pathways of white matter tracts in the brain.
Trial	A single instance or episode during which the participant is exposed to a specific stimulus or task condition.
True Positive Rate	The probability of correctly accepting the alternative hypothesis.
t-statistic	An inferential statistic used to make a decision on whether to reject or accept the alternative hypothesis in a t-test.
Version Control	A system that records changes to a file or set of files over time, allowing you to recall specific versions later.
Virtual machine	A software emulation of a computer system that runs an operating system and applications as if they were running on physical hardware.
Weighted and unweighted network	Types of graphs in which a weighted network assigns numerical values to edges, reflecting the strength of connections between nodes, while an unweighted network considers all connections as equal.
XML (Extensible Markup Language)	A flexible text format used to structure, store, and transport data, defined by tags and rules that are both human- and machine-readable.

INDEX

Robert Whelan and Hervé Lemaître (eds.), *Methods for Analyzing Large Neuroimaging Datasets*, Neuromethods, vol. 218, https://doi.org/10.1007/978-1-0716-4260-3, © The Editor(s) (if applicable) and The Author(s) 2025

N

O

P

R

S

T

V